Foundations of Personality

NATO ASI Series

Advanced Science Institutes Series

A Series presenting the results of activities sponsored by the NATO Science Committee, which aims at the dissemination of advanced scientific and technological knowledge, with a view to strengthening links between scientific communities.

The Series is published by an international board of publishers in conjunction with the NATO Scientific Affairs Division

A Life Sciences **B Physics**	Plenum Publishing Corporation London and New York
C Mathematical and Physical Sciences **D Behavioural and Social Sciences** **E Applied Sciences**	Kluwer Academic Publishers Dordrecht, Boston and London
F Computer and Systems Sciences **G Ecological Sciences** **H Cell Biology** **I Global Environmental Change**	Springer-Verlag Berlin, Heidelberg, New York, London, Paris and Tokyo

NATO-PCO-DATA BASE

The electronic index to the NATO ASI Series provides full bibliographical references (with keywords and/or abstracts) to more than 30000 contributions from international scientists published in all sections of the NATO ASI Series.
Access to the NATO-PCO-DATA BASE is possible in two ways:

– via online FILE 128 (NATO-PCO-DATA BASE) hosted by ESRIN, Via Galileo Galilei, I-00044 Frascati, Italy.

– via CD-ROM "NATO-PCO-DATA BASE" with user-friendly retrieval software in English, French and German (© WTV GmbH and DATAWARE Technologies Inc. 1989).

The CD-ROM can be ordered through any member of the Board of Publishers or through NATO-PCO, Overijse, Belgium.

Series D: Behavioural and Social Sciences - Vol. 72

Foundations of Personality

edited by

Joop Hettema
Tilburg University,
The Netherlands

and

Ian J. Deary
University of Edinburgh,
United Kingdom

Springer Science+Business Media, B.V.

Proceedings of the NATO Advanced Research Workshop on
Basic Issues in Personality: European-American Workshop on
Biological and Social Approaches to Individuality
Oisterwijk, The Netherlands
August 17-20, 1992

Library of Congress Cataloging-in-Publication Data

Foundations of personality / edited by Joop Hettema and Ian J. Deary.
p. cm. -- (NATO ASI series. Series D, Behavioural and social sciences ; vol. 72)
Papers presented at a workshop held in Oisterwijk, The Netherlands, in Aug. 1992.
Includes index.
ISBN 978-94-010-4725-8 ISBN 978-94-011-1660-2 (eBook)
DOI 10.1007/978-94-011-1660-2
1. Personality--Congresses. 2. Nature and nurture--Congresses.
I. Hettema, Joop. II. Deary, Ian J. III. Series: NATO ASI series. Series D. Behavioural and social sciences ; no. 72.
BF698.F654 1993
155.2--dc20 93-9801

ISBN 978-94-010-4725-8

Printed on acid-free paper

TABLE OF CONTENTS

PREFACE

A burning question that has occupied man throughout history is "where do I come from?" The roots of human individuality have always fascinated philosophers and scientists as well as the man in the street. An area especially occupying itself with this question is personality psychology. The explanations forwarded by personality psychologists put special emphasis on biological and social determinants. The relative contributions of biological and social determinants to personality as well as the relationships between the determinants are central issues in personality psychology. However, during the history of personality psychology, the attention paid to biological and social determinants of individuality has waxed and waned in opposition to each other. When emphasis on social factors was popular, interest in biological factors was minimal, and vice versa. Currently, there appears to be a 'coming together', the beginning of a dialogue on how to conceptualize the biological and social foundations, and how to best integrate the existing biological and social approaches to personality. Accordingly, this seems to be the right moment for bringing together those working from a biological perspective with those working from a social point of view, with the special aim to exchange views and develop ideas on how to proceed.

The initiative for an expert workshop to study the biological and social approaches to individuality was taken by the European Association of Personality Psychology (EAPP). The organization of a meeting of this scope requires dedicated teamwork and special funding. An international committee, consisting of Joop Hettema (The Netherlands, Director), Ian J. Deary (Scotland), Jan Strelau (Poland), and Gian Vittorio Caprara (Italy), took on the preparation of the workshop. Funding was acquired from NATO, who admitted the meeting as an Advanced Research Workshop in its International Scientific Exchange Programme. The workshop was held in Oisterwijk (The Netherlands) in august 1992.

Speakers for the ARW were invited with the special aim to obtain presentations from those working from a biological perspective as well as those working from a social point of view. Other speakers included those explicitly emphasizing both perspectives. The international nature of the workshop was stressed by inviting speakers from the USA and Canada as well as from several European countries. To obtain a clear view of the several positions advanced, a major intention of the workshop was to create an atmosphere of discussion. To that effect, for each paper a separate discussant was invited to prepare a commentary.

The character of several contributions -from speakers as well as discussants- was polemic. Discussions were vivid and many classical issues in personality were touched upon. For instance, while some of the discussants had a profound commitment with the dominant current trait approach, a sizable portion of the discussions was devoted to the Big Five against the background of the main issue.

Obviously, the workshop did not solve all the problems and controversies in personality psychology. However, on the central topic of the workshop -biological and social approaches to personality- a great deal of consensus became manifest. This permitted us to draw a rough outline of a new paradigm, integrating biological and social approaches to personality. The first chapter of this volume gives an overview. The book is organized in a way to facilitate reading. The order of presentation of the contributions roughly reflects a sequence from biological to social approaches. Each chapter is followed immediately by its commentary.

As stated earlier, a workshop like this requires special effort and financing. As editors we feel obliged to the NATO Science Committee for supplying the funds necessary for a successful workshop. We also want to express our gratitude to NWO, the Netherlands' Organization for Scientific Research, for supplying additional funds. And, finally, we are due thanks to Sandra Schruijer for organizational assistance, and to Francine Van Remunt and Rinus Verkooijen for carefully preparing the manuscripts of the workshop.

Joop Hettema
Tilburg University
The Netherlands

Ian J. Deary
University of Edinburgh
United Kingdom

CONTRIBUTORS

Alois Angleitner
Department of Psychology
University of Bielefeld
P.O. Box 8640
4800 Bielefeld 1
Germany

Thomas J. Bouchard, Jr.
Department of Psychology
University of Minnesota
Elliot Hall
Minneapolis, MN 55455
USA

David M. Buss,
Department of Psychology
University of Michigan
Ann Arbor, MI 48109-1346
USA

Gian Vittorio Caprara
Dipartimento di Psicologia
Università di Roma "La Sapienza"
Via Degli Apuli 8
00185 Roma
Italia

Ian J. Deary
Department of Psychology
University of Edinburgh
7 George Square
Edinburgh EH8 9JZ
United Kingdom

Boele de Raad
Department op Psychology
University of Groningen
Grote Kruisstraat 2/1
9712 TS Groningen
The Netherlands

Norman S. Endler
Department of Psychology
York University
4700 Keele Street
North York, Downsview
Ontario M3J 1P3
Canada

Hans J. Eysenck
Institute of Psychiatry
University of London
De Crespigny Park
Denmark Hill
London SE5 8AF
United Kingdom

Joop Hettema
Department of Pychology
Tilburg University
P.O. Box 90153
5000 LE Tilburg
The Netherlands

Willem K.B. Hofstee
University of Groningen
Department of Psychology
Grote Kruisstraat 2/1
9712 TS Groningen
The Netherlands

Douglas T. Kenrick
Department of Psychology
Arizona State University
Tempe, AZ 85287
U.S.A.

Randy J. Larsen
Department of Psychology
580 Union Drive
University of Michigan
Ann Arbor, MI 48109-1346
U.S.A.

John C. Loehlin
Department of Psychology
University of Texas
Austin, TX 78712
U.S.A.

Ivan Mervielde
Department of Psychology
University of Ghent
H. Dunantlaan 2
B-9000 Ghent
Belgium

Robert M. Stelmack
Department of General
Experimental Psychology
University of Ottawa
Monpetit Hall
Ottawa, Ontario KIN 6N5
Canada

Jan Strelau
Department of Psychology
University of Warsaw
Stawki 5/7
Warsaw
Poland

Guus L. Van Heck
Department of Psychology
Tilburg University
P.O. Box 90153
5000 LE Tilburg
The Netherlands

Marvin Zuckerman
Department of Psychology
University of Delaware
Newark, DE 19716
U.S.A.

BIOLOGICAL AND SOCIAL APPROACHES TO INDIVIDUALITY: TOWARDS A COMMON PARADIGM

Joop Hettema
Department of Psychology
Tilburg University
P.O. Box 90153
5000 LE Tilburg
The Netherlands

Ian J. Deary
Department of Psychology
University of Edinburgh
7 Geore Square
Edinburgh EH8 9JZ
United Kingdom

Abstract

This paper introduces and partly summarizes the different contributions of this volume. After a brief introduction of the biological and social approaches to personality, we argue that an integrative framework is to be preferred over competing approaches. A common framework is outlined for the joint study of biological and social determinants of individuality. The basis for this framework is sought in the (more distal or more proximal) relationships between several determinants of individuality on the one hand, and personality as appearing in daily life on the other. Different levels of explanation are forwarded, each delivering a specific contribution to personality as a whole. To study personality within this framework, specific conceptual units and specific measures are required at different levels of inquiry. Also, the definition of personality traits needs revision to include distal elements in addition to the commonly used proximal elements. Finally, some attention is paid to major research strategies within this framework, as well as some of the the problems to be faced.

Introduction

Personality psychology is polyparadigmatic by tradition. Until recently, when new textbooks came out, the discussion was almost invariably organized around presentations of psychoanalysis, learning, cognition and traits (this state of affairs has been described by Eysenck [this volume] as a 'Dutch auction'). Special mention is usually made of options such as neoanalysis, the self, humanistic approaches, social learning, needs and motives, self-regulation, constructionist and interactional approaches. Why is there this divergence of approaches, schools, models, theories, and paradigms?

First of all, as an object of study, personality is extremely complex; it offers no obvious single approach in terms of reductionistic explanation. If we want to explain why an individual acts as he does, a multitude of hypotheses readily present themselves. The choice of the hypothesis studied is usually not dictated by intrinsic considerations based on a clear conception of how the person really is, even if such a conception were readily attainable. Instead, ulterior motives often play a decisive role, such as the wish to classify persons into neat types or along a small number of dimensions, to place them in optimal or appropriate environments,

J. Hettema and I. J. Deary (eds.), Foundations of Personality, 1–14.

or to treat them and modify their behavior. Accordingly, several specialized paradigms have been developed, each with its own concepts, elements, processes, methodology and measures.

As is common practice in science in general, each paradigm tends to accentuate its own position, to posit itself as *the* approach to personality, while censuring or ignoring competing paradigms. For instance, psychoanalysts concentrated on their belief that early childhood experiences, pecially those of a sexual or violent nature, were powerful determinants of later personality. Learning theorists hypothesised that situational factors and reinforcement contingencies were major conditions governing individual behavior. Cognitive personality psychologists insisted that personal constructions of reality were important determinants of individual behavior. Trait theorists emphasised that individual behavior is governed by internal dispositions with the capacity to render many stimuli functionally equivalent and to produce consistent forms of behavior. Obviously lacking in personality psychology is a synthesis which might bring together the validated facts of different schools and paradigms, and that will be able to accept and integrate findings from different paradigms that are mutually consistent and reciprocally reinforcing. While many exponents of one school will concede the existence of others, there is rarely an attempt to marry the results of more than one school. Equally lacking from most broad contemporary accounts of personality is a proper sense of discrimination between those approaches which adopt an empirical scientific attitude and those which are essentially unfalsifiable modern myths (Eysenck, this volume).

Recent history reveals some first steps toward more than token integration. Currently, human personality is studied from different, yet converging perspectives. Two major developments have become visible during recent years. First of all, the social foundations of personality have been summarized in the framework of social behavior theory. Processes like social learning and social cognition, modeling, and vicarious reinforcement have been scrutinized as have subjective competence and self-efficacy. Closely associated with the social foundations of thought and action (Bandura, 1986) are several programs of training and treatment, such as modeling-based therapy and cognitive behavior modification.

Secondly, and more recently, personality psychologists, typically using one of the trait models of personality, have (re)turned to the search for the biological bases of personality (cf. Zuckerman 1991). The study of temperament and of physiological processes presumed to be relevant to individual differences in personality dimensions, has witnessed an upsurge. Drawing on modern evolutionary concepts and advances in genetic analysis, hereditary aspects of personality are currently studied using notions derived from ethology, sociobiology and behavior genetics. A major breakthrough has been obtained in recent research with twin studies employing self-report measures. The results of such studies (e.g. Bouchard, this volume) have indicated that several personality and attitudinal scale scores have considerable amounts of genetic influence.

The perspective looming here is one of dichotomizing rather than polypartitioning the field of personality. Biological and social approaches reflect different paradigms, each meant to explain individuality. From a biological point of view, man is a representative of the human species, predisposed to behave in certain ways by evolutionary laws and hereditary mechanisms. From a social point of view, man is a member of the community, subject to processes of learning and adjustment, specifying his behavior in daily life situations. Biological and social paradigms are both valid avenues to personality. However, in practice, they tend to be treated as competing paradigms, each claiming to provide *the* explanation of human individuality. Furthermore, both approaches tend to proceed without reference to each other's theories much of the time.

The aim of the NATO Advanced Research Workshop entitled 'Biological and Social Approaches to Individuality', from which this volume grew, was twofold. Its first objective was to scrutinize the biological and social approaches with a view to detecting commonalities and differences. Thus, it was attempted to obtain a clearer view of the boundaries between the two approaches. The second objective was to map out complementarities between the two approaches, with a view to drawing the outlines of a common framework to stimulate future research.

Two Disciplines of Personality

Biological and social approaches differ not only in their broad conceptions of the major determinants of personality, they tend also to emphasize different basic elements underlying the same type of behavior. As an example, consider the ways in which people select their environments.

Biological approaches emphasize heritable tendencies to behave in one way or another. Biologically oriented psychologists argue that the selection of environments is importantly governed by inborn tendencies; for instance, Plomin et al. (1977) proposed genotype-environment correlations to explain the extent to which children receive or create environments correlated with their genetic propensities. According to Eysenck (1981), introverts, who are normally more aroused, prefer environments with a relatively low arousal potential. Zuckerman (1979) assumes that sensation seekers tend to prefer environments with intense stimulation, whereas their counterparts protect themselves from overstimulation by avoiding those environments. In these examples, biologically prepared structures are the primary basis for selecting environments.

Socially oriented psychologists, on the other hand, assume that learning experiences are the major basis for interpreting and selecting environments. Mischel (1973) argued that learning experiences lead to individual differences in cognition, that, in turn, guide behavior. According to Mischel, people are confronted with a flood of potential stimuli in any particular environment. These stimuli must be perceived, selected, processed and interpreted before they can have any effect on behavior. In this process, cognitive person variables established through social learning play a major part. Examples of such variables are construction competencies, encoding strategies, outcome expectancies, subjective values and self-regulatory systems and plans. Actual behavior is seen as the end product of a sequence of cognitive transformations, involving some or all of the variables mentioned. Bandura's (1986) social learning approach focuses on the way that learning experiences lead to differences in perceived self-efficacy. These expectancies influence performance and persistence at tasks, the experience of stress, the selection of actions an individual undertakes, and the selection of situations allowing for such actions.

This example reveals profound differences in the way the biological and social approaches conceptualize a specific aspect of behavior, the choice of environments. Notwithstanding this comment, there is no necessary objection to both approaches being correct; social processes might act upon a range of possibilities narrowed by heritable propensities. The example indicates also the direction in which one must look for more basic paradigmatic differences between the two. Paradigmatically, the biological and social approaches exhibit connections with the socalled organismic and field positions in personality psychology.

Personality psychologists have always insisted that their subject should be studied from the vantage of the entire person functioning in his natural habitat (Hall and Lindzey, 1970). The wholeness of the individual and his environment can be broken down into two rather distinct forms. The *organismic position* stresses the interrelatedness of everything the individual does, the fact that each act he performs can be understood only against the background provided by his other acts. The organismic position describes persons as distinct elements, often using trait descriptions such as dominance or submissiveness, aggression or peacefulness, extraversion or introversion. The *field position* on the other hand is primarily concerned with the inextricable unity between a given behavioral act and the environmental context within which it occurs. The field position describes the acts a person performs, like dominant or submissive acts, aggressive or peaceful acts, extraverted or introverted acts. Those acts are primarily viewed in the context of the environment in which they occur.

In an even wider perspective, the two positions are connected with the two disciplines of scientific psychology: correlational and experimental. As Cronbach (1957) pointed out, the two disciplines exhibit differences in conceptualization, method, and scope, as well as in the major questions asked. The biological and social approaches to personality reflect the same types of differences. The biological approach tends to attribute behavior to the organism, to derive its process conceptions from physiology, to favor traits as major units of personality, and to use correlational methods. The social approach primarily attributes behavior to the environment, derives its process conceptions from (social) learning, favors cognitive units, and uses experimental methods to collect data. Obviously, biological and social approaches differ in major respects. Yet, they refer to the same object: personality.

How are they related?

Currently, several authors are notable for their attempts to reconcile the biological and social positions. For instance, Buss (this volume) starts from an evolutionary framework to explain species typical forms of social behavior and their individual differences. In his view, important adaptive problems, such as the retention of a mate, produce species-typical psychological mechanisms. The expression of these mechanisms may be strategically patterened by stable between-subject differences in the actual experience of the relevant adaptive problems in the social world. Kenrick (this volume) explicitly advocates a biosocial approach, based on an integration of the central features of the cognitive and evolutionary paradigms. Hettema (this volume) proposes two different behavioral systems, one primarily shaped through evolution, the other through ontogenetic experience. These proposed solutions all rest on the same basis: a definite conception of the foundations of human behavior. They also share the view that human individuality is multiply determined; both biological and social factors are major determinants of personality.

Levels of Explanation

Several attempts have been made to create order in the ways that biological and social determinants operate. All these attempts propose a conditional order in the operating of different determinants. According to Kenrick (1987; Kenrick et al., 1985; Hettema and Kenrick, 1989, 1992), environmental events (the focus of traditional learning models), cognitive representations (the focus of more recent social learning models), and physiological predispositions (the mainstay of the biological model), are all necessary, but not sufficient as explanations of individual differences. In Kenrick's view, biological approaches tend to pay most attention to distal explanations like evolution and behavior genetics. Learning concepts address behavior variations in more proximal terms, emphasizing conditioning and modeling. Cognitive approaches deal with the most proximal kind of explanations. However, more proximal determinants like learning are not independent from distal determinants like genetics. Genes influence the way we respond to the environment, but they do not construct an organism that is insensitive to outside pressures. Likewise, more proximal determinants are not independent from distal determinants. Modern learning theorists have found that learning is not independent of biological constraints, like biological preparedness and critical periods.

The functioning of distal and proximal determinants of individuality reveals several connections. Those connections are assumed here to be conditional rather than causal in a more absolute sense. More distal levels are necessary but not sufficient to explain individuality at a more proximal level. Zuckerman (1991; this volume) proposed the conditional order: Genetics - Neurology - Biochemistry - Physiology -Conditioning - Social behavior - Traits. In Zuckerman's view, when more distal determinants move, more proximal determinants must perforce also move. Thus, genes (the most distal determinants) have a definite priority, because they are there at the beginning. On the other hand, moves in more proximal determinants may disturb the functioning of more distal determinants.

In conclusion then, while the determinants proposed exhibit differences, many authors agree on the principle of distal-proximal determination, including the priority of the more distal biological over the more proximal social determinants involved. In our view, a (parsimonious) summary conditional ordering of the major determinants includes:

Genetics (1)

Evolution (2)

Physiology (3)

Learning (4)

Social behavior (5)

Personality traits (6)

This ordering reflects some major transitions from species characteristics to individual characteristics. The more basic levels are necessary but insufficient conditions for higher levels to materialize. Each level adds a major specification towards phenotypic individuality. However, the ordering should be seen as an aid to thinking about personality rather than a map of the reductionistic process. It is not an order, therefore, in the way that physics may be ordered below chemistry. For instance, while genes and physiology may readily be envisaged as entities, evolution and learning are more easily conceived of as mechanisms though, as we shall see, both of these mechanisms have putative psychological 'products'. Social behaviour is the output of the organism and lacks the clear descriptive units that, say, apply to genes and to physiological structures.

Traits have an uncertain position in the conditional order. They may be viewed as the perceived average of an individual's social behaviour over time, or they may be seen as more basic than social behaviour, acting in a way that constrains the form and content of social acts. Again, this reflects a difference between some biologically- and some socially-oriented personality psychologists. While most can agree that traits are stable descriptors of an individual's tendencies to behave in certain ways, their explanations of this observation differ. Those with a biological bent might see traits as more basic than social behaviour, in that they act as a constraint on social action and are relatively close to physiology. However, it is also possible, as in the above conditional ordering, to characterise traits as more superficial than social behaviour, in that they are the observed average of behavioural consistencies, and might even be constructed in the mind of the observer, with no necessary link to biological constraints and influences (Hampson, 1988).

The ordering departs from a normal reductionistic ordering in a further sense. In a typical reductionistic scheme one would expect the closest empirical links to exist between those phenomena that are adjacent to each other in the order. However, one of the better-established links in personality psychology is that between the highest and lowest levels. The evidence available supports a link between genes and traits (though, of course, there are other well-established links such as those between measured traits and social behaviour). This traits-genes link is stronger than any link to date between physiological mechanisms and traits (Brody, 1988; Zuckerman, this volume), even though physiology is nearer to traits in the ordering. However, this is likely to reflect the problems of technique more than it reflects an underlying problem with the hypothesis that traits have a physiological basis. Research on the dynamics of brain functioning is still in its infancy and measures of brain metabolism and of the functioning of individual neurotransmitter systems is still relatively crude.

The major implication of this provisional ordering of levels is that personality (i-ndividuality) is multiply determined. Organismic factors are necessary conditions underlying individuality. However, to be effective, each level requires specific environmental conditions to be realized in addition to organismic conditions. Different levels emphasize different environmental aspects. If the proper environmental conditions for a specific level are satisfied, individuality will develop according to the pre-existing organismic conditions and dispositions. If not, the pre-existing organismic conditions will be modified. Depending on the level of causality, they will either vanish through selection or be modified through environmental conditions.

For instance, at a distal level, if a species is exposed to radical changes in its living conditions, natural selection might cause changes to occur in the gene pool. Thus, evolution at level 2 proceeds through selection at level 1, genetics (Dobzhansky et al., 1977). Or, at a more proximal level, consistent exposure to painful and unavoidable shocks may extinguish avoidance behavior that was present previously, thus causing a condition of learned helplessness (Seligman and Maier, 1967) to be established. Explanations of learned helplessness in humans typically rely on expectations of future non-contingency (Abramson et al., 1978). Thus, cognitive expectations at level 5 may remove elements established at level 4, learning.

Thus far, we have been concerned mainly with the ability of biological and social concepts to explain features of human personality. However, is there an autonomous set of psychological concepts that can explain features of personality, or are biological and social determinants sufficient? In other words, do psychologists merely describe and order the phenomena of personality and then look to biological and/or social concepts to explain them?

Perhaps it is the hallmark of the psychometricians, the perhaps-unwitting disciples of the London School, that they are content to draw the maps of psychological phenomena such as personality and intelligence and look to biology for explanations (Sternberg, 1990; Howe, 1988; Deary and Matthews, in press). On the other hand, more socially oriented psychologists such as the social constructionists look to social processes involving interactions with other people to explain phenomena like traits (Hampson, 1988; Emler, 1990). The less scientific approaches to personality, such as the humanistic and the psychodynamic, have tended to incorporate psychological explanatory concepts in the field of personality. Freud's Oedipus complex, Adler's inferiority complex and Maslow's needs for loving and self-actualisation were attempts to explain human personality at the level of individual psychology, though none of these was particularly successful. Perhaps there is a recoverable agreement here between biologically- and socially-oriented personality psychologists, i.e. that the phenomena of personality require explanations at levels outwith that of the purely psychological.

But the psychological level is important. As stated earlier, all the above levels are conditional rather than deterministic. At each separate level, conditions are confronted with other conditions. The levels at which the different conditions converge is definitely psychological. For instance, if an individual is biologically impaired, e.g. through missing a limb or blindness, the challenge has to be dealt with at the psychological level. This can be done in several ways, ranging from the redistribution of energetical resources (level 3), the learning of compensatory patterns of behavior (level 4), to the developing of specific cognitions and expectancies (level 5). Thus, rather than replacing the psychological level of description, biological and social conditions are considered as inputs to be dealt with at that level.

Perspectives on Individuality

Clearly, individuality can be studied at different levels of explanation. Each level provides a separate perspective on individuality. From a *genetic perspective*, individuality is studied as a product of genotypic propensities. Using twin, family and adoption studies as their major tools, behavior genetics has been able to demonstrate a considerable influence of genotypes upon traits as measured by conventional personality measures (cf. Bouchard, this volume). In addition, the genetic perspective can throw more light on the effects of environments, either alone or in combination with genes.

From an *evolutionary perspective*, individuality is viewed primarily as a product of natural selection. Evolutionary personality psychology (Buss, 1991) emphasizes the survival value and reproductive potential of several types of social behavior. The ultimate aim of the evolutionary approach is to explain human nature as it has been shaped through millennia of evolution. Sexual reproduction is a major mechanism for evolution to be accomplished. Accordingly, within the evolutionary perspective, special attention is paid to generating and testing hypotheses on gender differences and mating strategies. Examples are the analyses of Buss and Kenrick (this volume).

The *physiological perspective* is specially concerned with individuality as reflected in biological processes. A major thesis of this perspective is that biological processes are necessary conditions for learning, the shaping of social behavior and, finally, traits. Physiological reactivity has an important effect upon the occurrence and course of learning processes. An example is a study by Eysenck and Rachman (1965), showing that emotional adjustment is related to differences in emotional arousal which cause neurotic individuals to condition anxiety to new situations more easily. It is likely that the genetic, physiological and trait perspectives will be brought closer by the advent of molecular genetic techniques and their application to polygenically detemined phenomena such as individual differences in personality traits (Deary, this volume).

The *learning perspective* emphasizes the acquisition of behavioral patterns on the basis of conditioning and the observation of others. However, it should be kept in mind that learning not only depends upon specific experiences, but also includes the genetic and physiological preparedness for learning to occur. Learning processes are major forces underlying the establishment of effective types of behavior in specific circumstances. Accordingly, learning abilities may be assumed to be major conditions for the shaping of personality (Wallace, 1966)

The *social behavior perspective* is especially concerned with the performance of learned patterns. This perspective emphasizes the ways in which social feedback and reinforcement affect the actual occurrence of specific types of behavior. As a consequence of social reinforcement, in specific situations the probability of occurrence of some behavioral patterns is enhanced, whereas the occurrence of other patterns is inhibited. Through feedback and reinforcement, specific patterns of behavior are prepared to provide the building blocks of phenotypic personality. Several mechanisms may be involved here. Amplification, as studied by Caprara (this volume) is an example.

Not only do social environments affect the person, the reverse is true as well. The reciprocal relationships or transactions between persons and situations are at the core of the interactional approach to personality. Transactions between persons and situations include mutual changes in persons and environments over time. Classical interactionism has specially emphasized social-cognitive aspects of the processes involved in person-situation transactions. Modern interactionism includes physiological systems and processes as well (Endler, this volume; Magnusson, 1990).

The *trait perspective,* finally, is concerned with stable differences among individuals that become visible in daily life. Most current trait approaches pay special attention to individual differences directly concerned with social relations. Their major aim is to explain and predict the behavior of individuals in socially important settings. Thus, they rely heavily on level 5, social behavior. However, as we have seen, individuals differ according to other levels as well, including their genetic make-up, species typical mechanisms, physiological resources and reactivity, learning abilities, and behaviors performed on social environments. Only part of these differences are summarized in trait terms as customarily used to describe individuals. However, a comprehensive trait perspective would include differences emanating at the other levels as well. The work of Eysenck (this volume) and Zuckerman (this volume) provides examples of a more comprehensive approach from the perspective of traits.

What units shall we employ?

Several decades ago, Allport (1958) raised the question which units we are to employ in the study of personality. Since that time, several candidates have been considered as *the* units of personality. Those units were often seen as competitors, mutually exclusive and with more or less explanatory power. We advocate a different approach.

A major question to be answered first is 'what is an explanation of behavior?'. In a recent paper, Simon (1992) has called attention to this basic question. In Simon's view, for psychology as for other sciences, it is essential to explain complex phenomena at several levels. Different levels should be conceived as complementary rather than competing. Only in this way will a comprehensive psychology become possible. Different levels of resolution use different constructs or units and emphasize different processes along which they operate. Units with optimal explanatory power at one level of inquiry may be less relevant at another level.

Explanation at different levels does not deny the possibility of reduction. As Simon (1992) proposes, higher level theories may use aggregates of the constructs at lower levels to provide parsimonious explanations of phenomena without explicit reference to the microconstructs. In terms of reductionism chemistry has been conspicuously successful in reducing: the variance in chemical phenomena to the interactions of certain molecular structures; the variance in molecules to their posession of different types of atom; and the variance in atoms to their possession of certain numbers of protons, neutrons and electrons. Do present trait models, for example, offer a periodic table or something more akin to a time when chemistry had four elements: air, earth, fire and water? This matters a great deal, because an accurate slicing-up of reality at a given level, i.e. the rendering of phenomema into valid basic units, is a prerequisite to successful reductionism.

However, it would be naive to assume that the units studied at one level exhibit a strict isomorphism with elements studied at another level (cf. Zuckerman, this volume). It is now clear that there was no straightforward explanation of neuroticism and extraversion waiting in the brain in terms of the autonomic nervous system and the reticular activating syst ems, respectively (Eysenck, 1967; Brody, 1988; Zuckerman, 1991). Further, there is no neat isomorphism between the various neurotransmitter systems and personality traits, though some slight associations exist (Zuckerman, this volume).

The present approach is focused on 'units of individualization'. It starts from the assumption that different levels are involved in individualization, and, consequently, may contribute to the explanation of personality. Different levels employ different explanatory constructs or units. For instance, at the genetic level, genes are the major units and heritability is the major process studied. With regard to personality, genes are particularly useful in explaining individual differences or traits. A major finding with this approach is that most personality traits exhibit comparable amounts of genetic involvement, although there are differences as well (cf. Bouchard, this volume). From a genetic point of view, the variance not explained by genetic mechanisms, can be conceived in different ways. Not only straightforward environmental explanations, but also several types of interaction between genotypes and environments deserve attention (Plomin et al., 1977; Plomin, 1986).

The evolutionary level of explanation focuses on species typical mechanisms shaped through evolution. Examples are competition for limited resources and for mates, mate preferences, preferential altruism directed toward kin, sanctions for crimes against the group, and the like (cf. Buss, this volume, for a more complete overview). An important issue with respect to the evolutionary level of explanation is the nature of the units to be employed. This issue refers to the question what elements are shaped by evolution. Are those elements content-laden and domain specific as claimed by sociobiologists (cf. Buss, this volume), or formal and stylistic as conceived by those studying temperament (Strelau, this volume)? In these views evolution has much to say about personality and its identifiable traits. Another possibility is raised by Hofstee (this volume), namely that personality traits might be evolutionarily neutral, having as much relevance to evolutionary selection as the colour of the front door has to the structure of a house.

At the physiological level, the functioning of different organ systems is studied. In psychology, the neurophysiological system deserves special attention. Major disciplines studying the neurophysiological level are neurology, biochemistry, and physiology. The units emphasized at this level include neuronal networks, hormones and energetical systems like arousal, effort, and activation. They provide the energetical basis for the cognitive system, the emotional system and learning processes (cf. Hockey et al., 1986). Some theorists conceive of energetics, including reactivity, speed and mobility as the major basis for temperament (cf. Strelau, this volume). Others have proposed specific tactical structures as an outflow of physiological mechanisms (Hettema, this volume). Still others have proposed connections between physiology and personality traits like neuroticism, introversion, and sensation seeking. However, the ways along which those relationships are conceived exhibit some major differences. Eysenck (this volume), for instance, assumes a rather direct connection, whereas Zuckerman (this volume) emphasizes learning and social behavior as major intermediaries.

At the next level, learning, simple as well as more complex behavioral structures are acquired that have the capacity to materialize specific goals. Learning produces habits and skills, that may be used to deal with the same or different situations in the future. Observational learning and modeling are powerful mechanisms to acquire complex patterns of social behavior and roles. Major units emphasized at this level include efficacy expectancies (Bandura, 1977), competencies for the construction of cognitions and of behavior, encoding strategies and personal constructs (Mischel, 1973).

The actual occurrence of specific behaviors in daily life situations is governed by cognitive structures, established on the basis of feedback experiences. At the level of social behavior, the major units emphasized are cognitive variables exemplified by stimulus and stimulus outcome expectations, subjective evaluations, systems and plans for selfregulation (Mischel, 1973). Several cognitive elements may be summarized in more comprehensive structures governing behavior, like behavior strategies (Hettema, this volume).

The Definition of Traits

Traits are the basic units of personality as it appears in daily life. However, traits may be conceived of in different ways. First of all, they may be conceptualised as the major elements used by naive observers to define other people (cf. Van Heck, this volume). In that case, their status is merely descriptive and they reflect a surface structure that may represent a relatively arbitrary social construction (Hampson, 1988). However, they may also be conceived of as the ultimate products of all the levels discussed earlier. In that case the deep structure of personality is emphasized, including genetic endowment, phylogeny, physiological structure, learning history and social experience. Needless to say, the present paper favors the latter approach. Only by assuming a deep structure can personality traits obtain the internal locus and causal primacy that characterises the belief that personality traits are valid descriptions, have predictive power and have biological bases (cf. Deary and Matthews, in press.).

Based on genetic and evolutionary propensities, traits obtain their final shape during ontogenetic development, when natural tendencies are confronted with learning conditions and social feedback. They include fluid as well as crystallized elements (Hettema, this volume), cognitive as well as affective elements, brought into harmony by systematic person-environment confrontations (Zuckerman, this volume). In our conception, personality includes a deep structure in addition to the surface structure exhibited with scales customarily used to assess personality. On the other hand, we realize that much research remains to be done before the deep structure of personality can be properly conceptualized and measured. Explanatory trait constructs cannot be defined on the basis of a mere taxonomy. To obtain a causal in addition to a descriptive status, the nomological network surrounding a trait should include relationships of verbal descriptions with genotypical peculiarities, evolutionary aspects, physiological bases and social-emotional as well as social-cognitive aspects (cf. Eysenck, Zuckerman, this volume). An important issue, deserving attention in the future, is the relationship between the deep structure of personality and the surface structure as currently approached with standard personality questionnaires.

Approaches to Personality

Thus far, biological and social approaches to personality have developed more or less independently. Both approaches have emphasized specific problems, developed specific concepts, and used specific methods. As stated earlier, they may even obtain the status of different disciplines. However, a comprehensive approach to personality should incorporate both approaches in addition to other approaches relevant for personality. In other words, there is an urgent need for a bridging of the gap between the two disciplines. Rather than the development of theoretical blends which act as stopping points between the two approaches, we advocate the study of personality at *different levels of inquiry*. Specialized researchers may be interested first and foremost in one or more of the levels defined, but they should keep an open eye for other levels as well. Furthermore, attempts to conceptualize and study personality at different levels simultaneously deserve especial support.

We propose that *each level studies its own validated units*. At the levels defined earlier the obvious units are: genes, species typical mechanisms, energetical resources, social skills, cognitions, and traits. In our view, a comprehensive deep structure of personality would contain information on all these levels separately but, in addition, it would be able to knit them together as a theoretically and explanatory coherent whole. For a proper explanation of a person's behavior, we would have to have information on his genetic endowment, species typical mechanisms, energetical resources, social skills, situation specific cognitions, and surface traits.

Connections among levels can be studied in different ways. Researchers can decide to start either at the bottom (the genetic level) or the top (the trait level). Both top-down and bottom-up approaches have been used in the past, although the top-down approach has been the most prominent. Several researchers have first developed a conception of traits and devised scales to measure them, before proceeding with attempts to ensure the construct validity of their measures. A disadvantage of this approach is that the results may remain confined to a descriptive trait level, without establishing connections with the other levels. Consequently, the top-down approach runs the risk of producing merely a surface structure, without any connections with deeper levels. More recently, researchers have attempted to start from the bottom and work their way up. Examples are the current genetic and evolutionary approaches. The problem with bottom-up approaches is that they may have difficulties connecting their results with the top level. How can they connect their findings with classical units employed in personality psychology, e.g. traits. Investigators favoring a bottom-up approach face the challenge of eventually translating their units into trait terms.

Connections among levels raise special problems, because they require the investigator to connect units of different types. To solve this problem, several investigators working at different levels have designed a common language. They have defined new middle level units (Buss and Cantor, 1989) with maximal utility at the levels scrutinized. Those units also enable them to proceed without immediate reference to the top (trait) level. An examples of new middle level units are evolutionary strategies (Buss, this volume), connecting level 2, evolution, with level 5, social behavior. A second example are behavioral strategies and tactics (Hettema, this volume), connecting level 3, physiology, with level 5, social behavior. A third example are cognitive social learning variables (Mischel, 1973), connecting level 4, learning, with level 5, social behavior. A fourth example are personal strivings (Emmons, 1986) and conditional dispositions (Wright and Mischel, 1987). Defined especially at level 5, social behavior, those units also exhibit connections with level 6, traits. Middle level units usually have a greater utility for studying the connections among specific levels. Thus, several relationships among levels can be conceptualized with middle level units.

Meanwhile, it is interesting to note that the more classical top-down approach is currently complemented with several bottom-up approaches, starting either from genetics, evolution, psychophysiology, learning or social behavior. The perspective offered by this state of affairs is one of two construction teams starting at both ends of a tunnel with the aim of meeting half-way. If either or both is successful, personality psychology will really have come of age.

CONCLUDING REMARKS

Human personality is complex and can be studied from several points of view. Biological and social approaches are major perspectives from which personality is currently studied. The results obtained with each are important in and of themselves. However, a more complete understanding can only be obtained if the connections between biological and social approaches are explained in detail. The present volume is dedicated to that purpose.

A first conclusion to be drawn is, that there is no such thing as *the* biological or *the* social approach to personality. Instead, there are several approaches within each. This result might lead to the conclusion that complexity is increased rather than diminished. However, complexity can be reduced considerably by an ordering into more distal and more proximal determinants of personality. An ordering as proposed here allows for the study of personality from one single point of view: the effects of more distal upon more proximal determinants.

What are the chances of improving our understanding of personality along the ways outlined here? First of all, recent work in behavior genetics has established connections between determinants postulated at the extreme ends of the distal-proximal continuum. Thus, the "range of indeterminacy" is narrowed down considerably. However, to obtain a more complete understanding of personality, the remaining between-level connections deserve our full attention. The research strategy following from this statement may be seen as a challenge to single-level studies, that are customary in current personality psychology.

While studying between-level connections, a simple isomorphism assumption cannot provide an easy way out. Instead, we have to pay full attention to the conceptual and measurement contingencies characterizing each of the levels. Only if we take account of those peculiarities, will it become possible to explain rather than merely describe an individual's personality.

References

Abramson, L.Y., Seligman, M.E.P., and Teasdale, J.D. (1978). Learned helplessness in humans: Critique and reformulation. *Journal of Abnormal Psychology, 87*, 49-74.

Allport, G.W. (1958). What units shall we employ? In: G. Lindzey (Ed) *The assessment of motives*. New York: Rinehart, pp. 239-260.

Bandura, A. (1977). Self-efficacy: Toward a unifying theory of behavioral change. *Psychological Review, 84*, 191-215.

Bandura, A. (1986). *Social foundations of thought and action: A social cognitive theory*. Englewood Cliffs, NJ: Prentice-Hall.

Brody, N. (1988) *Personality: In Search of Individuality*. New York: Academic Press.

Buss, D.M. (1991). Evolutionary personality psychology. *Annual Review of Psychology, 42*, 459-491.

Buss, D.M. and Cantor, N. (Eds)(1989). *Personality psychology. Recent trends and emerging directions*. New York: Springer Verlag.

Cronbach, L.J. (1957). The two disciplines of scientific psychology. *American Psychologist, 12*, 671-684.

Deary, I.J. and Matthews, G. (in press) Personality traits are alive and well. *The Psychologist*.

Dobzhansky, T., Ayala, F.J., Stebbins, G.L. and Valentine, J.W. (1977). *Evolution*. San Francisco: Freeman.

Emler, N. (1990) A social psychology of reputation. In W. Stroebe and M. Hewstone (Eds), *European Review of Social Psychology* (Vol. 1, 171-194). Chichester: Wiley.

Emmons, R.A. (1986). Personal strivings: An approach to personality and subjective well-being. *Journal of Personality and Social Psychology, 51*, 1058-1068.

Eysenck, H.J. (Ed.)(1981). *A model for personality*. Berlin: Springer Verlag.

Eysenck, H.J. and Rachman, S. (1965). *Causes and cures of neurosis*. London: Rutledge & Kegan Paul.

Hall, C.S. and Lindzey, G. (1970). *Theories of personality*. (Second edition). New York: Wiley.

Hampson, S. (1988) *The Construction of Personality: An Introduction (Second Edition)*. London: Routledge.

Hettema, P.J. and Kenrick, D.T. (1989). Biosocial interaction and individual adaptation. In: P.J. Hettema (Ed). *Personality and environment. Assessment of human adaptation*. Chichester: Wiley, pp. 3-29.

Hettema, P.J. and Kenrick, D.T. (1992) Models of person-situation interactions. In: G.V. Caprara and G.L. Van Heck (Eds.) *Modern personality psychology. Critical reviews and new directions*. New York: Harvester-Wheatsheaf, pp. 393-417.

Hockey, G.R., Gaillard, A.W., and Coles, M.G. (Eds)(1986) *Energetics and human information processing*. Dordrecht: Martinus Nijhof.

Howe, M.J.A (1988) Intelligence as an explanation. *British Journal of Psychology, 79*, 349-360.

Kenrick, D.T. (1987). Gender, genes, and the social environment: A biosocial interactonist perspective. In: P.C. Shaver and C. Hendrick (Eds). *Review of personality and social psychology*. Newbury Park CA: Sage.

Kenrick, D.T., Montello, D.R., and MacFarlane, S. (1985). Personality: social learning, social cognition, or sociobiology? In: R. Hogan and W.H. Jones (Eds.) *Perspectives in personality*. Volume I. Greenwich, CO: JAI Press, pp.201-234.

Magnusson, D. (1990). Personality development from an interactional perspective. In: L. Pervin (Ed.) *Handbook of personality theory and research.* New York: Guilford, pp. 194-222.
Mischel, W. (1973) Toward a cognitive social learning reconceptualization of personality. *Psychological Review, 80*, 252-283.
Plomin, R. (1986) Behavior genetic methods. *Journal of Personality, 54*, 226-261.
Plomin, R., DeFries, J.C., and Loehlin, J.C. (1977). Genotype-environment interaction and correlation in the analysis of human behavior. *Psychological Bulletin, 84*, 309-322.
Seligman, M.E.P. and Maier, S.F. (1967). Failure to escape traumatic shock. *Journal of Experimental Psychology, 74*, 1-9.
Simon, H.A. (1992). What is an "Explanation" of behavior? *Psychological Science, 3*, 150-161.
Sternberg, R.J. (1990) *Metaphors of Mind: Conceptions of the Nature of Intelligence.* Cambridge: Cambridge University Press.
Wallace, J. (1966) An abilities conception of personality: Some implications for personality measurement. *American Psychologist, 21*, 132-138.
Wright, J.C. and Mischel, W. (1987) A conditional approach to dispositional constructs: The local predictability of social behavior. *Journal of Personality and Social Psychology, 35*, 1159-1177.
Zuckerman, M. (1979) *Sensation seeking: Beyond the optimal level of arousal.* Hillsdale NJ: Erlbaum.
Zuckerman, M. (1991) *Psychobiology of Personality.* Cambridge: Cambridge University Press.

GENETIC AND ENVIRONMENTAL INFLUENCES ON ADULT PERSONALITY: EVALUATING THE EVIDENCE

Thomas J. Bouchard, Jr.
Department of Psychology
University of Minnesota
Elliot Hall,
Minneapolis, MN 55455, USA

Abstract

This paper is made up of two parts. In the first part we outline a frame of reference (the Big Five) for organizing the data from Behavior Genetic studies of adult personality and then briefly review methods (model fitting) used by behavior geneticists to analyze genetic and environmental influences on personality. We then proceed to summarize the major model fitting findings from twin, adoption, extended family, and twin reared apart studies. Part two critiques non-behavioral genetic approaches to the study of environmental influences on personality. With regard to part one we conclude that a) genetic factors account for about 40-50% of the variance in the Big Five personality traits as measured by standard instruments, and b) that non-additive variance contributes significantly to most of these traits (in the range .02 to .17). The hypothesis of a more modest heritability based on only additive variance (28-46%) and special monozygotic and sibling environments cannot, however, be completely ruled out. There are inconsistencies in the findings and failures to replicate the details of studies. We argue that this is due to the use of poor quality instruments, and that unlike scientists in other domains who push the limits of their instrumentation to test theoretical ideas, behavior geneticists have failed to implement the broad array advances in our understanding of the measurement of psychological traits that have occurred over the last forty years. With regard to part two we find that a) critiques of the twin method are largely unfounded, b) the methodologies and assumptions of the various environmental (non-behavior genetic) approaches to explaining variance in adult personality traits are largely untenable, c) the results of experimental studies have often been inappropriately extrapolated to domains to which they do not apply, and d) studies of a number of "so called" environmental variables (birth order, divorce, physical attractiveness, etc.) simply do not explain the phenomena they have been postulated to explain. We conclude that the non-behavior genetic approach to the study of environmental influences on personality has generated few if any credible findings. We further argue that studies of environmental influences on adult personality that do not impose genetic controls are uninterpretable with regard to explaining individual differences in adult personality.

J. Hettema and I. J. Deary (eds.), Foundations of Personality, 15–44.

Genetic and Environmental Influences on Adult Personality: The Behavior Genetic Approach

There are so many different ways of construing personality that it is often difficult to reconcile findings and arguments because the parties are simply not discussing the same concepts. Rather than attempting to reconcile these varying approaches standard textbooks tend to either devote a chapter to each approach or commit themselves to a particular point of view. Because of their need for large sample sizes Behavior Geneticists are committed, for the most part, to self-report measures with established psychometric properties. Even though personality traits can be measured reliably by self-report, and many measures have shown considerable validity, there is disagreement regarding how many personality traits should be measured and there is no widely agreed upon theory that points to a compelling resolution. The scales on personality inventories generally reflect the theoretical biases of the test constructor. The Myers-Briggs Type Indicator (MBTI) (Myers and McCaulley, 1985), for example, reflects the three major typological categories of Jungian psychology; Extraversion-Introversion, Sensation-Intuition, Thinking-Feeling, as well as a fourth trait, Judging-Perceiving. The scales of the Eysenck Personality Questionnaire (EPQ) reflect Eysenck's attempt to combine findings derived from a large number of psychometric studies of personality, experimental findings using personality variables and various brain-behavior relationships (Eysenck, 1992; cf. also Zuckerman, 1992). The scales of the EPQ represent what are often called the Big Three (Extraversion (E), Neuroticism (N) and Psychoticism (P)). The California Psychological Inventory (CPI), with eighteen scales, attempts to assess "folk concepts", not traits (Gough, 1968, 1987). The eleven scales of the Multidimensional Personality Questionnaire (MPQ), on the other hand, represent an iterated factor approach to the domain of self-report personality items that attempts to achieve relatively independent primary scales and yet still tap the Big Three (Tellegen and Waller, in press). The Neuroticism, Extraversion, Openness, Personality Inventory-Revised (NEO-PI-R) was designed to capture the so called Big Five model of personality (N, E, O and Agreeableness (A) and Conscientiousness (C)) (Costa and McCrae, 1992). The Big Five model is based on psychometric studies largely originating with the work of Tupes and Christal (1961).

THE BIG FIVE

The Big Five appear to be emerging as a dominant paradigm in personality research (cf. Goldberg, 1992: Goldberg, in press; John, 1990; Digman, 1990 for reviews), and we will discuss the behavior genetics findings within this frame of reference because a great deal of data can be organized under this rubric. This includes the meta-analyses of the twin literature by Nichols (1978), Eaves et al. (1989), and the recent reanalysis of a great deal of twin and family data by Loehlin (1992)[1]. In addition, our own data on twins reared apart can be easily incorporated into this scheme. The use of this frame of reference should not be taken to mean that we subscribe to the view that it is the best scientific scheme for characterizing personality (cf. Hough, 1992; Tellegen and Waller, in press).

Table 1. The Big Five Factors (bold), Sample Bipolar Scales, the Six NEO-PI-R facet-scales, the California Psychological Inventory Regression Equations for Predicting the Big Five and the One or Two Highest Loading Multidimensional Personality Questionnaires Scales

Extraversion

introverted	extroverted
unenergetic	energetic
timid	bold

Warmth, Gregariousness, Assertiveness, Activity, Excitement seeking, Positive emotions
Extraversion = Dominance + Self acceptance - Self control
Social closeness, Social potency

Neuroticism

angry	calm
nervous	at ease
emotional	unemotional

Anxiety, Angry hostility, Depression, Self-consciousness, Impulsiveness, Vulnerability
Adjustment= Well being + Work orientation - Anxiety
Stress reaction, Well being

Conscientiousness

disorganized	organized
irresponsible	responsible
careless	through

Competence, Order, Dutifulness, Achievement striving, Self-discipline, Deliberation
Conscientiousness= Responsibility + Achievement vis Conformance - Flexibility
Control, Achievement

Agreeableness

cold	warm
selfish	unselfish
distrustful	trustful

Trust, Straightforwardness, Altruism, Compliance, Modesty, Tendermindedness
Agreeableness = Socialization + Tolerance - Narcissism
Aggression

Openness

intelligent	unintelligent
reflective	unreflective
creative	uncreative

Fantasy, Aesthetics, Feelings, Actions, Ideas, Values
Culture=Empathy + Achievement via independence + Creativity
Absorption, Harm avoidance

Table 1 shows the major dimensions of the Big Five and attempts to orient the reader with respect to this approach to personality description. It lists a) sample bipolar scales from Goldberg (1992), b) the six NEO-PI-R sub-scales for each trait (Costa and McCrae, 1992), c) the CPI equations that attempt to capture each of the big five traits (Gough, 1992), d) and the one or two MPQ scales that tend to load highest on the relevant Big Five factors (Tellegen and Waller, in press). We need only note that the MBTI Intuition, Feeling, and Judgment scales map against the Big Five Openness, Agreeableness and Conscientiousness scales and that the EPQ Psychoticism scale does not fit neatly into the scheme but rather splits across Agreeableness and Conscientiousness. It is not surprising therefore that Openness and Psychoticism are contested traits in each of these schemes.

McCrae and Costa (1990) provide an overview of personality in adulthood and evidence of the validity of the Big Five.

Quantitative Genetic Models

The definition of quantitative genetic influence on a personality trait (or any trait for that matter) is unambiguous and follows from a well verified theory of inheritance, a theory that has been validated across the plant and animal kingdoms. The operational methods make use of kinships in which degrees of genetic and environmental communality are known (Jinks and Fulker, 1970; Eaves et al., 1989; Plomin et al., 1990, Neale and Cardon, 1992). This is not to assert that all genetic mechanisms are known and identified. Far from it, new and important (in both the sense of being scientifically interesting and in explaining a considerable amount of variance) genetic mechanisms are constantly being discovered. Fragile-X syndrome is one of the most recent examples. This mutable mutation differs dramatically from ordinary Mendelian mechanisms and is now known to be the most common cause of mental retardation after Down's syndrome. It was not discovered until 1969 (Lubs, 1969), and its mode of action is still being unraveled (Bouchard, in press; Rennie, 1992). In addition we still do not know the genetic mechanisms underlying the major psychoses.

In quantitative genetic models environmental influences are often evaluated as residuals confounded with error of measurement. This is, of course, not a necessary feature of such models. Environmental indexes can be built in. The difficulty is in properly conceptualizing and measuring environmental influences. As Plomin and Bergeman (1991) have shown, self-reported environmental measures gathered from twins may reflect genetic factors inherent in their personalities and perceptual systems. As we will show later in this chapter objective characteristics of the environment, such as parental education and SES, also reflect genetic influence when measured in biological families (Scarr, 1992; Bouchard et al., 1990).

We begin by briefly sketching the logic of the basic twin and adoption designs that provide much of the data that we plan to review. Path analysis is a powerful and relatively straightforward tool for illustrating the logic underlying behavior genetic methodology (Loehlin, 1989, 1992). The diagram in the upper left corner of Figure 1 shows the path diagram for monozygotic twins reared together (MZTs).

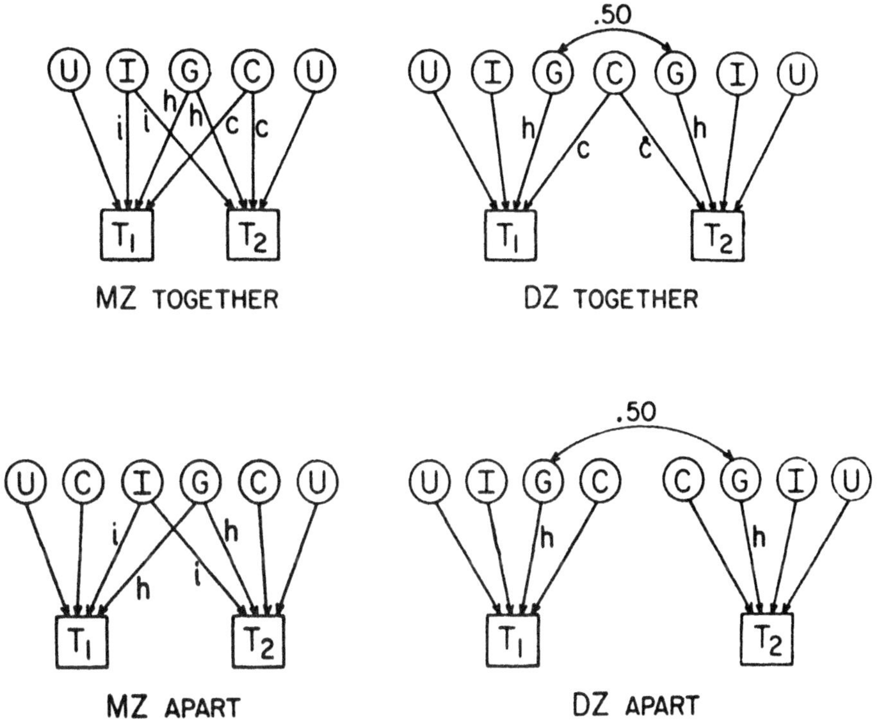

Figure 1. Alternative MZ-DZ Twin Models: Nonadditive Genetic Variance

Items in circles indicate underlying latent variables, and items in boxes indicate measurable phenotypes (scores) for the kinships indicated. The symbols on the directed paths, *h, i, and c* (standardized partial regression coefficients between latent and observed variables) represent the effects of additive genetic effects, non-additive or epistatic genetic effects, and shared environmental effects. The phenotype of each twin is influenced both by their genotype and by their environment. That is, the single-headed arrows denote *theoretical causal* influences, with the lower case letters representing the degree to which the phenotype is a function of the latent causal entities. The expected correlation between any two observables can be derived using the tracing rules of path analysis. There are three paths influencing the correlation between MZTs, a path which reflects common environmental influences, an additive genetic path, and an epistatic path (non-additive genetic). The coefficient of genetic relationship is set at 1.00 for MZ twins, and therefore the *i* and *h* arrows flow from common latent traits. For DZ twins (and siblings) the coefficient of genetic relationship is .50 as shown by the correlation of .50 between the Gs in the upper right hand diagram in Figure 1. For DZ twins *i* is considered negligible. They are not shown, but the coefficients of genetic relationship would be .25 for half-siblings, and .00 for unrelated individuals reared together. The MZT correlation ($r_{MZT} = h^2 + i^2 + c^2$) confounds three sources of variance. The DZT correlation ($r_{DZT} = .5h^2 + c^2$) confounds two sources of variance. If the twins are truly reared apart in environments in which trait relevant features of the environment are uncorrelated the MZA correlation does not contain a *c* component ($r_{MZA} = h^2 + i^2$) and confounds two sources of variance. The comparable DZA correlation is ($r_{DZA} = .5h^2$). A similar logic underlies the models for biological and adoptive families, but these models also require assumptions about the similarity between childhood and adult genotypes of parents in the biological families (cf. Loehlin, 1992, p. 33).

These theoretical models, as well as models that also incorporate ordinary adoptees and biological families can, with certain assumptions, be solved for, and the fit of the data (correlations or mean squares) to the models can be evaluated statistically. We will first review the overall findings based on the numerous twin studies of personality.

META-ANALYSES OF PUBLISHED TWIN CORRELATIONS

Table 2 presents the personality data for MZ and DZ twins for the Big Five taken from the meta-analysis of the twin literature by Nichols (1978).

Table 2. Mean Intraclass Correlations for Monozygotic and Dizygotic Twins Organized According to the Big Five Personality Factors

Trait	Number of Studies	r^{MZ}	r^{DZ}	$r^{MZ} - r^{DZ}$
Extraversion	30	.52	.25	.27
Neuroticism	23	.51	.22	.29
Conscientiousness[1]	12	.44	.24	.20
Agreeableness[2]	6	.49	.23	.26
Openness[3]	7	.43	.17	.26

1. Weighted mean of Conformity and Flexibility. 2. Socialization.
3. Masculinity-Femininity

Table 3 is a meta-analysis of the twin literature before 1976 taken from Eaves et al. (1989). The remarkable similarity of the results should not be surprising as both reviews cover essentially the same literature. These results, or results very similar to them, will be incorporated into most of the model fitting to be discussed subsequently. They are presented for the purpose of providing a complete picture. They constitute the starting point of most analyses and generally provided the largest sized samples and the most power.

Table 3. Mean Intraclass Correlations for Monozygotic and Dizygotic Twins Organized According to the Big-Three Personality Factors.

Trait	Number of Studies	r^{MZ}	r^{DZ}	$r^{MZ} - r^{DZ}$
Extraversion	36	.53	.24	.29
Neuroticism	22	.44	.22	.22
Psychoticism	15	.46	.23	.23

MODEL FITTING OF TWIN DATA

Eaves et al. (1989) fit models to the mean squares of data gathered on adult twins in four large studies, one conducted in London, one in the U.S. (Loehlin and Nichols, 1976, The National Merit Scholarship Twins) one conducted in Sweden (Floderus-Myrhed et al., 1980) and one conducted in Australia (Martin and Jardine, 1986). We summarize what is actually a rather complex analysis in Table 4.

The London study made use of an early form of the Eysenck Personality Inventory and was based on modest sample sizes (MZ males = 70; MZ females = 233; DZ males = 47; DZ females = 125; O-Sex DZ = 67). A simple additive model fits the data best. Nevertheless, as the authors point out, this is not a statistically powerful study, and substantial amounts of non-additve variance could remain undetected.

The U.S. study (based on adolescent CPI data scored for Extraversion and Neuroticism; MZ males = 179; MZ females = 267; DZ males = 111; DZ females = 176) yields fits for both sexes that do not differ from each other and are similar to the London data.

The Swedish study made use of brief versions (nine item scales and allowing for a minimum of six responses) of the Eysenck Extraversion and Neuroticism scales and very large samples. The sample sizes vary from cohort to cohort (MZ males = 510-1053; MZ females = 606-1240; DZ males = 817-1752; DZ females = 1040-1798). The 1946-1958 cohort is by far the largest. There are clear differences as we move from older cohorts (less heritable) to younger cohorts (more heritable). It is, however, dangerous to draw strong conclusions from tests that contain as few as six items and a maximum of nine items. As Eaves et al. (1989) point out we could be looking at changes in gene expression with age, interaction of genetic and environmental effects with secular change, or simply selective mortality. Nevertheless, these large samples allowed for the detection of differential sex and cohort effects across traits and non-additive genetic effects on Extraversion.

The Australian twin study made use of the entire EPQ and large samples (MZ males = 567; MZ females = 1233; DZ males = 352; DZ females = 751; O-sex DZ = 907). This study does not show heterogeneity by sex for Extraversion as did the Swedish study, but the Neuroticism results are sex specific. Both the Swedish and the Australian data suggest significant non-additive genetic variance for Extraversion.

Table 4. Summary Findings of Genetic and Environmental Influence on Extraversion, Neuroticism and Psychoticism from Four Twin Studies Analyzed and Reported by Eaves, Eysenck and Martin, 1989.

	Source of Variance			
Trait Additive	Non-additive genetic	Common genetic	Idiosyncratic family environmental	environmental and error
The London Study				
Extraversion	.49	.00	.00	.51
Neuroticism	.40	.00	.00	.60
Psychoticism	.47	.00	.00	.53
The U.S. Study				
Extraversion	.39	.00	.00	.61
Neuroticism	.48	.00	.00	.52
The Swedish Study				
Males 1926-1935				
Extraversion	.43	.02	.00	.55
Neuroticism	.38	.00	.00	.62
Females 1926-1935				
Extraversion	.06	.43	.00	.50
Neuroticism	.40	.00	.00	.60
Males 1936-1945				
Extraversion	.35	.03	.00	.62
Neuroticism	.39	.00	.00	.61
Females 1936-1945				
Extraversion	.17	.32	.00	.51
Neuroticism	.51	.00	.00	.49
Males 1946-1958				
Extraversion	.23	.28	.00	.49
Neuroticism	.53	.00	.00	.47
Females 1946-1958				
Extraversion	.51	.06	.00	.43
Neuroticism	.61	.00	.00	.39
The Australian Study				
Females				
Extraversion	.21	.32	.00	.47
Neuroticism	.51	.00	.00	.49
Psychoticism	.37	.00	.00	.63
Males				
Extraversion	.21	.32	.00	.47
Neuroticism	.46	.00	.00	.54
Psychoticism	.45	.00	.00	.55

From Eaves et al. (1989).

FITTING MODELS TO ADOPTEES AND EXTENDED FAMILIES

Eaves et al. (1989) also fit models to P, E and N data from 33 different kinship groups (including the twins previously reported on). They report the narrow and broad heritabilities shown in Table 5.

Table 5. Broad and Narrow Heritabilities for Extraversion, Neuroticism and Psychoticism Estimated From Models Fitting Thirty-three Kinship Groups

Trait	Additive genetic (narrow h2)	Non-additive genetic	Broad heritability
Extraversion	.48	.04	.52
Neuroticism	.12	.30	.42
Psychoticism	.28	.21	.49

From Eaves et al. (1989).

As the authors readily admit, the model fit to this data ignores the possibility of a specially high environmental correlation for MZ twins (see Loehlin's model below). They point out that in the absence of separated MZ twins they cannot resolve these alternative hypotheses. *The estimate of .30 for non-additive genetic variance for Neuroticism is inconsistent with all other twins data previously analyzed by this group* (see table 4). This analysis did not include twins reared apart. Loehlin's (1992) analysis (see table 7 below), which includes a very different data set and twins reared apart, does suggest non-additive genetic variance for most of the Big Five. We will return to this point when we discuss the twin reared apart data.

A META-ANALYSIS OF THE TWIN, ADOPTION AND FAMILY DATA

Table 6 is the result of model-fitting to a meta-analysis of twin, twin reared apart (Swedish and American but not Finnish), adoption and family studies of data organized according to the Big Five scheme by Loehlin (1992).

The bottom line in the Loehlin analysis is that, even with MZA data, it is difficult to decide between models that require special twin and sibling environment terms and models of non-additive genetic variance. Common family environmental influences are modest under either model with Agreeableness showing the most influence under both (approximately 10%). Extraversion clearly shows a sizable non-additive genetic effect, consistent with the large Australian study, if one prefers that model, but not consistent with the results of the analysis in Table 5. It should be noted that the Loehlin analysis incorporated the U.S. twin study and therefore that study is not an independent replication.

Table 6. Summary: Estimates from Fitting Simple Models to Correlation Data from the Big Five.

	Alternate Models					
	Additive genes Special MZ and Sibling environments			Additive genes Epistasis Equal environments		
Big Five Factor	Additive genetic Environ.	Special MZ Environ.	Sibling and DZ	Additive genetic	Non-additive genetic	Common family Environ.
Extraversion	.36	.15	.00	.32	.17	.02
Neuroticism	.31	.17	.05	.27	.14	.07
Conscientiousness	.28	.17	.04	.22	.16	.07
Agreeableness	.28	.19	.09	.24	.11	.11
Openness	.46	.05	.05	.43	.02	.06

From Loehlin (1992).

A Closer Look at the Minnesota MZA Data

Table 7 compares the MZA data used by Loehlin from the Swedish and the Minnesota studies as well as additional data from the Minnesota study.

The additional data sets consist of a) Big Five estimates, using the Gough formulas shown in Table 1 for the CPI administered to the Minnesota Study of Twins Reared Apart (MISTRA) sample (age and sex corrected) and also scored on the sample studied by Horn et al. (1976), b) updated MISTRA sample and a much larger sample of MZT and DZT twins using the MPQ, and c) a weighted composite of a and b. We will focus on the weighted composite. In many senses it can be considered a semi-independent cross-validation sample. The MZT and DZT MPQ sample has been increased in size enormously over that used by Loehlin. The use of the same Gough scoring algorithm for the CPI on both the reared apart and reared together twin samples makes a unique contribution to the estimates. The other data sets give a good picture of the variation expected as a result of the use of different tests and varying sample sizes. The MISTRA MZA data suggest an average heritability of about .50 and a common family environmental effect of zero (r_{MZT} not reliably greater than r_{MZA}). For three of the five traits the DZA and DZT correlations suggest non-additive genetic influence (DZ correlation less than half the MZ correlation). The exception is Openness, and this is consistent with the Loehlin data.

Table 7. Correlations for MZA, DZA, MZT and DZT on the Big Five Factors foor the Swedish Study and Various Samples of the Minnesota Study

	Minnesota Study - MPO used by Loehlin, 1992				Minnesota Study - MPO Lager Sample updated sample			
	MZA	MZT	DZA	DZT	MZA	MZT	DZA	DZT
N (pairs)	44	217	27	114	52	553	33	459
Extraversion	.34	.63	-.07	.18	.40	.53	-.13	.16
Neuroticism	.61	.54	.29	.41	.53	.46	.41	.17
Conscientiousness	.57	.58	.04	.25	.61	.59	-.04	.38
Agreeableness	.46	.43	.06	.14	.32	.41	.09	.18
Openness	.61	.49	.21	.41	.61	.40	.23	.17
Mean	.52	.53	.11	.28	.49	.48	.11	.21

	Minnesota and California Study - CPI scored with Gough equations				Minnesota Study - MPO and CPI Sample weighted means			
	MZA	MZT	DZA	DZT	MZA	MZT	DZA	DZT
N (pairs)	61	99	42	99				
Extraversion	.60	.56	.04	.22	.51	.53	-.03	.17
Neuroticism	.55	.50	.16	.05	.54	.47	.27	.15
Conscientiousness	.41	.46	.20	.19	.50	.57	.09	.35
Agreeableness	.67	.48	.11	.16	.51	.42	.10	.18
Openness	.59	.61	.37	.21	.60	.43	.31	.18
Mean	.56	.52	.18	.17	.53	.48	.15	.20

	Swedish Study used by Loehlin, 1992			
	MZA	MZT	DZA	DZT
N (pairs)	95	151	218	204
Extraversion	.30	.54	.04	.06
Neuroticism	.25	.41	.28	.24
Conscientiousness	.25	.39	.16	.13
Agreeableness	.18	.37	.09	.32
Openness	.43	.51	.23	.14
Mean	.28	.44	.16	.18

A Note on the Finnish and Swedish MZA Data

The Finnish MZA data (Langinvainio et al., 1984) are sufficiently deviant, statistically, from other samples that Loehlin has chosen not to incorporate them into any of his models (Loehlin, 1989, 1992). This data are perplexing and it would be desirable if these cases were seen personally and assessed with better instruments rather than by mail. A comment by Rose et al. (1990) also suggests that better scales would probably yield better results.

> We offer some previously unpublished data from personality studies of Finnish MZAs. We have MMPI data on 75 adult Finnish twin pairs, 28 male and 47 female, varying in age at separation from birth to 28 years. The MMPI items administered to these twins form eight of Wiggins' Content Scales (1966). Eighteen of these MZ pairs, separated by age 8, can be categorized as MZAs. These early-separated MZ twin pairs show significant (p>.05) resemblance for five of the eight Wiggins' scales: their intraclass rs range from .41 to .74, and correlations for two of the other three scales achieve p=.06! (p. 775).

The Swedish MZA data, shown in Table 7, suggest an average heritability of .28. The MZA vs. MZT contrast suggests a common family environmental influence of about .16. This heritability estimate is lower than all other estimates and the common family environment estimate is larger than all other estimates except for the Loehlin model which fits a special MZ and Sib environment. Pedersen et al. (1988) assert that their lower heritability estimates are due to their lower MZA correlations and proceed to argue that "Differences between the current study and earlier MZA studies may lie in the method of ascertainment. Both the SATSA (Swedish Adoption/Twin Study of Aging) and the Finnish MZA were identified from population-based information, whereas other studies of MZA typically relied on identification by third parties or response to media appeals. Pairs may have come to the investigator's (and each other's) attention because of their remarkable similarity" (p. 955). We hope to be able to test this hypothesis directly for the MISTRA data in the future when we have enough cases to group twins according to how they located each other. A number of twin pairs came into the study because we or someone we knew were involved in the search. These twins had no knowledge of their zygosity or degree of similarity and all have participated in the study. Prior to carrying out such an analysis we would like to suggest a possible alternative hypothesis, namely that the differences reflect *in part* the quality of the measures in the Scandinavian studies. Their instruments are very inferior to ours. We also note that their zygosity diagnosis is entirely by questionnaire and their data collected by mail. Our zygosities are determined by blood type and the data collected under our supervision. Their Extraversion and Neuroticism scales are 9 items long and they required responses to only six items in order to generate a score. Their Agreeableness and Conscientiousness scales were 10 items long and the authors do not report how many items needed to be answered in order to score the scale. The Openness scale, however, was 25 items long and yielded the highest heritabilities (Bergeman et al., in press)! Average scale length on the MPQ is 24.7 items. All the regular CPI scales are much longer. Taken together these factors can reduce phenotypic correlations somewhat (cf. Nichols, 1978). As part of another study we scored the CPI on a number of shorter scales (with no overlapping items), namely the 11 theoretical scales (average scale length = 14.3) derived by Hase and Goldberg (1967). Note that these scales are still considerably longer than the Scandinavian scales and there are no missing data for the reared apart twins and very little missing data for the reared together twins. The average correlation across the 11 scales for the MZA, MZT, DZA and DZT twins were, .38, .41., .08 and .12, not remarkably different from the SATSA data. If we shortened our scales further and thus threw in a little random error our correlations would all be lower than theirs. The cohort from which their Extraversion and Neuroticism data for the MZA, DZA, MZT and DZT samples was largely drawn, the 1926-45 cohort, yields lower heritabilities than the 1946-58 cohort (Floderus-Myrhed et al., 1980) (see table 4). As pointed out by Eaves et al. (1989) there could be a variety of reasons for this difference. Their DZA correlations (based on a much larger sample than the MZAs, N=218) also leave us puzzled as they imply MZA correlations of at least .56 for Neuroticism and .32 for Conscientiousness. Their DZ data, in disagreement with the MZ data, implies almost no common family environmental influence (except for Agreeableness and this is because the DZT correlation is much larger than the MZT correlation!). We would argue that the relative messiness of the Swedish data relative to the broad heritabilities (sum of additive and non-additive effects in table 7) reported by Loehlin (1992) is due, in part, to the use of poor scales and in part to sampling peculiarities. The variation from scale to scale and inconsistencies between various classes of twins is much more striking when the results of individual studies are examined. It would seem reasonable to argue that the study underestimates the heritability of personality if one is interested in generalizing to somewhat younger samples measured with standard instruments. It is of considerable interest to note that a sub-sample of their MZA twins (n = 45) was assessed with a cognitive battery and the first principal component, a highly reliable measure which reflects 4 hours of testing, yielded exactly the same results as the MISTRA study, namely an MZA correlation of .78 (Pedersen et al., in press).

Treating the Minnesota data in the same manner we can look at it in at least two ways. First, we might assume that the DZA data are unbiased (we must be cautious here as the sample size is very small) in terms of case finding (at least relative to the MZAs) and the difference between the DZA and the DZT (.05) sample estimates common family environment. This value is quite close to the estimate of .066 in the Loehlin analysis. Let's use .06. If we subtract this from our MZTs we get an estimate of h^2 of .42, precisely the same as Loehlin! This would suggest a bias of about .11, for these traits, in the Minnesota MZA data. As an alternative we might assume that the large Swedish DZA sample yields a correct estimate (the Swedish DZA and DZT and Minnesota DZT samples all yield figures between .15 and .20). The Loehlin analysis, however, underestimates h^2 relative to what it would have been had it been gathered with a reliable standard instrument (it is based on a mishmash). Let's be conservative and add .03. The Loehlin figure depends heavily on mailout data and we know that the error rate in zygosity determination reduces the heritability somewhat[2]. Let's be conservative and add .03. This would suggest that the Minnesota data has a bias of about .05 for these scales and that the heritability of ordinary personality scales that assess the Big Five is about .48.

THE GENERALITY OF THESE FINDINGS

We do not have the space to review the matter here, but heritabilities of about .40 -.50 are ubiquitous for virtually all traits measured with ordinary psychological tests. For psychological interests see Lykken et al. (in press); Moloney et al. (1991); Nichols (1978); Waller et al. (in press). For data on Work Values see Keller et al. (1992). For data on Religiosity see Waller et al. (1990). For Special Mental Abilities see Bouchard et al. (1990) and Nichols (1978). Intelligence is an exception as it yields higher heritabilities (Bouchard, in press; Bouchard et al., 1990; Pedersen et al., in press). This difference is due, in part, to the use of superior instruments to measure IQ relative to most other measures of individual differences. Most personality and interest inventories require about five minutes to measure an individual trait[3]. The measurement of special mental abilities generally takes a few minutes more. Good IQ tests on the other hand require about an hour and the brief ones require about 20-30 minutes.

Where Should Behavior Geneticists Go From Here?

We would like to argue that behavior geneticists have not been good scientists in a fundamental sense. We have failed to implement what Lubinski and Dawis (1992) call "normal science". That is we have failed to implement our knowledge about validity and precision of measurement. It is all well and good to carry out a behavior genetic analysis of some number of personality traits using a standard instrument. But it is well known that few personality instruments approach the reliability and precision that they could or should! The fact is that most instruments satisfice. That is, they are just good enough. They are good enough to collect data from students in order to develop the necessary psychometrics. They seldom use more than one format (true or false). They are often much too brief because it is important to "get everything you need in one hour". This latter point is related to the data gathering problem (the world of student and patients revolves around time spans of 45 to 59 minutes) and the fact that instruments that require too much time to complete will not sell well. In most studies the population is sampled at only one point in time because it is too difficult to persuade participants to complete inventories on more than one occasion. In the behavioral sciences we have, to a large extent, taken a stance comparable to engineers. They use what works and is readily available at the right price in order to get "something" done. They simply overbuild by a large order of magnitude in order to avoid structural failure and let it go at that. The scientific approach is quite different. Scientific studies are designed to yield answers within the known limits of measurement error of the scientific technology that can be brought to bear on a problem. Physicists, for example, do not limit their questions to those that can be answered by the use of off the shelf measuring devices. If a theoretical question is important enough they build instruments with technology capable of generating the precision necessary to answer the question. What would this mean for a behavior geneticist interested in genetic and environmental influences on personality? We know that a wide variety of confounding factors interfere with the precision of our measures (i.e., method variance, construct irrelevancies, systematic bias, ambient noise, accidental sampling of a participant at a bad time, etc. etc. (cf. Cronbach et al., 1972)). What if we tried to take these artifacts and sources of error into account systematically via measurement? Imagine the results of a twin and/or adoption study of adults in which the participants completed five different, five scale (Big Five?) personality instruments on ten occasions (each instrument completed twice over the course of twelve months). The five instruments would use entirely different response formats (i.e., True-False, Multiple self-rating of trait descriptions, Adjective ratings or check list, Like-Indifferent-Dislike format, Forced choice) and contain a sufficient number of items to reach a high level of reliability (>.90). In addition these scales would have been built using an iterative procedure designed to yield convergent and divergent validity as well as extrinsic validation profiles (Lubinski et al., 1983; Lubinski and Dawis, 1992). These requirements may appear strict, but they simply flow from the known limits of our measuring tools. Loehlins (1992) powerful summary review of the twin and adoption data on personality presented earlier in this paper required the combining of data gathered in different studies using at least 12 different measures as indexes of each of the Big Five factors. The correlations between many of these scales, purporting the measure the same factor, are often in the .50 range or less. Indeed, we would, in many instances, have chosen different correlation coefficients, but the justification for our choices would have been no more persuasive than those of Loehlin. It is not widely appreciated that measures that correlate as highly as .90 can still display different patterns of external correlations demonstrating that they are poor substitutes for each other. Take a realistic example, consider the case of a correlation of .70 between test 1 and test 2. A correlation of .70 would be considered an excellent level convergent validity in almost any context. If test 1 has a correlation of .50 with an external criterion (.50 would be considered a very good validity coefficient), the upper and lower limits of the correlation of test 2 with the same external criterion are .97 to -.27. (cf. Lubinski and Dawis, 1992, Table 2). Not a single one of these suggestions is novel. The design simply requires the implementation of common knowledge or what should be "normal science".

SUMMARY

The Loehlin (1992) analysis (Table 7), which includes non-additive genetic variance is, in our opinion, by and large the very best that can be achieved with the quality of the data that has been gathered to date. Plausible adjustments to the Loehlin analysis as well as to the Swedish and Minnesota data suggest a broad heritability of 40-50% for ordinary personality scales administered to ordinary adult samples. Nevertheless, the striking inconsistencies in results from study to study are an embarrassment to the field. The parameters that are being estimated are, of course, not constants and may vary from population to population and over time. There is, however, considerable agreement among investigators (in private if not in print) that the inconsistencies are much larger than might be expected on such a basis. The Loehlin analysis strongly implicates non-additive genetic variance and some common family environmental influence for four of the Big Five. Both of these findings conflict with major data sets. In addition, the hypothesis of a special MZ and sibling environmental effect cannot be conclusively refuted. We recommend that these issues be attacked using the procedures of "normal science" a practice woefully lacking in Psychology as a discipline.

Non-Behavior Genetic Approaches to Environmental Influences on Adult Personality: What have they shown?

It would be impossible to review even a small part of the "environmental" literature. Our task is, however, simplified somewhat by the appearance of a recent review of a segment of this literature by L.W. Hoffman (1991). Our critique will be specific to a few of the domains of research cited in her paper. It is important to state at the outset that it is not our intention to engage in a study by study critique. We are not arguing that the factual results of particular studies are incorrect (they may or may not be), but rather that the interpretation of the results of many studies, both by Hoffman and just as often by the original authors, is seriously flawed. Hoffmans paper was written to deal with family environmental influences on personality from the point of view of accounting for sibling differences. As she puts it "one specific aim of this article is to clarify how developmental researchers conceptualize environmental influences. The second is to discuss developmental research findings that indicate why environmental influences even those from within the family, are not the same for different family members" (p. 187). Her analysis, however, extends well beyond these goals and so does our critique. It is our contention that, a) to the extent that Hoffmans opinions are representative of those of developmental researchers at large (we have no reason to think that this is not true as she is a distinguished investigator in her own right and her paper appeared in a distinguished journal and was most likely reviewed by her peers in the field) their methods of conceptualizing environmental influences on adult outcomes are largely invalid, and b) Hoffman largely fails to present plausible evidence for the influence of environmental factors on adult personality.

CORRELATIONS ARE NOT CAUSES

There are numerous studies in developmental psychology that draw causal inferences regarding environmental influences on adult psychological characteristics from a) correlations between parental traits (i.e., child rearing patterns) and traits of adult offspring, and b) from correlations between characteristics of the environment (parental education, socioeconomic status, etc.,) and adult offspring characteristics. More often, however, the correlations are between the factors cited and a childs characteristics. In the latter case the unspoken inference drawn is that the effect is permanent and will exhibit itself in adulthood. It is impossible to repeat sufficiently often the dictum that such phenotypic correlations, when obtained on biological family members, are completely confounded by genetic influences. Interpretation of such correlations in terms of environmental causation presumes the theory they are purporting to test. Scarr (1985, 1992) has preached this point to developmentalists for over twenty years but, as we will show, to very little avail. What change is occurring is largely occurring because developmentalists being trained in behavior genetics are slowly infiltrating the field. A recent and particularly flagrant example of this flawed approach to understanding environmental influences on behavior can be found in Tomlinson-Keasey and Little (1990). This study uses the Terman longitudinal data and structural modeling to demonstrate parental influences on educational attainment, intellectual skill and personal adjustment. Bouchard et al. (in press) present a critique of this study with respect to the causal analysis of intellectual skill. They point out that the causal model implemented in that study presumes that parental educational levels are solely environmental measures. This assumption is made in the face of overwhelming evidence that parental educational levels are in fact genetically influenced to a significant degree. In addition they show virtually no impact on children's IQ when assessed in the context of adoption studies. Tomlinson-Keasey and Little do not even mention the possibility of a genetic interpretation of the data even though Barbara Burks, Termans most brilliant student, demonstrated the importance of adoption studies in this context over fifty years ago (Burks, 1938). This problem haunts most of Hoffmans interpretation of the environmental literature.

CAUSAL MODELS MUST BE SPECIFIED AND EVALUATED

In addition to the use of within family studies to draw causal inferences, Hoffman repeatedly presumes causal models and causal chains that if specified as path diagrams would most likely result in no explanation whatsoever. Consider the argument, repeatedly cited by Hoffman throughout her paper, that differences in physical attractiveness elicit differences in treatment from caregivers and others. This "model" assumes, a) strong consistencies in the behavior of those providing the "influence" (i.e., attractiveness elicits, across many different individuals, the same or functionally equivalent behaviors that over time shape personality traits), b) strong agreement among caregivers and others on what constitutes physical attractiveness from age to age, c) consistent treatment based on attractiveness regardless of other characteristics of the child, d) consistency in attractiveness of the child from age to age, and e) a real influence (an *e* effect in a path model). The data for a complete path diagram are not available, but the numerical estimates necessary to make the diagram a plausible explanation are so unlikely that the burden of proof on Hoffman would be enormous even if she were allowed to use confounded within-family data. Perhaps the most frustrating feature of Hoffman's review is her repeated use of "significant findings" without meaningful cumulation. After fifty years of research on environmental influences on personality it would be nice to see a meta-analysis of replicated findings even if they made use of confounded with-family data. Finally, in repeated instances a particular environmental process is cited as a possible causal factor when in fact there is no outcome evidence demonstrating the environmental process had any influence whatsoever. The existence of a process is taken as evidence of a cause without linking it to an outcome. The purported effects of differential socialization by birth order is a classic example and discussed below. This approach may sound strange, but it is a consequence of a peculiar and narrow view of how research should be carried out. Consider the following quote from Hoffman (1991):

> Research in developmental psychology typically examines combinations of "environment" and "person" variables that are directly measured and controls on other variables in predicting child outcomes. *The focus is more on the process of influence rather than the outcome. An outcome, such as a personality trait, cannot tell one the environmental influences* (emphasis added). Different scores on a personality inventory for a trait do not mean that common environmental experiences did not go into that outcome, but only that total environmental package was not the same. Because there are so many aspects of the environment that interact in affecting personality, it is not likely that even a common experience that marks all children in some way will result in the same outcome. (p. 192).

While plausible on it's face this approach fails all the fundamental tests that characterize empirical science. If it were followed strictly it is unlikely that any hypotheses regarding environmental causation would be falsified. If one believes that aspects of the environment interact with the person in affecting personality then one is obliged to demonstrate such an influence, not simply assume it. This is a hypothesis about a genotype by environment interaction and methods are available for studying such phenomena, methods developed by behavior geneticists (Plomin et al., 1977). The study of processes of influence without concern for outcomes is a totally unacceptable procedure if an extrapolation to outcomes is envisaged. The existence of a systematic influence process, i.e., child rearing pattern, may say something about the person exercising the process (parent) and it may control the expression of some behaviors while it is in force, but it may also have nothing to do with the eventual or final development of a child's personality. The study of outcomes is mandatory if one is to distinguish between these possibilities. The study of differential outcomes when similar groups are subjected to different environmental treatments or different groups are subject to the same treatment is a fundamental scientific procedure and is the only way in which one can examine systematic interactions.

EXPERIMENTAL RESULTS SHOULD NOT BE OVER INTERPRETED OR OVER GENERALIZED

Experimental procedures are widely used by developmentalists to illustrate that specific manipulations influence particular traits. The experimental approach is held in such high esteem in contemporary psychology that it is important to demonstrate that the results of experimental studies are often cited in support of arguments that they are incapable of supporting. Experimental manipulations during a brief experiment often temporarily influence the expression of a behavior or trait (change behavior or a score on a psychological test). Such evidence is not proof, nor even persuasive evidence, that a similar manipulation caused the trait in the first place, nor that the change in behavior or the trait will maintain itself once the manipulation is removed. Even manipulations or reinforcement sequences maintained for long periods of time need not have enduring effects (Breland and Breland, 1961). All such manipulations must be shown to be linked to the outcome behavior, and this must be done in a context where genetic influences have been excluded as plausible alternatives. The lack of correlation between a child rearing behavior pattern and a childs behavior observed within biological families is sometimes sufficient to falsify such hypotheses. For example, experimental studies have repeatedly demonstrated that attractive people are treated differently from unattractive people (Hartup, 1983; Langlois, 1986; Lernez et al., 1987; R. M. Lerner, 1987; Sorell and Nowak, 1981). These studies (all cited by Hoffman, see below), most often based on brief and contrived circumstances, are cited as evidence for environmental influence on personality. They are also often cited as the cause of the similarity in personality between monozygotic twins reared apart. We show below that the appropriate correlations do not exist, consequently such claims are unfounded.

This general developmental point of view can be characterized as the doctrine of environmental specificity. Under this view (a legacy of behaviorism and the philosophy of meliorism that pervades the study of child development (Charlesworth, 1992)) nearly every variation in features of the environment is assumed to be important and to influence the organism. In the domain of Psychopathology there is a comparable model, namely the "spun glass theory of mind" (Schofield, 1964). According to the "spun glass theory" every affront, no matter how minor, breaks something! The alternative point of view might be called the doctrine of organismic robustness. Bouchard et al. (in press) drawing heavily on Scarr (1992) have argued that there is compelling empirical and theoretical evidence that human beings have evolved to survive in "an average expected environment" that has wide bounds. As Scarr (1992) has argued "Fortunately, evolution has not left development of the human species, nor any other, at the easy mercy of variations in their environments. We are robust and able to adapt to wide-ranging circumstances - a lesson that seems lost on many ethnocentric psychologists. If we were so vulnerable as to be led off the normal developmental track by slight variations in our parenting, we should not long have survived" (p. 16). We are in full agreement with this line of argument and believe it is entirely consistent with the findings from behavior genetics. Bouchard et al. (1990) present the plausible conjecture that genes drive behavior and that behavior determines the environments which we experience. Under such a model the effective environment is decoupled from the control of parents and others who believe they have major influence on the development of the child. The effective environment is the one that is available to the child and most compatible with his/her genotype.

The Environmental Critique of the Twin Method: How Valid?

Hoffman begins her paper with a critique of the twin literature. The critique is sufficiently in error to deserve discussion. She asserts that the equal environmental assumption of the ordinary twin method is simply not tenable and argues that MZ and DZ twins do not have equally similar environments. Consequently, she asserts, the twin method overemphasizes genetic influences. Loehlin's (1992) analysis of all the kin correlations and reported in Table 6 demonstrates that we are not yet in a position to completely reject the possibility of special MZ and Sib. environments, thus Hoffman may be correct to some degree. If, however, she is correct it is for the wrong reasons. Her analysis is seriously flawed. For example, she argues that MZ twins look "exactly alike" (an erroneous assumption on her part) and consequently, since there is "abundant evidence that adults, peers, parents and siblings, respond differently to different appearances" (p. 188) this evidence supposedly demonstrates that the greater similarity of MZ than DZ twins in personality can be explained environmentally. Hoffman does not seem to realize that the demonstration of differential responsiveness to appearances of DZ twins or the existence of environmental similarities for MZ twins does not by itself constitute evidence that the equal environment assumption is violated. The equal environment assumption is required only for trait relevant features of the environment; features of the environment that have causal status. Causal status must be demonstrated, not assumed. To make this point crystal clear consider the hypothetical argument that MZA twins are alike because they are 100% concordant for the outside color of the homes they grew up in. Would anyone ascribe their similarity in personality to that fact? Obviously not. The reason is that the outside color of ones home is not believed to be a trait relevant environmental variable. It is absolutely mandatory that Hoffman demonstrate that the differential treatments she cites have a causal influence on the traits whose similarity she is trying to explain. This is a very difficult task. Loehlin and Nichols (1976) showed that similarity of treatment of MZ twins by their mothers could not explain more than a trivial portion of the twin differences, differences which must be environmental in origin. These results were also replicated in the DZ sample. Bouchard and McGue (1990) used the same methodology to determine if difference in self reported child rearing practices might be related to the MZA twin differences in personality. The results were essentially negative and were replicated using a different method, namely correlating the child rearing scores with personality traits in the entire sample of adoptees (both MZA and DZA twins).

Hoffman cites Bronfenbrenner (1986) as demonstrating that "degree of similarity among twins [MZA twins] is considerably reduced when they are reared, not only in different families, but also in different community environments" (p. 188). Bronfenbrenner's arguments have to do with IQ and are based on the work of Taylor (1980). Taylor's analysis was systematically refuted shortly after his book appeared (Bouchard, 1983). None of the so called environmental similarities purported by Taylor to explain the twin similarities in IQ survived cross validation when a different IQ test was used! To our knowledge no critic of the twin literature has replied to the Bouchard analysis and it is not cited by Hoffman. She must be aware of this study as she cites Bouchard et al. (1990) in which this study is discussed. Regarding the 1990 paper she asserts that "very high estimates of genetic influence and low estimates of influence from the shared family environment were reported. However, in this study, the age of separation varied, and some monozygotic twins had maintained contact during childhood. Furthermore, some had been together for 20 years before assessment" (p. 188). There is a literature on the influence of contact between twins in adulthood as a source of similarity in personality and alcohol consumption. The effects are complex, sometimes contact is a cause and sometimes a consequence (Rose et al., 1990). The problem must be worked out in each instance. In the 1990 paper we demonstrated that IQ similarity was unrelated to various measures of contact. We have also shown that similarity in vocational interests is unrelated to contact (Moloney et al., 1991). The Swedish study of reared apart twins has also failed to find any effect of contact on IQ (Pedersen et al., in press; Pedersen et al., 1985). It also failed to find any influence of age at separation and degree of separation on twin personality resemblance. These researchers did find some influence of "time prior to separation" on Neuroticism and Impulsivity (Pedersen et al., 1988). They report that "twins who had their first reunion shortly after separation (regardless of age at separation) were more similar for Neuroticism and Impulsivity than twins who first met their partner later in life or who never met" (p. 956). They do not report the magnitude of the effect, but if contact is an important variable one wonders why unrelated siblings reared together when studied as adults have so little similarity. There is zero similarity in IQ (Bouchard et al., 1991). For various personality traits assessed in adolescence the average correlation for unrelated individuals reared together is .07 (Scarr et al., 1981). This is a point at which verbal arguments appear to explain away a finding whereas quantitative analysis reveals the ephemerality of the argument. Hoffman also cites Farber (1981) on this issue, but does not cite the refutation of Farber's pseudoanalyis by Bouchard (1982a). For further discussion of this issue with respect to MZA twins see Bouchard (1982b, 1987, in press).

FAILURE TO DISTINGUISH BETWEEN DISTAL CAUSES AND PROXIMAL CONSEQUENCES

Hoffman (1991) construes personality as reflecting a particular behavior or set of behaviors, often but not necessarily always, measured in a particular situational context. This is the perspective most often taken by investigators with a strong experimental orientation. It is not surprising therefore that Hoffman accepts Mischel's (1973) view of personality. She argues that; "if one conceives of personality as a result of multiple interacting influences (Mischel, 1973) then these environmental variations can be important in accounting for differences among siblings in personality" (p. 191). Notice that this "interaction view" assumes away the question that the behavior geneticist is asking. The behavior geneticist agrees that transactions with the environment are necessary for an organisms to grow and develop. Whether interactions are a cause (in a distal sense) of individual differences is a different and empirical question. Food, for example, is necessary for the development of stature. Within a population of well nourished individuals there is great variation in stature and it is almost entirely genetic in origin. Different individuals in the population eat quite different foods in different amounts, but all end up consuming the necessary nutrients for proper growth. One could argue that "stature is the result of multiple interacting influences" but that would not vitiate the conclusion that most of the variation among individuals owes its origin to the fact that they carry different genes. The "so called" multiple interacting influences in this case are functionally equivalent and do not contribute to variation among individuals. What must be explained is variation in a character from individual to individual in a population. The fact that different individuals consume (experience) different foods in different amounts is not in itself proof that this variation is a cause of differences in stature among them, it may simply be a consequence. Short people by and large consume less food than tall people over the course of their development. This variation is not, however, the fundamental cause of their differences in stature. Such experiential differences, if we can call them that, are a consequence of genetic differences. If an investigator proposes that in the long run multiple interacting influences are an important determinant of variation from person to person on a trait then this fact must be demonstrated using an appropriate design (generally longitudinal and adoption) and measured outcomes. It cannot simply be asserted.

BIRTH ORDER AND PERSONALITY

Hoffman cites Lasko (1954) among others, with respect to the possible influence of birth order on personality. Novice mothers, it is argued, behave differently with respect to first born children relative to later born children if only because they have more experience when the later born children arrive. Lasko showed that there are consistencies in mothers methods of handling children of different birth rank and in their policies of child rearing across children. We need make only a few points here. It is, as we have argued, necessary for someone to shown that the variations in child rearing across birth orders is sufficiently related to adult personality to explain some of the variance. Causation cannot simply be assumed. No causal evidence is presented by either Lasko or Hoffman. Indeed it is of some interest to note that the most comprehensive review of the birth order literature, an entire book by Ernst and Angst (1983), is not even cited by Hoffman. Let's briefly look at this literature.

Serious questioning of the popular belief in the effect of birth order on personality began almost as soon as personality tests were invented. Stagner and Katzoff (1936) in a study of 430 men who completed one of the very first personality inventories, the Bernreuter Personality Inventory, concluded:

> "The fact that the findings of the present study are largely negative should not be surprising to anyone who has worked with personality problems. The surprising point is that so many presumably thoughtful psychologists have emphasized the importance of birth order in determining personality" (p. 345).

Schooler (1972) in a review of the literature over thirty years later could only echo the same conclusion.

Finally Ernst and Angst (1983) in their massive review of this literature emphatically point out that just because a behavioral manipulation can be demonstrated it does not follow there need be an effect.

> *Differential socialization* (emphasis in original) by birth order, on the other hand, has been well established at least for firstborns in comparison to second borns and at infancy and preschool age.... These differences in socialization, however, do not seem to leave indelible traces that can be predicted (p. 187).
>
> Whenever studies of representative adult samples were carried out with an unobjectionable method, birth order differences were nil. Whenever it was possible to estimate the amount of variance explained by birth order it was negligible (p. 186).

BIRTH ORDER AND IQ

Hoffman also refers to the often cited work of Zajonc (1983) with respect to birth order differences in IQ. It is worth noting here that the evidence in support of this theory has always been highly equivocal (cf. Bouchard and Segal, 1985) and these cautions are not cited. In addition, it is now known that the mathematical basis of the Zajonc confluence model is in error. When large data sets, i.e., the Wisconsin Longitudinal study, are properly analyzed there is no evidence whatsoever for the model (Retherford and Sewell, 1991).

Hoffman argues that behavior geneticists report that the influence of ordinal position on personality traits is small and therefore dismiss it. She goes on to assert that this is an inappropriate stance because the effect is indirect via the environment and "personality outcomes are affected by a multiplicity of interacting environmental influences, and any given one is unlikely to explain much variance" (p. 194). Apart from another ex cathedra assertion regarding the "multiplicity of environmental influences" the argument that the effect is small would ordinarily be acceptable (behavior geneticists use the term polyenvironmental effects in contrast to polygenic effects to characterize numerous small influences). We, however, hesitate to accept such an argument when the domain is known to be plagued with artifacts and when there is no substantive evidence whatsoever of causal influence.

PHYSICAL APPEARANCE

In line with many critics of the behavior genetics literature Hoffman cites a wide variety of studies that purport to show among other things, "that more attractive children elicit more positive responses and evaluations from parents, other adults, and peers". She goes on to assert that, "All these can be differences among siblings and lead to different environmental experiences" (p. 196). We are again told "these can be differences", however, no evidence whatsoever is presented to show that these treatments have any enduring influence on personality traits. What are the facts? Fortunately, Feingold (1992) has prepared a meta-analysis of the literature on the relationship of physical attractiveness, as independently rated by others, and a variety of measures of personality and cognition. There is almost nothing to report (recall that these correlations must be squared in order to be interpreted as amount of variance accounted for and most of the traits are correlated and as a consequence the variances accounted for by each variable do not add up independently). The sample size, median and mean correlations for each trait are as follows; Sociability (N=1,710, .00, .04), Dominance (N=2,858, .04, .07), General Mental Health (N=2,597, .02, .05), Self-esteem (N=4,942, .04, .06), Internal locus of control (N=3,683, .00, .02), Freedom from loneliness (N=430, .04 .15), Freedom from general social anxiety (N=1,155, .06 ,.09), Freedom from heterosexual anxiety (N=1,539, .19, .22), Freedom from public self-consciousness (N=578, -.20, -.18), Freedom from self-absorption (N=746, .00, -.08), Freedom from manipulativeness (N=252, .03, -.01), Social skills (N=1,050, .25, .23), Popularity (N=982, .04, .08), Intelligence (N=3,497, .00, -.04) and Grades (N=3,445, .07, .02). It is well worth emphasizing that, while there are few dispositional differences between physically attractive and physically unattractive people, the experimental literature on this topic clearly demonstrates that there is a powerful physical attractiveness stereotype. People believe that there are such differences (Feingold, 1992). Feingold (1992) also concludes that "good-looking people are not what we think" and we would add that "the influence of attractiveness on stable personality traits is unlikely to be what some people thought". These data are, of course, not definitive regarding the influence of differential treatment due to physical attractiveness in childhood and adolescence because they do not fulfill the model specification cited earlier. They do, however, throw considerable doubt on the idea that physical attractiveness is an important determinant of personality. There is no doubt, however, that physical attractiveness influences social behavior and attitudes. The correlations for Social skills and Freedom from heterosexual anxiety and Freedom from public self-conciousness are higher than for other traits. In addition the comparable figures for Noncoital sexual experience are (N = 1,167, .16, .13) and for Global sexual experience (N=1,896, .18, .18) (Feingold, 1992). These findings suggest that physical attractiveness has a very narrow and specific influence on behavior and attitudes. This suggestion is confirmed by the fact that various measures of sexual activity yield quite low correlations.

DIVORCE

Studies of the children of divorce, much like studies of the children of alcoholics, are a growth industry. These studies are of importance for the practitioner as they yield information that is of value regarding what factors to focus on in counseling. With regard to the explication of underlying causes they are of almost no value. Contrary to Hoffman's view, and that of most of the field, divorce is not simply an environmental factor independent of genetic influences. There are now two studies that demonstrate a significant genetic component to divorce (McGue and Lykken, in press; Turkheimer et al., 1992). If this effect is mediated by stable individual differences, as is very likely, then they will be transmitted differentially to offspring and we would expect differences between children in the same family in their personality characteristics due to genetic factors, not solely to the environmental factor of divorce, if there is any such influence. The recent cross-national study (Great Britain and the Unites States) by Cherlin et al. (1991) concludes that "Overall, the evidence suggests that much of the effect of divorce on children can be predicted by conditions that existed well before the separation occurred" (p. 1388). While these authors provide an environmental explanation, the results are entirely consistent with a genetic one. Hoffman on the other hand concludes "these studies show that each child experiences the divorce through his or her level of understanding and the needs and anxieties that are ascendant at the particular age. In this sense, then, the same divorce may be experienced by siblings of different ages as a different environmental influence. Divorce may seem like an experience that is shared by siblings but its interpretations and impact are not shared" (p. 194-195). Again all we have is the claim of the existence of experiences and the speculation that they may make a difference, without evidence of influence on outcomes or appropriate controls for genetic influence.

SOUND OF A NEONATE'S CRY

A final example, one that postulates a long causal chain between childhood events and possible adult outcomes, is the following; "Differences as subtle as the sound of the neonate's cry have been shown to evoke different responses in both parents and unrelated adults (Lamb and Bornstein, 1986; Malatesta et al., 1989; Wiesenfeld and Malatesta, 1982). Experimental studies have also shown that parental responses to the same stimuli, such as recorded infant cries, are not homogeneous. They can be affected by, for example, whether the parent is an experienced parent or a new parent (Lounsbury and Bates, 1982)." What are we to make of such arguments? Is this a serious claim about early environmental influence on latter personality? If so we simply cannot take it seriously without a great deal more evidence. None is presented.

BROAD SOCIETAL INFLUENCES

There is one domain of influence cited by Hoffman as an important mediator of influence that we agree with. It is the influence of broad societal factors such as the depression. Elder's work on this topic, while subject to some criticism from a behavior genetic point of view, is on the whole superb (Elder, 1974; Elder et al., 1986). We need only add that this work is far from mainstream developmental psychology and that it makes use of longitudinal studies which assess adult outcomes. These studies, however, would have been far more informative had they included adoptive families as is being done in the Colorado Adoption Project (Plomin and DeFries, 1985; Plomin et al., 1988) or twins as was done by Wilson (1983).

Summary

Hoffman's paper is entitled "The influence of the family environment on personality: Accounting for sibling differences." Nevertheless, she argues near the end of her paper that "The outcomes examined in developmental psychology research on the family environment are rarely personality traits: they are more often coping skills, competency, and moral internalization, and these are not the variables examined in the behavioral genetics work" (p. 198). She then goes on to assert that on these alternative measures it is not known that siblings are not similar. If Hoffman is not interested in personality then one wonders why she wrote her paper. If we do not know if siblings are alike or different on the "outcomes examined in developmental psychology research" then there is no sense in arguing about how they should be explained. Data will have to be gathered using the proper designs.

It is not possible to respond to every point made by Hoffman in a brief summary review. We can only assert that Hoffman commits a number of the errors outlined by Meehl (1978; 1990) in his critiques of soft psychology and that what she presents as evidence for environmental influences on personality, is at best, a series of plausibility arguments. In many cases, as we have shown, the models required are clearly much to weak to be convincing. In other cases sufficient data have already been gathered using proper methodology to refute the arguments presented.

What are the implications of Hoffman's review? With respect to the work of behavior geneticists she concludes that, "Research on the environmental processes that lead to sibling similarities and differences can be an important new direction for unraveling the complexities of social development. The real value of this recent research from behavior genetics does not lie in its implications for reinterpreting existing research in developmental psychology, but rather in its impetus to research on the environmental precursors of sibling similarities and differences" (p. 199).

Our analysis of her arguments leads to a different conclusion. Behavior genetics research has demonstrated that samples of adoptees and their parents are mandatory in order to demonstrate environmental influences. The behavior genetic findings in the realm of personality are now sufficiently robust that no study of environmental influences should be carried out without "genetic" controls. Contrary to Hoffman we must conclude that the lessons of behavior genetic research have been completely lost on some developmentalists. Correlational research in developmental psychology that does not impose genetic controls is uninterpretable with respect to the question under discussion, the shaping of adult personality. In addition, without the concurrent study of outcomes, process research has virtually no bearing on the explanation of adult individual differences in personality. Variation in most of the factors cited by Hoffman as possibly important environmental determinants of personality may well be "will of the wisps" with respect to adult traits (abilities, personality traits and interests). Some of these processes are undoubtedly necessary if only to keep a child's behavior within acceptable bounds during the developmental years, but a reasonable competing hypothesis is that most of them (child rearing patterns, teaching styles, family structures) are functionally equivalent and variation in their implementation by caregivers is, within wide bounds, irrelevant with respect to the determination of adult status.

Author's note

This research was supported by the Pioneer Fund, the Koch Charitable Foundation and the Seaver Institute. I would like to thank Yoon-Mi Hur for carrying out the analysis of the CPI data. Reprint requests should be directed to: Thomas J. Bouchard, Jr., Department of Psychology, Elliott Hall, 75 East River Rd., University of Minnesota, Minneapolis, MN., 55455-0304.

Footnotes

1. Loehlin (1992) is the first person to organize the behavior genetic data according to this scheme and this paper was re-organized after the appearance of his book.
2. We could, in theory, also correct for missclassification of other relatives as well. Mother offspring genetic relationships are virtually certain but father offspring relationships are not and estimates of unrelatedness seldom go below 5%. Also important and seldom recognized is the data coding error rate which is higher than most people believe (Hunter and Schmidt, 1991). All of these errors reduce the heritability in an extended family design. We would assert that zygosity errors are virtually absent in our MZA study and coding errors have been held to a minimum.
3. While preparing this section we received a brochure from Psychological Assessment Resources, Inc. from whom we purchase the Stroop test. The brochure touts the *Personality Assessment Inventory, a comprehensive* 344 item inventory of adult psychopathology. It contains 22 scales and supposedly can be completed in 40-50 minutes. Thus all it takes is 1.8 to 2.3 minutes to complete each 15.6 item scale (average length).

References

Bergeman, C.S., Chipuer, H.M., Plomin, R., Pedersen, N.L., McClearn, G.E., Nesselroade, J.R., Costa, P.T., Jr. and McCrae, R.R. (in press). Genetic and environmental effects on openness to experience, agreeableness, and conscientiousness: An adoption/twin study. *Journal of Personality*.

Bouchard, T.J. Jr., (1982a). Review of Identical twins reared apart: A reanalysis. *Contemporary Psychology, 27*, 190-191.

Bouchard, T.J. Jr., (1982b). Review of The intelligence controversy. *American Journal of Psychology, 95*, 346-349.

Bouchard, T.J. Jr., (1983). Do environmental similarities explain the similarity in intelligence of identical twins reared apart? *Intelligence, 7*, 175-184.

Bouchard, T.J. Jr., (1987). The hereditarian research program: Triumphs and tribulations. In S. Modgil and C. Modgil (Eds.), *Arthur Jensen: Consensus and controversy* (pp. 55-75). London: Falmer International.

Bouchard, T.J. Jr., (in press). The genetic architecture of human intelligence. In P.A. Vernon (Ed.), *Biological approaches to the study of human intelligence*. Norwood, NJ: Ablex.

Bouchard, T.J. Jr., Lykken, D.T., Tellegen, A. and McGue, M. (in press). Genes, drives, environment and experience: EPD theory revised. In C. Benbow and D. Lubinski (Eds.). *From psychometrics to giftedness: Essays in honor of Julian C. Stanley*. Baltimore: John Hopkins University Press.

Bouchard, T.J. Jr., Lykken, D.T., McGue, M., Segal, N.L. and Tellegen, A. (1990). Sources of human psychological differences: The Minnesota study of twins reared apart. *Science, 250*, 223-228.

Bouchard, T.J. Jr., Lykken, D.T., McGue, M., Segal, N.L. and Tellegen, A. (1991). IQ and heredity. *Science, 252*, 191-192.

Bouchard, T.J. Jr. and McGue, M. (1990). Genetic and rearing environmental influences on adult personality: An analysis of adopted twins reared apart. *Journal of Personality, 58*, 263-292.

Bouchard, T.J. Jr. and Segal, N. (1985). Environment and IQ. In B.J. Wolman (Ed.), *Handbook of intelligence: Theories, measurements, and applications* (pp. 391-464). New York: Wiley, 391-464.

Bouchard, T.J. Jr., Segal, N.L. and Lykken, D.T. (1990). Genetic and environmental influences on special mental abilities in a sample of twins reared apart. *Acta Geneticae Medicae et Gemellologiae, 39*, 193-206.

Breland, K. and Breland, M. (1961). The misbehavior of organisms. *American Psychologist, 16*, 681-644.

Bronfenbrenner, U. (1986). Ecology of the family as a context for human development: Research perspectives. *Developmental Psychology, 22*, 723-742.

Burks, B.S. (1938). On the relative contributions of nature and nurture to average group differences in intelligence. *Proceedings of the National Academy of Sciences, 24*, 276-282

Charlesworth, W.R. (1992). Darwin and developmental psychology: Past and present. *Developmental Psychology, 28*, 5-16.

Cherlin, A.J., Furstenberg, F.F. Jr., Chase-Lansdale, P.L., Kiernan, K.E., Robins, P.K., Morrison, D.R. and Teitler, J.O. (1991). Longitudinal studies of effects of divorce on children in Great Britain and the United States. *Science, 252*, 1386-1389.

Costa, P.T. Jr. and McCrae, R.R. (1992). Four ways five factors are basic. *Personality and Individual Differences, 13*, 653-665.

Cronbach, L.J., Gleser, G.C., Nanda, H. and Rajaratnam, N. (1972). *The dependability of behavioral measurements: Theory of generalizability of scores and profiles.* New York: Wiley.

Digman, J.M. (1990). Personality structure: Emergence of the five-factor model. In M.R. Rosenzweig and L.W. Porter (Eds.), *Annual Review of Psychology* (Vol 41, pp. 417-440). Palo Alto, CA: Annual Reviews.

Eaves, L.J., Eysenck, H.J. and Martin, N.G. (1989). *Genes, culture and personality.* New York: Academic Press.

Elder, G.H. Jr. (1974). *Children of the great depression.* Chicago: University of Chicago Press.

Elder, G.H. Jr., Caspi, A. and Nyguyen, T. (1986). Resourceful and vulnerable children: Family influence in hard times. In R.K. Silbereisen, K, Eyferth and G. Rudinger (Eds.). *Development as action in context* (pp. 167-186). New York: Springer-Verlag.

Ernst, C. and Angst, J. (1983). *Birth order: Its influence on personality.* New York: Springer-Verlag.

Eysenck, H.J. (1992). Four ways five factors are *not* basic. *Personality and Individual Differences, 13*, 667-673.

Farber, S.L. (1981). *Identical twins reared apart: A reanalysis.* New York: Basic Books.

Feingold, A. (1992). Good-looking people are not what we think. *Psychological Bulletin, 111*, 304-341.

Floderus-Myrhed, B., Pedersen, N. and Rasmuson, I. (1980). Assessment of heritability for personality, based on a short-form of the Eysenck Personality Inventory. *Behavior Genetics, 10*, 153-162.

Goldberg, L.R. (1992). The development of markers for the big-five factor structure. *Psychological Assessment, 4*, 26-42.

Goldberg, L.R. (in press). What the hell took so long? Donald Fiske and the Big-Five factor structure. In P.E. Shrout and S.T. Fiske (Eds.), *Advances in personality research, methods, and theory; A festschrift honoring Donald W. Fiske*. New York: Erlbaum.

Gough, H.G. (1968). An interpreter's syllabus for the California Psychological Inventory. In P. McReynolds (Ed.), *Advances in psychological assessment* (Vol 1). Palo Alto: Science and Behavior Books.

Gough, H.G. (1987). *California Psychological Inventory administrator's guide*. Palo Alto: Consulting Psychologists Press.

Gough, H. (1992). Personal communication to Thomas J. Bouchard, Jr.

Hartup, W.W. (1983). Peer relationships. In P.H. Mussen (Ed.), *Handbook of child psychology: Socialization, personality and social development* (4th ed., pp. 102-196). New York: Wiley.

Hase, H.D. and Goldberg, L. (1967). Comparative validity of different strategies of constructing personality inventory scales. *Psychological Bulletin, 67*, 231-248.

Hewitt, J.K. (1984). Normal components of personality variation. *Journal of Personality and Social Psychology, 47*, 671-675.

Hoffman, L.W. (1991). The influence of the family environment on personality: Accounting for sibling differences. *Psychological Bulletin, 110*, 187-203.

Horn, J.M., Plomin, R. and Rosenman, R. (1976). Heritability of personality traits in adult male twins. *Behavior Genetics, 6*, 17-30.

Hough, L.M. (1992). The "Big Five" personality variables-construct confusion: Description versus prediction. *Human Performance, 5*, 139-155.

Hunter, J.E. and Schmidt, F.L. (1990). *Methods of meta-analysis: Correcting error and bias in research findings*. Newbury Park: Sage.

Jinks, J.L. and Fulker, D.W. (1970). Comparison of the biometrical genetical, MAVA, and classical approaches to the analysis of human behavior. *Psychological Bulletin, 73*, 311-349.

John, O.P. (1990). The "Big-Five" factor taxonomy: Dimensions of personality in the natural language and in questionnaires. In L.A. Pervin (Ed.), *Handbook of personality theory and research* (pp. 66-100). New York: Guilford Press.

Keller, L.M., Bouchard, T.J. Jr., Arvey, R.D., Segal, N.L. and Dawis, R.V. (1992). Work values: Genetic and environmental influences. *Journal of Applied Psychology, 77*, 79-88.

Lamb, M.E. and Bornstein, M.H. (1986). *Development in infancy* (2nd ed.). New York: Random House.

Langinvainio, H., Kaprio, J., Koskenvuo, M. and Lonnqvist, J. (1984). Finnish twins reared apart. III: Personality factors. *Acta Geneticae Medicae et Gemellologiae, 33*, 259-264.

Langlois, J.H. (1986). From the eye of the beholder to behavioral reality: Development of social behaviors and social relations as a function of physical attractiveness. In C.P. Herman, M.P. Zanna and E.T. Higgins (Eds.), *Ontario symposium on personality and social psychology* (Vol. 3, pp. 23-51). Hillsdale, NJ: Erlbaum.

Lasko, J.K. (1954). Parent behavior toward first and second children. *Genetic Psychology Monographs, 49*, 97-137.

Lerner, R.M. (1987). A life-span perspective for early adolescence. In R.M. Lerner and T.T. Foch (Eds.), *Biological-psychosocial interactions in early adolescence* (pp. 9-34). Hillsdale, NJ: Erlbaum.

Lernerz, K., Kucher, J.S., East, P.L., Lerner, J.V. and Lerner, R.M. (1987). Early adolescents' physical organismic characteristics and psychosocial functioning: Findings from the Pennsylvania early adolescent transition study (PEATS). In R.M. Lerner and T.T. Foch (Eds.), *Biological-psychosocial interactions in early adolescence* (pp. 225-248). Hillsdale, NJ: Erlbaum.

Loehlin, J.C. (1989). Partitioning environmental and genetic contributions to behavioral development. *American Psychologist, 44*, 1285-1292.

Loehlin, J.C. (1992). *Genes and environment in personality development.* Newbury Park, CA: Sage.

Loehlin, J.C. and Nichols, R.C. (1976). *Heredity, environment, and personality.* Austin: University of Texas Press.

Lounsbury, M.L. and Bates, J.E. (1982). The cries of infants of differing levels of perceived temperamental difficulties: Acoustic properties and effects on listeners. *Child Development, 53*, 677-686.

Lubinski, D. and Dawis, R.V. (1992). Aptitudes, skills, and proficiencies. In M. Dunnette and L.M. Hough (Eds.). *Handbook of Industrial and Organizational Psychology* (Vol 2). Palo Alto: Consulting Psychologist Press.

Lubinski, D., Tellegen, A. and Butcher, J.N. (1983). Masculinity, femininity, and androgyny: Viewed and assessed as distinct concepts. *Journal of Personality and Social Psychology, 44*, 428-439.

Lubs, H.A. (1969). A marker X chromosome. *American Journal of Human Genetics, 21*, 231-244.

Lykken, D.T., Bouchard, T.J. Jr., McGue, M. and Tellegen, A. (in press). How might genes determine interests? A twin study. *Journal of Applied Psychology.*

Malatesta, C.Z., Culver, C., Tesman, J.R. and Shepard, B. (1989). The development of emotion expression during the first two years of life. *Monographs of the Society for Research in Child Development, 54*, (1-2, Serial No. 219).

Martin, N. and Jardine, R. (1986). Eysencks contribution to behaviour genetics. In S. Modgil and C. Modgil (Eds.), *Hans Eysenck: Consensus and controversy.* London: Falmer International.

McCrae, R.R. and Costa, P.T. Jr. (1990). *Personality in adulthood.* New York: Guilford.

McGue, M. and Lykken, D.T. (in press). Genetic influence on risk of divorce. *Psychological Science.*

Meehl. P.E. (1978). Theoretical risks and tabular asterisks: Sir Karl, Sir Ronald, and the slow progress of soft psychology. *Journal of Consulting and Clinical Psychology, 46*, 806-834.

Meehl, P.E. (1990). Why summaries of research on psychological theories are often uninterpretable. *Psychological Reports, 66*, 195-244.

Mischel, W. (1973). Towards a cognitive social learning reconceptualization of personality. *Psychological Review, 80*, 252-283.

Moloney, D.P., Bouchard, T.J. Jr. and Segal, N.L. (1991). A genetic and environmental analysis of the vocational interests of monozygotic and dizygotic twins reared apart. *Journal of Vocational Behavior, 39*, 76-109.

Myers, I.B. and McCaulley, M.H. (1985). *Manual: A guide to the development and use of the Myers-Briggs Type Indicator.* Palo Alto: Consulting Psychologists Press.

Neale, M.C., Cardon, L.R. and Associates (1992). *Methodology for genetic studies of twins and families*. Dordrecht: Kluwer Academic Publishers.

Nichols, R.C. (1978). Twin studies of ability, personality and interests. *Homo, 29*, 158-173.

Pedersen, N.L., McClearn, G.E., Plomin, R. and Friberg, L. (1985). Separated fraternal twins: Resemblance for cognitive abilities. *Behavior Genetics, 15*, 407-419.

Pedersen, N.L., Plomin, R., Nesselroade, J.R. and McClearn, G.E. (in press). A quantitative genetic analysis of cognitive abilities during the second half of the life span. *Psychological Science*.

Pedersen, N.L., Plomin, R., McClearn, G.E. and Friberg, L. (1988). Neuroticism, extraversion, and related traits in adult twins reared apart and reared together. *Journal of Personality and Social Psychology, 55*, 950-957.

Plomin, R. and Bergeman, C.S. (1991). The nature of nurture: Genetic influences on "environmental" measures. *Behavior and Brain Sciences, 14*, 373-427.

Plomin, R. and DeFries, J.C. (1985). *Origins of individual differences in infancy: The Colorado adoption project*. San Diego: Academic Press.

Plomin, R. DeFries, J.C. and Fulker, D.W. (1988). *Nature and nurture during infancy and early childhood*. New York: Cambridge University Press.

Plomin, R., DeFries, J.C. and Loehlin, J.C. (1977). Genotype-environment interaction and correlation in the analysis of human behavior. *Psychological Bulletin, 84*, 309-322.

Plomin, R., DeFries, J.C. and McClearn, G.E. (1990). *Behavioral genetics: A primer* (2nd ed). New York: Freeman.

Rennie, J. (1992). Mutable mutations. *Scientific American*, May, 34.

Retherford, R.D. and Sewell, W.H. (1991). Birth order and intelligence: Further tests of the confluence model. *American Sociological Review, 56*, 141-158.

Rose, R., Kaprio, J. Williams, C.J., Viken, R. and Obremski, K. (1990). Social contact and sibling similarity: Facts, issues, and red herrings. *Behavior Genetics, 20*, 763-778.

Scarr, S. (1985). Constructing psychology: Making facts and fables for our times. *American Psychologist, 40*, 499-512.

Scarr, S. (1992). Developmental theories for the 1990s: Development and individual differences. *Child Development, 63*, 1-19.

Scarr, S., Webber, P.L., Weinberg, R.A. and Wittig, M.A. (1981). Personality resemblance among adolescents and their parents in biologically-related and adoptive families. *Journal of Personality and Social Psychology, 40*, 885-898.

Schofield, W. (1964). *Psychotherapy: The purchase of friendship*. Englewood Cliffs, N. J.: Prentice-Hall.

Schooler, C. (1972). Birth order effects: Not here, not now! *Psychological Bulletin, 78*, 161-175.

Sorell, G. T. and Nowak, C.A. (1981). The role of physical attractiveness as a contributor to individual development. In R.M. Lerner and N.A. Bush-Rossnagle (Eds.), *Individuals as producers of their development* (pp. 389-446). San Diego, CA: Academic Press.

Stagner, R. and Katzoff, E.T. (1936). Personality as related to birth order and family size. *Journal of Applied Psychology, 20*, 340-346.

Taylor, H.F. (1980). *The IQ game: A methodological inquiry into the heredity environment controversy*. New Brunswick, NJ: Rutgers University Press.

Tellegen, A. and Waller, N.G. (in press). Exploring personality through test construction: Development of the Multidimensional Personality Questionnaire. In S.R. Briggs and J.M. Cheek (Eds.), *Personality measures: Development and evaluation* (Vol 1). Greenwich, CN: JAI Press.

Tomlinson-Keasey, C. and Little, T.D. (1990). Predicting educational attainment, occupational achievement, intellectual skill and personal adjustment among gifted men and women. *Journal of Educational Psychology, 82*, 442-455.

Tupes, E.C. and Christal, R.E. (1961). Recurrent personality factors based on trait ratings. Technical report ASD-TR-61-97. Lackland Air Force Base, TX: USA. Air Force.

Turkheimer, E., Lovett, G., Robinette, C.D. and Gottesman, I.I. (1992). The heritability of divorce: New data and theoretical implications. Paper delivered to the Behavior Genetics Association Annual Meetings, Boulder Colorado, July 4, 1992.

Waller, N.G., Kojetin, B.A., Bouchard, T.J. Jr., Lykken, D.T. and Tellegen, A. (1990). Genetic and environmental influences on religious interests, attitudes, and values: A study of twins reared apart and together. *Psychological Science, 1*, 138-142.

Waller, N.G., Lykken, D.T. and Tellegen, A. (in press). Occupational interests, leisure time interests, and personality: Three domains or one? Findings from the Minnesota Twin Registry. In R. Dawis, D. Lubinski (Eds.), *Assessing individual differences in human behavior: New concepts and findings*. Minneapolis: University of Minnesota Press.

Wiesenfeld, A.R. and Malatesta, C.Z. (1982). Infant distress. In L.W. Hoffman, R.J. Gandelman and H.R. Shiffman (Eds.), *Parenting, its causes and consequences* (pp. 123-139). Hillsdale, N. J.: Erlbaum.

Wiggins, J.S. (1966). Substantive dimensions of self-report in the MMPI item pool. *Psychological Monographs, 80*, (Whole No. 630).

Wilson, R.A. (1983). The Louisville twin study: Developmental synchronies in behavior. *Child Development, 54*, 298-316.

Zajonc, R.B. (1983). Validating the confluence model. *Psychological Bulletin, 93*, 457-480.

Zuckerman, M. (1992). What is a basic factor and which factors are basic? Turtles all the way down. *Personality and Individual Differences, 13*, 675-681.

BEHAVIORAL GENETICS: VARIABLES, MECHANISMS AND DISORDERS

Commentary on the chapter by T.J. Bouchard

Ian J. Deary
Department of Psychology,
University of Edinburgh,
7 George Square,
Edinburgh EH8 9JZ, U.K.

First, Catch Your Variables

The issue I should like to discuss first is the validity of the variables that are to hand for the behaviour geneticist who wishes to study personality.

The genetic study of personality and cognitive ability has become more interesting as it has become less controversial. Prior to recent times researchers in behavioural genetics may be said to have had good reasons for being defensive. For a start, the variables they were studying were themselves a matter for considerable controversy. It can not have been easy to conduct research on the heritability of intelligence when the disputes concerning the stucture of human intellect pitted models emphasising the general factor against those which entertained the possibility of up to 120 separate abilities. Similarly, when it was popular to deny the existence of personality traits, and when there was little consensus within differential psychology concerning the key personality traits, researchers had an apparent credibility problem with their variables. I should like to think that even zeitgeists are sensitive to empirical data and that the ease with which one may discuss heritability nowadays has come about because the evidence is better.

The currently accepted structure of intellect has converged on a model that achieves integration of older models which used to be seen as exclusive alternatives rather than complementary accounts. Although such a structure was inherent in the results of hierarchical factor analytic studies of ability, it is the arrival of the more explicitly testable models allowed by confirmatory factor analysis that has convinced most researchers that there is a place for g, group factors, primary mental abilities and more specific abilities within a unified model of human mental ability (Gustaffson, 1984; Undheim and Gustafsson, 1987).

Something similar has happened in the field of personality trait research. It is not such a long time ago that the existence of personality traits was called into question (Mischel, 1968) and yet, in the last year or so, there have been many reports insisting not only on the existence of traits, but suggesting that there is an emerging consensus as to the type and number of traits that exist in a full description of human personality differences. As Bouchard indicates, this consensus has converged on the so-called Big Five personality dimensions, the best known version of which has arisen from the work of McCrae and Costa (Digman, 1990; Angleitner, 1991; McCrae 1989; Wiggins and Trapnell, 1991). However, not all of these traits enjoy an equal level of consensus. There is hardly any dispute about the existence or importance of

J. Hettema and I. J. Deary (eds.), Foundations of Personality, 45–53.

Extraversion and Neuroticism, but the traits beyond these two dimensions still cause some controversy within differential psychology, with Eysenck (1992a,b) advocating Psychoticism as the third of the 'Gigantic Three' and Costa and McCrae (1992a,b) supporting Conscientiousness, Agreeableness and Openness as the trio that complete the Big Five. Eysenck has suggested that A and C might be primary level traits associated with the broader dimension of P - McCrae and Costa put A and C on a par with N and E - and one can discern a not dissimilar hierarchical compromise in the results of Zuckerman et al. (1991). Zuckerman (1991) adds a further interesting suggestion to this discussion by stating that it is the narrower personality traits that will provide better starting point for boiolgical investigations of personality than the broader dimensions beloved of Eysenck. In this author's opinion N and E have more substantial validity than other dimensions. Given that P, A and C all have substantial heritabilities in Bouchard's summary, one is led to doubt whether behaviour genetic data will help to resolve this live psychometric argument.

Variables Derived from Psychopathology

The research reviewed by Bouchard almost exclusively concerns personality traits describing personality in the population at large. This emphasis on research on dimensions distinguishes differential psychology from more medically oriented research which tends to concentrate on medical syndromes. Within the medical arena even personality research is conducted within a 'syndrome' model, with the DSM-IIIR classification recognising 11 types of personality disorder. Therefore, which direction should future behaviour and molecular genetic research take?; type or trait. There are good reasons for excluding neither approach. Personality traits have considerable validity, as we have seen with the emergence of the five factor model, especially E and N. However, personality disorders have a large impact on the lives of those suffering from them. Some of the abnormal personality variants might be related to psychiatric illnesses; schizoid personality disorder is sometimes included in the 'schizophrenic spectrum', and those diagnosed as schizoid in childhood tend to develop schizoid personality disorder or schizophrenia later in life (Wolff et al., 1991).

Whereas differential psychologists have spent decades refining scales to measure the key dimensions of normal personality, psychiatric researchers have worried about nosology and have only recently taken on the issues of measurement. Additions and subtractions to the list of personality disorders tends to be done by committees more concerned with clinical impressions than psychometric data. This is starting to change: psychologists such as Claridge and Hare (Hare et al., 1991) have investigated potential psychometric approaches to schizotypal and anti-social personality disorders, respectively. Within psychiatry, too, one of the issues of the day is whether the typological scheme should be replaced by a dimensional scheme (Blackburn, 1988). Such a change might rid the DSM-III Axis two of a number of chronic problems such as co-morbidity and multiple diagnoses, overlapping criteria of personality disorders and Axis I diagnoses, and questionable validity (Widiger and Shea, 1991).

Early attempts at a factor analytic approach to abnormal personality suggested that the large number of dimensions might be replaced by as few as four dimensions, viz., social deviance, submissiveness, obsessional-schizoid characteristics and hysteria (Presly and Walton, 1973). More recent attempts to render personality disorders dimensional within a patient group suggested that three dimensions might suffice: social involvement (affiliation), assertion (dominance), and anxious rumination (Widiger et al., 1987). The resemblance between these three factors and some basic dimensions agreed to underlie normal personality is striking and emphasises the need for research aimed at integrating the two efforts. In fact, investigations by Livesley and Schroeder (1990) on the DSM-IIIR cluster A personality disorder diagnoses have established that the factorial structure of these personality features is similar in the general population and in a clinical sample. Further research has confrmed that personality disorders are not clinical exotica and that they are found in a significant proportion of people who do not present to psychiatrists (Zimmerman and Coryell, 1989; Casey and Tyrer, 1990). Another reason that clinicians might be warming to dimensions is that they provide a scheme with higher reliability than one based on categories (Heuman and Morey, 1990).

There is some evidence from traditional behaviour genetic research methods to indicate that some aspects of abnormal personality are partly heritable (McGuffin and Thapar, 1992). This is true for at least antisocial personality disorder, anxious or avoidant personalities, obsessionality and schizoid-schizotypal disorders. However, McGuffin and Thapar argue that it is not possible to state to what degree these simply reflect the heritabilities of related personality dimensions and psychiatric illnesses. Therefore, the genetic study of personality and its disorders would benefit from the integration of the psychometric and the psychiatric approaches.

Molecular Genetics: Getting at Mechanisms?

Why do behaviour geneticists estimate heritabilities?; not to get to the point where they may say that a number of personality traits are about 50% heritable, surely. I rather like the line that Robert Plomin has taken in emphasising that behaviour genetic research provides information about the effects of the environment on personality. The effects of the environment will be raised in a later section, but at present it will be suggested that estimates of heritability are important in that they allow further research into the mechanisms through which genes have an impact on the phenotype. Therefore, if it is known that a trait has substantial heritability, one may ask whether a molecular genetic approach might be the next stage in personality research.

Advances in recombinant DNA technology have led to massive advances in our understanding of the human genome (Baron and Rainer, 1988). The human genome, by degrees, is being mapped and sequenced and, in advance of delivering a complete account of the human genome, the new technology has already pinpointed the molecular genetic deficiency involved in some forms of human disease. The present brief account will be limited to a few conditions that have a substantial psychological component. Most will know that the genetic problem that leads to the autosomal dominant dementing disorder Huntingdon's chorea was recently located on chromosome 4. Therefore, premorbid testing for this genetic variant allows diagnosis of the disorder to be more than 97% accurate. Although the region of the chromosome has been narrowly identified, the gene product and the pathogenesis of the disorder are still unknown. The second most prevalent cause of mental subnormality, fragile X syndrome, has recently proved tractable to molecular genetic research (York Moore, 1992). It is now known that the molecular lesion in fragile X syndrome is a multiply repeated CGG sequence of DNA base pairs on the X chromosome.

More is known about the genetic disorder underlying some forms of familial Alzheimer's dementia. Affected members of some families have been shown to have a mutation on the region of chromosome 21 that codes for the amyloid precursor protein (Hardy and Allsop, 1991). This is a transmembrane protein in normal individuals which contains a recognised enzyme sequence. In affected individuals there is an abnormality in post-translational cleavage that leads to the release of amyloid protein subunits. These form amyloid deposits which some consider to be the key stage in the formation of plaques and neurofibrillary tangles in neuronal tissue, pathognomonic signs of Alzheimer's disease. Familial Alzheimer's disease highlights the power of molecular genetic research for advancing biological psychology. Previously, behaviour genetic research using twins had established that some forms of Alzheimer's disease appeared to have a genetic component (Deary and Whalley, 1988). Molecular genetic research pinpointed the problem at the level of specific DNA base pairs and, in a very short time, genetic research became part of the molecular biological approach to the disorder, allowing the mechanisms of the disease to be understood.

About 15 years ago most of the biological research in psychological disorders focussed on schizophrenia and depression, with very little attention afforded to Alzheimer's type dementia. The recent molecular advances in Alzheimer's disease have outstripped those in the other psychiatric disorders and the original much-publicised molecular genetic 'discoveries' in schizophrenia and major depressive disorder provide a good warning to those who might think that the molecular genetic path will prove smooth or straight.

One major difficulty in conducting molecular genetic research is deciding which regions of which chromosomes to examine in attempting to locate genes that are important bases of disorders or traits. In 1988 Bassett et al., reported the occurrence of schizophrenia in a Chinese proband and a maternal uncle. Apart from their psychopathological features, they shared facial features and other minor physical abnormalities that distinguished them from other members of the family. They also shared a partial trisomy of chromosome 5; an extra segment of chromosome 5 was inserted into chromosome 1. The proband's mother had a balanced translocation; no triplicated genes and no psychosis. Therefore, the translocation site on chromosome 5 became a candidate region for schizophrenia predisposition; it is probably some understatement to say that this was like knowing that a given bucketful of hay in the haystack contained the needle.

Later in the same year Sherrington et al. (1988) reported that a dominant allele in the chromosome 5q 11-13 region was inherited with the same pattern as schizophrenia in 5 Icelandic and 2 English families. Among the 7 families there were 39 cases of schizophrenia (all sub-types), 5 cases of schizoid personality and 10 other psychiatric diagnoses (including manic-depressive illness). Concordance was improved if schizoid personality was included as a 'case'; the lod score was 6.49, i.e. the disease and the marker were closely linked (a lod score of 3 is usually taken as evidence of significant linkage). For a short time it seemed as if one of the genetic bases of this major psychiatric disroder had been located. However, several studies in the USA, UK and in Sweden have failed to replicate the finding and, further, have furnished strong evidence against a linkage between schizophrenia and the candidate region on chromosome 5 (see Crowe et al., 1991). These disappointing results were emerging at about the same time as the announcement that the previously trumpeted molecular genetic basis of bipolar affective disorder in an Amish pedigree had failed to hold up when more members of the group were included (Kelsoe et al., 1989). These failures led Owen (1992) recently to ask whether schizophrenia poses such great barriers to molecular geneticists that their careers will be wasted in this area.

Why have these failures in schizophrenia and affective disorder followed the solid successes in Huntingdon's chorea, Alzheimer's dementia, fragile X syndrome and cystic fibrosis? In fact, there are special problems in the key psychiatric disorders which will apply to the key personality dimensions also. The inheritance patterns of the disorders are not known; therefore, multiple genetic models tend to be tested. Different estimates of penetrance are usually included in the models. The definition of an affected phenotype is usually uncertain (one schizophrenia subtype, any schizophrenia, all psychoses, schizophrenia spectrum including relevant personality disorders?) and the models tend to be run using different definitions. All this makes type I errors likely.

These cautionary notes notwithstanding, I remain convinced that molecular genetic research will soon start to make an impact on the psychology of personality traits and cognitive abilities. There are certain necessary conditions before success will come from this enterprise. First, the phenotypic traits have to be characterised clearly; it may be that relatively narrow personality traits such as impulsiveness and spatial ability, respectively, will prove more tractable than, say, extraversion and general intelligence. The traits must have an established genetic basis: traits such as perfect pitch, dyslexia and handedness are the subject of molecular genetic efforts at present, without positive results to date. Most demandingly, it is necessary for the molecular geneticist to have some idea of where to search on the chromosome; therefore candidate regions will have to be suggested by those conducting research on, say, neurotransmitter correlates of personality. This desideratum will not persist forever; the day will come when blind searches of the genome will take place to find the basis of any heritable trait. This would demand an effort that is beyond the scope of present knowledge and technology by several orders of magnitude; nevertheless, there is no longer any problem in principle of locating any phenotypic trait to its putative genetic bases. This should be cause for excitement for biologically-oriented psychologists; the path from psychological traits to the molecular genetic mechanisms underlying their expression is becoming ever more accessible.

Personality traits, having continuous distributions, are assumed to be inherited in a polygenic fashion. If molecular genetic approaches are to make an impact, therefore, it must be established that the method can detect multiple gene loci, each with a limited effect on the phenotype. There is promise that this is now possible; the much-cited example of success in this area is the demonstration that six quantitative trait loci control fruit mass in the tomato, whereas five contribute to pH of the fruit and four have an impact on liquid soluble concentrations (Paterson et al., 1988). In other words, there is now no barrier to conducting multivariate research to discover the small effects of several gene loci on human traits; as Plomin (1990) stated, "behaviour is not too complex for molecular biology."

Environmental Effects and Development

It is worth emphasising what behaviour genetics can do other than offer heritability estimates of human psychological traits. Of course, the aportioning of environmental variance in personality to between and within family effects has consistently delivered the message that family upbringing has less of an effect than most would have predicted and that individual environmental variance is much more important. There is also the fact that the effect of genes on personality persisted when the effect of shared environment was removed in the study of about 14000 Finnish co-twins. Perhaps, though, one may ask for a fuller discussion of the fact that there is the persistent finding that the correlations between sibs and dizygotic twins are often much less than half that of monozygotic twins. This is a feature of the data sets presented in Bouchard's paper, and is one that is not generally seen in studies of cognitive ability. Certainly, non-additive genetic variance has been proposed, and would fit with known genetic mechanisms (Pedersen et al., 1988). In addition, Eaves et al. (1989) propose an alternative or complementary explanation based upon competitive social interaction between non-identical twin pair members and siblings based on their genotypes. This interesting suggestion deserves futher attention and perhaps might lend itself to experimentally based research. At least in the case of extraversion it is rather easy to envisage a situation within a family in which it is difficult for all siblings to pursue successful strategies as extraverts. Another area that might be highlighted more is the contribution that behaviour genetic research designs can make to the study of personality change over time. For instance, Plomin and Nesselroade (1990) have suggested that there is a genetic contribution to personality change during childhood, but much less so in change during adulthood.

Structural Equation Models

Perhaps a short comment on the statitistical techniques involved in behaviour genetic genetic research is appropriate. Although structural equation models can be powerful, it must be realised that there are limitations to these techniques. The models being tested are often only a subset of those that might be tested. Various criteria may be used to estimate the goodness of fit of models, and none of these is definitive. For instance, not all readers of papers that present structural equation models appear to appreciate the apparent paradox of the size of the samples involved when using the chi square test; it becomes relatively easier to reject a model as the sample size increases. Another example of the wrinkles in these methods is that indices of fit for models can be quite different when using method of least squares versus maximum likelihood methods. Perhaps the greatest problem with these techniques is that structural equation modeling is not generally understood by the majority of psychological researchers. This means that most of our colleagues are not in a position to be sufficiently impressed or sceptical of the results of behaviour genetic studies. It will be easier to communicate the substantial progress in understanding the heritability of personality traits when the communication of model fitting procedures is better or when the scientific public make more efforts to keep abreast of modern multivariate techniques. Bouchard presents impressive data and communicates them clearly, but there is a very considerable education effort required before the non-psychometric/behaviour genetic researcher will have the knowledge to appreciate the ineluctable nature of his results.

Conclusion

Plomin (1990) stated recently that there is much more general acceptance of the fact that there is a genetic contribution to individual differences in human behaviour patterns, including personality and cognitive ability. In this article, that fact is taken for granted and some additional questions are asked. In particular, the problem of the variables that are to be studied was raised. Although there is consensus to some extent in differential psychology, there are interesting variables from psychiatry that have attracted attention and their relationships with the main personality dimensions are only partly understood. The future for genetic studies of behaviour lies in molecular genetic research, perhaps based on the initial findings gleaned from traditional behaviour genetic methods which use twin, family and adoption designs. A major consideration for the psychologist interested in the genetic contribution to behaviour traits will be in confronting the expertise that is needed to understand the publications in this area. The biology of molecular genetics and the statistical techniques are abstruse to the average psychologist, even those who are inclined to be sympathetic to genetic research. Much of the burden of education must lie with those psychologists conducting research in behaviour genetics; they have important facts about human nature to communicate and they must create a sophisticated audience to receive them.

References

Angleitner, A. (1991). Personality psychology: trends and developments. *European Journal of Personality, 2*, 121-142.

Baron, M. and Rainer, J.D. (1988). Molecular genetics of human disease: implications for modern psychiatric research and practice. *British Journal of Psychiatry, 152*, 741-753.

Bassett, A., McGilvray, B.C., Jones, B.D. and Pantzar, J.T. (1988). Partial trisomy chromosome 5 co-segregating with schizophrenia. *Lancet, i*, 799-801.

Blackburn, R. (1988). On moral judgements and personality disorders. *British Journal of Psychiatry, 153*, 505-512.

Casey, P.R. and Tyrer, P. (1990). Personality disorder and psychiatric illness in general practice. *British Journal of Psychiatry, 156*, 261-265.

Costa, P.T. and McCrae, R.R. (1992a). Four ways the five factors are basic. *Personality and Individual Differences, 13*, 645-652.

Costa, P.T. and McCrae, R.R. (1992b). Reply to Eysenck. *Personality and Individual Differences, 13*, 861-865.

Crowe, R.R., Black, D.W., Wesner, R., Andreasen, N.C., Cookman, A. and Roby, J. (1991). Lack of linkage to chromosome 5q11-13 markers in six schizophrenia pedigrees. *Archives of General Psychiatry, 48*, 357-361.

Deary, I.J. and Whalley, L.J. (1988). Recent research on the causes of Alzheimer's disease. *British Medical Journal, 297*, 807-810.

Digman, J.M. (1990). Personality structure: emergence of the five factor model. *Annual Review of Psychology, 41*, 417-440.

Eaves, L.J., Eysenck, H.J. and Martin, N.G. (1989). *Genes, Culture and Personality*. London: Academic Press.

Eysenck, H.J. (1992a). Four ways the five factors are not basic. *Personality and Individual Differences, 13*, 667-674.

Eysenck, H.J. (1992b). A reply to Costa and McCrae. P or A and C - the role of theory. *Personality and Individual Differences, 13*, 867-868.

Gustafsson, J.E. (1984). A unifying model for the structure of mental abilities. *Intelligence, 8*, 179-203.

Hardy, J. and Allsop, D. (1991). Amyloid deposition as the central event in the aetiology of Alzheimer's disease. *Trends in Pharmacological Sciences, 12*, 383-388.

Hare, R.D., Hart, S.D. and Harpur, T.J. (1991). Psychopathy and the DSM-IV criteria for antisocial personality disorder. *Journal of Abnormal Psychology, 100*, 391-398.

Heumann, K.A. and Morey L.C. (1990). Reliability of categorical and dimensional judgments of personality disorder. *American Journal of Psychiatry, 147*, 498-500.

Kelsoe, J.R., Ginns, E.I., Egeland, J.A., Gerhard, D.S., Goldstein, A.S., Bale, A.M., Pauls, D.L., Long, R.T., Kidd, K.K., Conte, G., Housman, D.E. and Paul, S.M. (1989). Re-evaluation of the linkage relationship between chromosome 11p loci and the gene for bipolar affective disorder in the Old Order Amish. *Nature, 342*, 238-243.

Livesley, W.J. and Schroeder, M.L. (1990). Dimensions of personality disorder: the DSM-IIIR cluster A diagnoses. *The Journal of Nervous and Mental Disease, 178*, 627-635.

McCrae, R.R. (1989). Why I advocate the five factor model: joint factor analyses of the NEO-PI with other instruments. In D.M. Buss and N. Cantor (Eds) *Personality Psychology: Recent Trends and Emerging Directions* (pp 237-245). New York: Springer.

McGuffin, P. and Thapar, A. (1992). The genetics of personality disorder. *British Journal of Psychiatry, 160*, 12-23.

Owen, M.J. (1992). Will schizophrenia become a graveyard for molecular geneticists? *Psychological Medicine, 22*, 289-293.

Paterson, A.H., Lander, E.S., Hewitt, J.D., Peterson, S., Lincoln S.E. and Tanksley, S.D. (1988). Resolution of quantitative traits into Mendelian factors by using a complete linkage map of restriction fragment length polymorphisms. *Nature, 335*, 721-726.

Pedersen, N.L., Plomin, R., McClearn, G.E. and Friberg, L. (1988). Neuroticism, extraversion and related traits in adult twins reared apart and together. *Journal of Personality and Social Psychology, 55*, 950-957.

Plomin, R. (1990). The role of inheritance in behaviour. *Science, 248*, 183-188.

Plomin, R. and Nesselroade, J.R. (1990). Behaviour genetics and personality change. *Journal of Personality, 58*, 191-200.

Presly, A.S. and Walton, H.J. (1973). Dimensions of abnormal personality. *British Journal of Psychiatry, 122*, 269-276.

Sherrington, R., Brynjolfsson, J., Pertursson, H., Potter, M., Dudleston, K., Barraclough, B., Wasmuth, J., Dobbs, M. and Gurling, H. (1988). Localisation of a susceptibility locus for schizophrenia on chromosome 5. *Nature, 336*, 164-166.

Undheim, J.O. and Gustafsson, J.E. (1987). The hierarchical organisation of cognitive abilities: restoring general intelligence through the use of linear structural relations. *Multivariate Behavioural Research, 22*, 149-171.

Widiger, T.A. and Shea, T. (1991). Differentiation of Axis I and Axis II disorders. *Journal of Abnormal Psychology, 100*, 399-406.

Widiger, T.A., Trull, T.J., Hurt, S.W., Clarkin, J. and Frances, A. (1987). A multidimensional scaling of the DSM-III personality disorders. *Archives of General Psychiatry, 44*, 557-563.

Wiggins, J.S. and Trapnell, P.D. (1991). Personality structure: the return of the big five. In S.R. Briggs, R. Hogan and W.H. Jones (Eds) *Handbook of Personality Psychology*. Orlando, FL: Academic Press.

Wolff, S., Towenshend, R., McGuire, R.J. and Weeks, D.J. (1991). 'Schizoid' personality in childhood and adult life II: adult adjustment and the continuity with schizotypal personality disorder. *British Journal of Psychiatry, 159*, 620-629.

York Moore, D.W. (1992). New developments in the fragile X syndrome. *British medical Journal, 305*, 208.

Zimmerman, M. and Coryell. W. (1989). DSM-III personality disorder diagnoses in a non-patient sample. *Archives of General Psychiatry, 46*, 682-689.
Zuckerman, M. (1991). *Psychobiology of Personality*. Cambridge: Cambridge University Press.
Zuckerman, M., Kuhlman, D.M., Thornquist, M. and Kiers, H. (1991). Five (or three) robust questionnaire scale factors of personality without culture. *Personality and Individual Differences, 12*, 929-941.

FROM DNA TO SOCIAL BEHAVIOUR: CONDITIONS FOR A PARADIGM OF PERSONALITY RESEARCH

Hans J. Eysenck
Institute of Psychiatry,
University of London
De Crespigny Park
Denmark Hill
London SE5 8AF, United Kingdom

Abstract

Is is suggested that in order to agree on a paradigm of personality, it is necessary to have agreed criteria. It is argued that factor analysis is not sufficient to provide a proper answer, and that genetic and psychophysiological antecedents and experimental and social outcomes of behavior characteristic of the major dimensions of personality are needed to complete any proper theory of personality. These constitute the distal and proximal causes and effects of such behaviors, and give meaning and relevance to the otherwise purely abstract traits "discovered" by correlational studies.

Beyond Psychometrics: The Need for Theory

It is only too well-known that paradigms, essential to progressive science, are woefully missing in the social sciences (Barnes, 1982); the study of personality is no exception. Yet I have suggested that such a paradigm is not only possible, but is in process of being created (Eysenck, 1983a). This view is based on a fundamental assumption which will probably receive at least lip service from most serious research workers, namely that man is a biosocial animal (Eysenck, 1980a,b). In other words, our behaviour is determined in part by *biological causes*, in part by *social learning*. It follows that any meaningful theory of personality must incorporate *both* biological and social determinants. Yet while few would probably deny an anxiom of this kind, in actual practice most psychologists interested in personality and individual differences generally have concentrated almost exclusively on the social side; indeed, most have got stuck at the preliminary level of taxonomy, using factor analysis (or some related technique, such as multi-dimensional scaling) to sort questionnaire answers or ratings into primary and second-order factors. Yet experts in the field of taxonomy have shown that such calculations cannot decide between alternative hypotheses in the absence of *causal theories* (Sokal and Sneath, 1963).

I have elsewhere argued for the importance of theories in this connection (Eysenck, 1985), the general importance of the study of individual differences in a scientific psychology (Eysenck, 1984, 1985), and the relevance of studying the way individual differences are based on biological foundation, and in turn affect social behaviour (Eysenck, 1983b). But the clearest

J. Hettema and I. J. Deary (eds.), Foundations of Personality, 55–66.

proof for the importance of biological factors is the demonstration that genetic factors are responsible for the major part of the personality phenotype differences that have been studied (Eaves et al., 1989; Eysenck, 1983c). The evidence on this point is now overwhelming, with many studies using up to 15,000 twins, and agreeing on the major findings, namely that genetic variance accounts for more phenotypic variance than any other factor, or even all other factors combined, and that environmental factors are overwhelmingly within-family; in other words, the usual theories of personality development, which postulate between-family factors, are wrong.

PEN versus Big Five

We thus have at the one end DNA, producing individual differences which emerge finally as personality traits and types, leading in turn to differential social behaviours. Clearly we must postulate some intermediary biological factors (physiological, neurological, hormonal, enzymatic or what not) which link the DNA with the phenotypic expression of personality. Only by formulating and testing theories seeking to explain this link can we hope to arrive at a paradigm of personality. I have elsewhere discussed the claims of three major groups, to provide a taxonomic paradigm of personality phenotypes - the Cattell 16 PF, the "Big Five", and the PEN system (Eysenck, 1991a). It will be clear to readers that agreement can only be reached by going beyond simple psychometric techniques, and by relying on specific theories having both physiological and behavioural consequences which admit of being tested.

Consider a simple point. My PEN system postulates a major factor of *psychoticism* (Eysenck, 1992), while the Big Five postulate the existence of two major factors called 'agreeableness' and 'conscientiousness'. P is highly correlated with A and C (negatively), the multiple R amounting to .85 when disattenuated. We can model this statistical finding in two ways, as illustrated in Fig. 1. We may say that A and C are primary factors forming part of the group which defines P (a), or we can say that psychoticism is a compound of A and C (b). Within psychometrics there is no test to decide between these two hypotheses, and consequently no proper taxonomic conclusions can be arrived at which are binding on proponents of the two hypotheses. This is a serious deficit which can only be eliminated by referring to a wider, more inclusive schema.

The Big Five taxonomy is based on a hypothesis accepted as axiomatic, but essentially lacking any empirical support, namely the view that frequency of adjectives relating to personality qualities will faithfully match the actual importance of the traits involved (Goldberg, 1982, 1990). It is certainly possible that this is approximately true, but in the absence of empirical proof it cannot be assumed, and made the basis of a paradigm. Neither is it obvious why adjectives rather than adverbs, verbs or nouns (De Raad, 1992) should be used. De Raad's attempt to use verb and noun structurers shows that these cannot be made to support anything like the Big Five structure. His analysis of the adjective structure illustrates the great problems faced in extending this method to another language. The subjectivity of interpretation of the resulting factors can best be shown by asking readers to name a factor characterized by the following adjectives: superficial, cringing, hypocritical, fanatical, soft, narrow-minded, silly, prudish and characterless, and opposed by critical, sharp and militant. Does this factor make any psychological sense? I have asked several psychologists to name it; none had any confidence in his interpretations, and none agreed with anyone else's - or De Raad's interpretation! The obvious failure to discover a proper duplicate of the Big Five in a related language should be contrasted with the high duplicability of the PEN system over many cultures (Barrett and Eysenck, 1984).

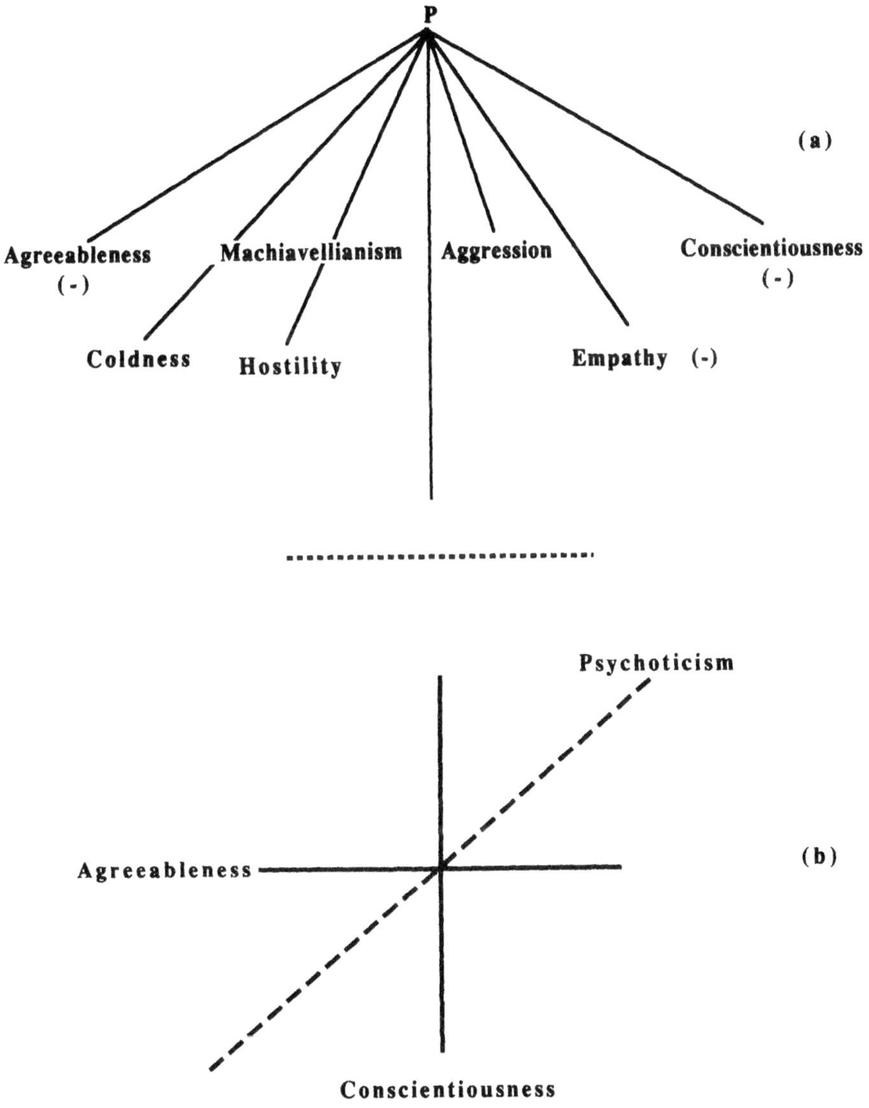

Figure 1. Alternative models of the relation between psychoticism (P), Agreeableness (A) and Conscientiousness (C).

It is often assumed (or even stated directly) that there is unanimity about the interpretations of the "Big Five" factors. That is clearly not so. Factor O is interpreted as "openness" by McCrae (1992), as "intellect" by Goldberg and other early adherents of the Big Five. Yet openness only correlates .46 with Goldberg's intellect, i.e. showing only 21% of the variance in common (McCrae, 1992). Even the theoretically unconnected Sensation-Seeking Scale correlates more highly with openness (r = .53) in McCrae's sample. The Hogan Personality Inventory Scale of Intellectance only correlates .21 with openness in the same study. As Goldberg and Rosolack (1992) have pointed out, "the disagreement about the specific nature of Factor V is somewhat of a scientific embarrassment". Such disagreement must throw considerable doubt on the psychological reality and meaning of the factor.

Digman (1990) comments on the lack of accord with respect to the meaning of all the Big Five factors. There is considerable agreement on E and N, but little on the other three. Readers may like to consult his Table 1, and his comments, to see how varied are the interpretations of the 3 factors in question. Clearly there is not only a variety of interpretations, but a complete failure to furnish us with a *criterion* to decide which interpretation (if any!) is correct. Vague, fuzzy and subjective interpretations are incapable of being tested, hence they lack the first requirement of *testability* which is needed for any theory to be considered scientific.

Nomological Networks: The case of Psychoticism

I have elsewhere argued the case for the importance of the concept of psychoticism (Eysenck, 1992). Fig. 2 shows in diagramatic form what is claimed. The abscissa shows a quantitative continuum ranging from schizophrenia through schizo-affective disorders onto affective disorders to unipolar depression, schizoid behaviour, psychopathy, aggressive, and hostile behaviour to impulsivity and criminality. At the opposite end are low-P traits like altruism, socialization, empathy, unconventionality, and conformity (plus agreeableness and conscientiousness, if you like). The normal curve roughly describes the distribution of this trait of psychoticism, and P_A suggests the probability of a person at a given point on the continuum, suffering from the psychopathologies indicated at the extreme right.

Note that this model is built on hundreds if not thousands of careful observations and experiments going back well over 100 years. These demonstrate, as I have shown elsewhere (Eysenck, 1992), the following essential points without which the scheme would not be viable. (1) Different diagnostic psychotic illnesses are not clearly separated into disease entities, but form a continuum without precise boundaries. (2) Functional psychoses are not clearly separated from psychopathic, schizoid and other odd and eccentic behaviours, but form a continuum with them. (3) These behaviours, in turn, shade into average types of behaviour, and these in turn into low-P activities. (4) The P scale correlates in predictable ways with different groupings within the continuum, i.e. very ill psychotics have higher P scores than less ill psychotics; psychopaths and criminals have higher P scores than normals, etc. (Apparently contradictory data find a ready explanation, but to go into such byways would take us too far in this context). (5) The data obey the *proportionality criterion*, i.e. a given test or measure which, following theory, should differentiate schizophrenics from normals, also correlates with P within a normal, or within a psychotic group. In other words, if T1 differentiates significantly between psychotics and normals, it also correlates with P within either group. Many such tests have been discussed in my presentation of the theory, but an example may be useful.

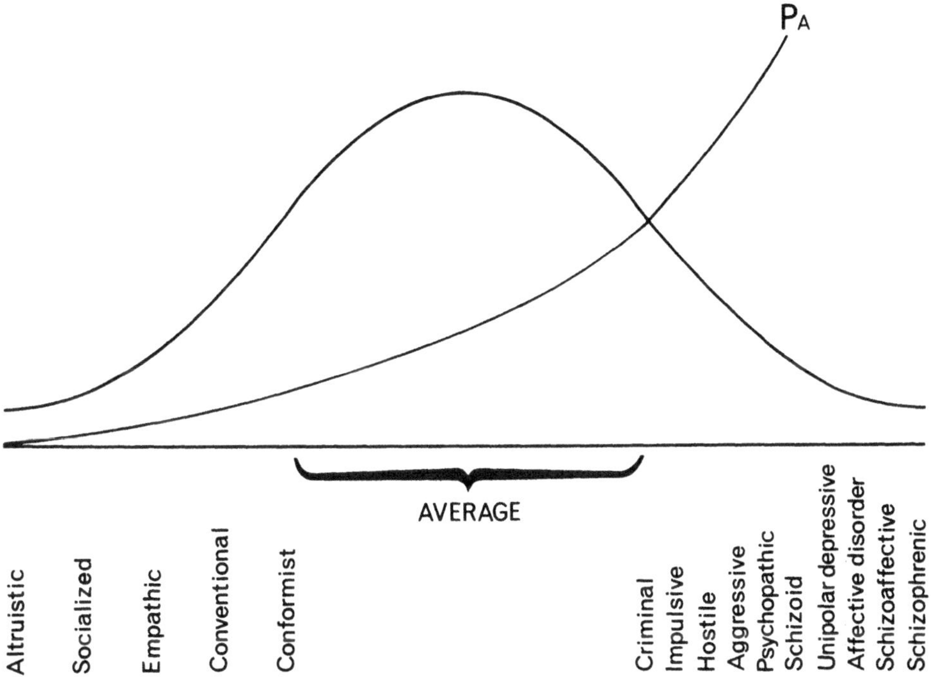

Figure 2. The psychoticism continuum. PA indicates the increasing probability of psychotic breakdown with increasing P score. (From Eysenck, 1992).

Consider Claridge's (1985) theory of psychosis, using the "phenomenon of reversed covariation". He suggested that psychosis did not involve a simple shift in, say, emotional arousal but represents instead a much more complex dissociation of CNS activity. He argued that in the schizophrenic, physiological mechanisms that are normally congruent in their activity and thereby maintain integreted CNS function become uncoupled and dissociated. He concentrated on two aspects of central nervous system functioning which he considered to be particularly involved in this uncoupling process; *emotional arousal* and a mechanism concerned with the regulation of *sensory input*, including variations in perceptual sensitivity and the broadening and narrowing of attention. He and his colleagues found strong evidence for this theory in comparing schizophrenics and normals but he also found that normal, high P subjects demonstrated the phenomenon of reversed covaration as well, when compared with low P scorers. Similarly, close (normal) relatives of schizophrenics, and normals given LSD, showed reversed covariation (Claridge, 1985; Claridge and Chappa, 1973; Claridge et al., 1983; Eysenck, 1993a give a more detailed discussion of the argument).

Let us consider for a moment the way in which the argument can be used to *interpret* the nature of a given factor more objectively than can be done by simply looking at items having high loadings, and subjectively interpreting the factor. In the first place, P is based on a theoretical conception (Eysenck, 1952), having much empirical support. In the second place, the proportionality criterion supports experimentally the interpretation of the factor. Big Five factors like A and C lack both these advantages; they are discovered by exploratory factor analysis, are subjectively interpreted by simple inspection of high-loading items, and completely lack the nomological network which alone gives scientific meaning to a concept (Cronbach and Meehl, 1955; Garber and Strassberg, 1991). Advocacy of the Big Five throws away all the advantages of proper construct validity in favour of an extreme subjectivism.

Another example may be useful. In it I have attempted to relate DNA through psychoticism to creativity and genius. All the parts of the theory have firm empirical support, but of course the very lengthy statement of theory and evidence in my monograph has to be cut to the very bones for the purpose of this article. We start with the twin observations that there appears to exist a relationship between psychopathology and genius (Richards, 1981), and that close relatives of schizophrenics (although not schizophrenics themselves) are unusually creative (Eysenck, 1983d). I suggested that creativity was related to psychoticism (Eysenck and Eysenck, 1976), and hence to genius, and both hypotheses have received empirical support (Eysenck, 1993b).

We thus have the beginnings of a link from personality through creativity to genius (defined in terms of creative and socially valued production). It is also known that P has strong genetic antecedents (Eaves et al., 1989); what lies between DNA and personality? Creative people and high P scorers are characterized by a trend towards associative overinclusiveness (as are schizophrenics); this may constitute the essential feature of creativity, i.e. an unusually wide supply of associative ideas. What may lie at the base of overinclusiveness? One possible cause may be an absence of *latent inhibition* (Lubow, 1989), a theory according to which schizophrenics fail to note experienced associations which would inhibit the calling up of a wide set of associations to certain experimentally manipulated stimuli. High P scorers also have been found to have little latent inhibition (Lubow, 1989; Lubow et al., 1992; Baruch et al., 1988), and hence to resemble psychotics and normal "psychosis prone" individuals. We may add, at least provisionally, latent inhibition as a link between DNA and P. Note that already Bleuler talked about the "loosening of associated threads" in schizophrenia, and Lubow (1989) went on to portray such behaviour in terms of a failure to attend to appropriate context information.

Lubow (1989) has presented a strong argument, based on extensive experimental work on animals and humans, linking schizophrenia and latent inhibition with dopaminergic effects, and all this may provide our last missing link between DNA and personality. We now have a complete chain from DNA through dopaminergic functioning to latent inhibition, psychoticism, overinclusiveness, creativity, and genius. Obviously other factors also play an important part (Eysenck, 1993b), but it would take us too far to discuss these factors (e.g. intelligence, social class, special abilities, etc.) here; the major links in the chain are as indicated.

This all-too-brief sketch of a theory must suffice; it will indicate ways of testing experimentally such interpretations of P as psychoticism, and using its psychophysiological underpinnings as an aid in accounting for its important social effects. Comparison with A and C will immediately reveal the complete lack of a nomological network of these concepts. Psychometrically the high (negative) correlation between P, on the one hand, and A and C, on the other, might allow us to say that genius is caused by disagreeableness and lack of conscientiousness! It will be obvious why such an inversion will not work; the possible correlational associations do not suggest any causal relation, and are completely unrelated to any theory related to creativity. The Big Five theory is drawing promissory notes on a non-existing bank account; factors like agreeableness, conscientiousness and openness simply do not have even the beginnings of the nomological network that many hundreds of empirical studies have constructed around concepts like P, E and N. This is not to say that no other dimensions of personality exist, or will ever be found; it would be nonsensical to make any such prediction. All that is being maintained is that no such dimensions have been suggested, or have received the necessary theoretical underpinning and experimental psychological and physiological support needed. It is on these grounds that I venture to argue that Fig. 1a is a better, more useful and more fruitful representation than Fig. 1b, even though psychometrically there may be little to choose between them.

Nomological Networks: The case of Extraversion

The advantages of having a proper theory, and a nomological network underpinning the construct validity of a given concept in this field lies in the introduction of *causal* relations into the taxonomic debate. Consider extraversion. A central aspect of extraverted behaviour is sociability. Critics of trait descriptions of personality have often criticized this approach by saying that it is tautological - people vary in their social behaviour from being very sociable to being very unsociable; on this variability we base the concept of a trait of sociability, only to go on to explain the observed behaviour in terms of the trait derived from it! But such a criticism does not apply to the concept of extraversion. Extraverted behaviour is explained in terms of low cortical arousability (Eysenck, 1967), i.e. a largely inherited property of the cortex (possibly mediated by the ascending reticular activating system). Extraverts have an ARAS which is relatively inactive, and hence requires strong sensory imput to enable the cortex to achieve a proper level of cortical arousal. Social contacts are a good way to achieve this greater arousal, and hence extraverts are sociable. Of course such a theory must also explain why extraverts demonstrate the other traits which, correlating together, define the concept; this is what is meant by a nomological network. I have elsewhere discussed at length the evidence for the biological theories linking P, E and N with psychophysiological, hormonal, neurological and other biological determinants (Eysenck, 1990, 1993a). It is clear from the empirical evidence that we are beginning to get a proper understanding of the biological foundations of P, E and N (Zuckerman, 1991); there is as yet not a whisper of any such understanding for the components of the Big Five other than E and N. Note that it is not claimed that more than a beginning has been made in building this bridge between DNA and personality; there are still many anomalies, and these will require much work along the lines of ordinary science to iron out. But we are beginning to understand the reasons why some experimental configurations, electrode placements, stimulus intensities and other details of procedure are more successful than others in producing results favourable or unfavourable to our theories, and we are beginning to control these variables in line with our improved understanding. This is an important step forward, indicating that our theorising is ultimately on the right lines.

Forging links between DNA and personality is only the first stage, we also have to forge similar links between personality and social behaviour. As an example, take sexual behaviour (Eysenck, 1976). What would the arousal theory of extraversion predict in this field? I made the following predictions (Eysenck, 1971), based on the theory that extraverts would require *stronger* stimulation than introverts to maintain a reasonable level of cortical arousal:

1) Extraverts will have intercourse *earlier* than introverts.
2) Extraverts will have intercourse *more frequently* than introverts.
3) Extraverts will have intercourse with *more different partners than introverts.*
4) Extraverts will have intercourse *in more different positions* than introverts.
5) Extraverts will indulge in *more varied* sexual behaviour outside intercourse than introverts.
6) Extraverts will indulge in *longer pre-coital love-play* than introverts.

The evidence (Eysenck, 1976) strongly supports these deductions, as well as others concerning P and N; much of sexual behaviour is clearly related to personality, in accordance with laws which can be traced back to biological causes. This type of argument extends the link from DNA and personality to social behaviour, and of course sexual behaviour is not the only type of social behaviour covered; the system has been extended to criminal behaviour (Eysenck, 1977; Eysenck and Gudjonsson, 1989), education (Eysenck, 1978), smoking and disease (Eysenck, 1991b), sport (Coleman, 1980; Eysenck et al., 1982) and many other social areas. We thus have a chain from DNA through psychophysiological mechanisms to personality and finally social activities of considerable importance. It follows that for P, E and N there exists a nomological network validating the theoretical constructs over and above the simple psychometric validation afforded by factor analysis and multidimensional scaling (Eysenck and Eysenck, 1985, Eysenck, 1993b).

We may say that *personality* is the central concept in our theory. The *distal* cause is DNA, the *proximal* cause is the physiological, hormonal, neurological set of intermediaries linking DNA to behaviour, and interacting with environmental factors. On the other side, representing behaviours consequent upon personality factors, are social events on the *distal* side, as argued above, and experimental studies on the proximal side. Consider one example. According to Walker's action decrement theory, learning a given material produces a period of *consolidation* during which *reproduction* of the material learned is inhibited; when consolidation is complete, reproduction is at its optimum level. This theory has produced both confirmatory and contradictory results in efforts to replicate the original studies. Howarth and Eysenck (1968) argued that the higher arousal of introverts would lead to a longer period of consolidation, involving *poorer* recall immediately after learning, but better recall after consolidation had been completed. Exactly the opposite course was predicted for extraverts. The results are shown in Fig. 3; clearly the theory is strongly supported by the data.

Note the *reciprocal* relation between personality study and experimental psychology, each helping and aiding the other. Walker's theory, taking its origin in experimental psychology, suggests an important way to explicate and test a theory of personality. But note that if we test Walker's theory *without* regard to individual differences, we obtain contradictory results, some studies supporting the theory, others contradicting it. Clearly introverts show the predicted action decrement effects, extraverts do not, for reasons given, hence the make-up of the sample tested will decide whether Walker's theory is supported or not - it will be if introverts predominate, it will not if extraverts do! If there are equal numbers, forgetting and consolidation will just about balance out, and although *both* are present, *neither* will become apparent in the results! Clearly Cronbach (1957) was right in insisting on the need for both disciplines of a scientific psychology to be combined to make a unitary science.

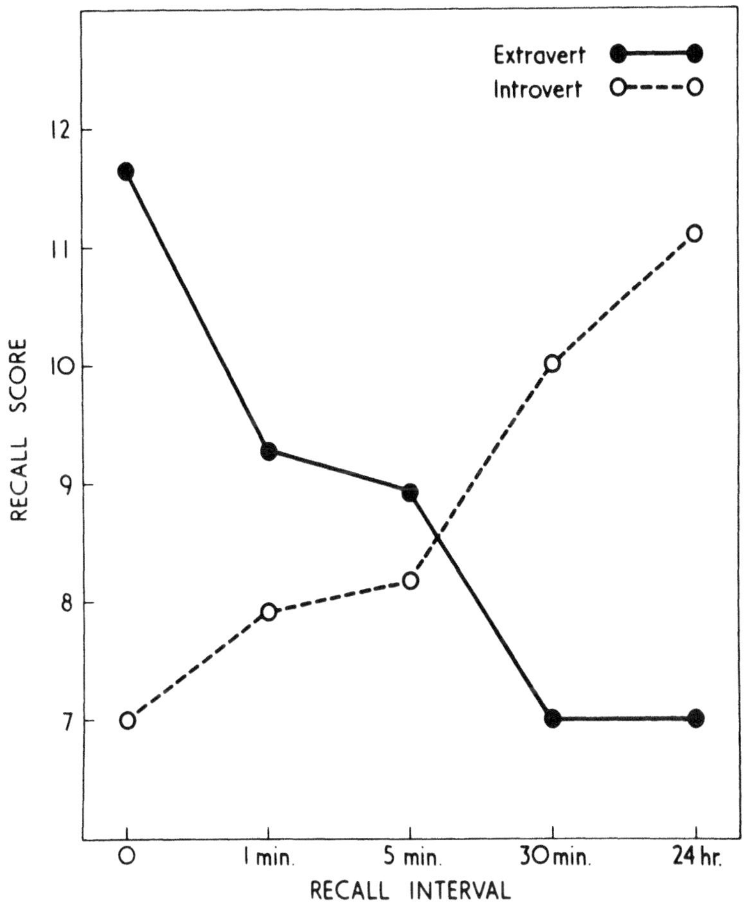

Figure 3. Recall scores of introverts and extraverts at different recall intervals (From Howarth and Eysenck, 1968).

I have always argued that what is true of experimental psychology (where many other examples could be given - see Eysenck and Eysenck, 1985) is equally true of social psychology, educational psychology, industrial psychology, and indeed any kind of psychology; by failing to take individual differences into account we make proper testing of theories in all these branches of psychology impossible, and throw a large part of true variance on to the error variance heap. This neglect, although fairly universal, is a denial of scientific method, and accounts for a large part of the "failure to replicate" and the failure to find strong main factor variance. Any nomological network in psychology must include individual differences, and vice versa any theory of individual differences must include these proximal and distal consequences to become viable.

It is the absence of a nomological network such as would mediate such experimental predictions to be made, and tested, that suggests to me that factors like O, A and C are not at the same level as P, E and N; what is missing are both the proximal causes and proximal consequences that are essential for any nomological networks around the concept of "personality". A purely descriptive, factorially derived taxonomy will not do; much, much more is required if the criteria I am suggesting are to be satisfied.

Concluding Remarks

Factor analysis and psychometrics generally, would seem to be a good servant, but a bad master. Cronbach (1957), in his Presidential Address to APA, pointed out that the existence of two separate disciplines of scientific psychology, experimental and correlational, made unification difficult if not impossible. No paradigm is possible in personality research if we fail to transcend the barrier between these two disciplines, and attempt the construction of a unified science incorporating both. The PEN system has from the beginning attempted to produce such a relationship, based on theoretical analysis of known facts, and suggesting empirical progress to hitherto unknown facts, social, experimental, and psychophysiological. Nothing short of such a wide-ranging approach will do; looking narrowly at only one element, such as the psychometric, is not enough.

Psychologists have always been ready to neglect and disregard important sources of knowledge. Skinner, to take but one example, advised leaving psychophysiological processes alone, and treating them as "black box" events, looking at input and output relations but not worrying about what happens in between. As a result his system has been embarrassingly narrow, leaving out of account most of human behaviour, and lacking in proper causal relations mediated through reasonably well understood physiological channels. In a similar way many personality theorists refuse to look at important empirical findings dependent on biological discoveries, whether genetic, psychophysiological, hormonal, or what not. Exclusive reliance on psychometric manipulations of highly restricted data may give us suggestions, but can never lead to the promised land of proper paradigms. Only by taking a much wider view are we likely to achieve this aim.

References

Barnes, B. (1982) *T.S. Kuka and social science*. London: Macmillan.

Barrett, P. and Eysenck, S.B.G. (1984) The assessment of personality factors across 25 countries. *Personality and Individual Differences, 5*, 615-632.

Baruch, I., Hemsley, D.R. and Gray, J.E. (1988) Latent inhibition and "psychotic proneness" in normal subjects. *Personality and Individual Differences, 9*, 777-784.

Claridge, G.S. (1985) *Origins of mental illness*. Oxford: Basil Blackwell.

Claridge, G.S. and Chappa, H.J. (1973) Psychoticism: A study of its biological nature in normal subjects. *British Journal of Social and Clinical Psychology, 12*, 175-187.

Claridge, G., Robinson, D.L. and Birchall, P. (1983) Characteristics of schizophrenics' and neurotics' relations. *Personality and Individual Differences, 4*, 651-664.

Coleman, J.A. (1980) Personality and stress in the shooting sports. *Journal of Psychosomatic Research, 24*, 287-296.

Cronbach, L.J. (1957) The two disciplines of scientific psychology. *American Psychologist, 12*, 671-684.

Cronbach, L.J. and Meehl, P. (1955) Construct validity in psychological tests. *Psychological Bulletin, 52*, 281-302.

De Raad, B. (1992) The replicability of the Big Five personality dimensions in three word-classes of the Dutch language. *European Journal of Personality, 6*, 15-29.

Digman, J.M. (1990) Personality structure: Emergence of the Big-Five model. *Annual Review of Psychology, 41*, 417-440.

Eaves, L.J., Eysenck, H.J. and Martin, N. (1989) *Genes, culture and personality: An Empirical Approach.* New York: Academic Press.

Eysenck, H.J. (1952) *The scientific study of personality.* London: Routledge Kegan Paul.

Eysenck, H.J. (1967) *The biological basis of personality.* Springfield: C.C. Thomas.

Eysenck, H.J. (1971) Personality and social adjustment. *British Journal of Psychiatry, 118*, 553-808.

Eysenck, H.J. (1976) *Sex and personality.* London: Open Books.

Eysenck, H.J. (1977) *Crime and personality.* (3rd Ed.) London: Paladin.

Eysenck, H.J. (1978) The development of personality and its relation to learning. In: S. Murray (Ed.), *Melbourne Studies in Education*, 134-181. Melbourne University Press.

Eysenck, H.J. (1980) Man as a biosocial animal: Comments on the sociobiology debate. *Political Psychology, 2*, 43-51.

Eysenck, H.J. (1980b) The biosocial model of man and the unification of psychology. In: A.J. Chapman and P.M. Jones (Eds.), *Models of Man.* London: British Psychological Society.

Eysenck, H.J. (1983a) Is there a paradigm in personality research? *Journal of Research in Personality, 17*, 369-397.

Eysenck, H.J. (1983b) The social application of Pavlovian theories. *The Pavlovian Journal of Biological Science, 18*, 117-125.

Eysenck, H.J. (1983c) A biometrical-genetical analysis of impulsive and sensation-seeking behaviour. In: M. Zuckerman (Ed.), *Biological Bases of Sensation-Seeking, Impulsivity, and Anxiety.* Hillsdale, N.J., Erlbaum.

Eysenck, H.J. (1983d) The roots of creativity: cognitive ability or personality trait? *Roeper Review, 5*, 10-12.

Eysenck, H.J. (1984) The place of individual differences in a scientific psychology. *Annals of Theoretical Psychology, 1*, 233-286.

Eysenck, H.J. (1985) The place of theory in a world of facts. *Annals of Theoretical Psychology, 3*, 17-72.

Eysenck, H.J. (1990) Biological dimensions of personality. In: L.A. Pervin (Ed.), *Handbook of Personality: Theory and Research*, 244-276.

Eysenck, H.J. (1991a) Dimensions of personality: 16, 5 or 3? Criteria for a taxonomic paradigm. *Personality and Individual Differences, 12*, 773-790.

Eysenck, H.J. (1991b) *Smoking, personality and stress: Psychosocial factors in the prevention of cancer and coronary heart disease.* New York: Plenum Press.

Eysenck, H.J. (1992) The definition and measurement of psychoticism. *Personality and Individual Differences, 13*, in press.

Eysenck, H.J. (1993a) Personality: Biological Foundations. In: P.A. Vernon (Ed.), *Handbook of the neuropsychology of individual differences.* New York: Academic Press.

Eysenck, H.J. (1993b) Creativity and personality: Suggestions for a theory. *Psychological Inquiry*, in press.

Eysenck, H.J. and Eysenck, S.B.G. (1976) *Psychoticism as a dimension of personality.* London: Hodder and Stoughton.

Eysenck, H.J. and Eysenck, M.W. (1985) *Personality and individual differences: A natural science approach.* New York: Plenum Press.

Eysenck, H.J. and Gudjonsson, G. (1989) *The causes and cures of criminality.* New York: Plenum Press.

Eysenck, H.J., Nias, D.K.B. and Cox, D.N. (1982) Sport and personality. *Advances in Behaviour Research and Therapy, 1*, 1-56.

Garber, J. and Strassberg, Z. (1991) Construct validity: History and application to developmental psychopathology. In: W.M. Grove and D. Ciocetti (Eds.), *Personality and Psychopathology*, 219-258. Minneapolis: University of Minnesota Press.

Goldberg, L.R. (1982) From ace to Zambia: some explorations in the language of personality. In: C.D. Spielberger and J.N. Butcher (Eds.), *Advances in Personality Assessment, 1*, 203-234. Hillsdale, N.J., Erlbaum.

Goldberg, L.R. (1990) An alternative "description of personality"; the Big Five factor structure. *Journal of Personality and Social Psychology, 59*, 1216-1229.

Goldberg, L.R. and Rosolack, T.K. (1992) Pitting the Big-Five Structure against its competitors: An empirical comparison with Eysenck's P-E-N model. In: C.F. Halverson, G.A. Kohnstamm, and R.P. Martin (Eds.). *The Developing Structure of Temperament and Personality from Infancy to Adulthood.* New York: Erlbaum.

Howarth, E. and Eysenck, H.J. (1968) Extraversion, arousal, and paired-associate recall. *Experimental Research in Personality, 3*, 114-116.

Lubow, R.E. (1989) *Latent inhibition and conditional attention theory.* Cambridge: Cambridge University Press.

Lubow, R.E., Ingberg-Sacks, Y., Zabstein-Orda, N. and Gewirtz, J.C. (1992) Latent inhibition in low and high "psychotic-prone" normal subjects. *Personality and Individual Differences, 13*, 563-572.

McCrae, R.R. (1992) Openness to experience as a basic dimension of personality. Presented at the Symposium *"Intelligence and Openness in the Big Five Personality Structure"*, at the APA Convention in Washington, DC.

Richards, R.L. (1981) Relationship between creativity and psychopathology: An evaluation and interpretation of the evidence. *Genetic Psychology Monographs, 103*, 261-324.

Sokal, R. and Sneath, P. (1963) *Principles of numerical taxonomy.* London: Freeman.

Zuckerman, M. (1991) *Psychobiology of personality.* Cambridge: Cambridge University Press.

THE GENES AND P
Commentary on the chapter by H. J. Eysenck

John C. Loehlin
Department of Psychology
University of Texas
Austin, TX 78712, USA

Let me begin by saying that I am entirely in sympathy with Professor Eysenck's view that a full understanding of personality will include its physiological and biochemical underpinnings and its genetic and environmental etiology. I'm not so sure I agree with him that this is the only way to go. It seems to me that there is still plenty of room for specialists, such as experts in the description and measurement of phenotypic personality - people who say, "I'll describe the regularities, let those other guys explain them." Nevertheless, it is clear that Eysenck's approach is a challenging one, that it is a route taken by a number of other participants at this conference, and that his ideas deserve our serious attention.

Psychoticism as a Personality Dimension

In his paper, Eysenck offers as the primary example of his approach the personality dimension which he calls Psychoticism. This is a dimension which, in the normal range of the population, contrasts relatively agreeable, empathic and well-socialized people with relatively cold, hostile, and impulsive ones. Beyond the normal range, according to Eysenck, it orders a variety of psychopathic, borderline, and psychotic conditions, from antisocial personality to schizophrenia.

As Eysenck notes, Big Five advocates would split this dimension in the normal range into two, Agreeableness and Conscientiousness - one stretching from the warm and empathic to the cold and hostile, the other from the conscientious and well-socialized to the impulsive and rebellious. And of course one could subdivide these still further: coldness and hostility are by no means identical; nor are impulsivity and rebelliousness. Eysenck suggests that as long as we remain on the purely psychometric level we cannot decide among these alternatives. Actually, this is not quite true: From a psychometric standpoint, the more finely differentiated our predictors, other things equal, the better we can hope to predict. A Big Five should do better than a Big Three, and a Big Fifteen better than either. But this is not what primarily concerns Eysenck. He argues that the decisive reason for choosing one set of dimensions rather than another is the "nomological network" of relationships into which it enters, in particular, the alignment of the dimensions with underlying biological facts. In these terms, the reason for preferring Psychoticism as a dimension to its components impulsivity and hostility is the presumption that Psychoticism reflects an underlying biological unity. Furthermore, it is

J. Hettema and I. J. Deary (eds.), Foundations of Personality, 67–72.

hypothesized that this biological dimension runs from the normal into the abnormal range, and in the latter discriminates among various disorders: ordinary criminals on the average have less Psychoticism than psychopaths, who have less of it than schizoids, who have less of it than unipolar depressives, who have less than bipolars, who have less than schizoaffectives, who have less than schizophrenics.

Eysenck suggests a number of approaches to evaluating such a theory, but I will focus on just one - genetics. If we suppose that there is a single genetic dimension underlying the observed grouping of traits and ordering of pathological conditions, certain testable consequences flow from this.

Is P a Single Genetic Dimension?

The most natural approach, in terms of Eysenck's model, is to assume that there is a single, continuous dimension, presumably influenced by many genes, that forms the normally-distributed underlying variable P.

Then we can ask several questions. First, is P, as measured, heritable? To answer this, we obtain the correlations or covariances of P scores among various classes of relatives: identical and fraternal twins, adopted children and their biological and adoptive parents and siblings, members of ordinary families, identical twins who have been reared apart, members of the families of pairs of identical twins, and so on. Then we fit models to the obtained covariances or correlations, based on various assumptions about the genetic and environmental determination of the resemblances, and see which ones fit. Normally, these models contain free parameters concerning the strengths of additive and nonadditive genetic effects and shared and unshared family environment, and a model that fits the data also provides us with estimates of these parameters.

P, as measured by Eysenck's scales in normal populations, has a substantial degree of genetic determination and not much shared family environmental variance (Eaves et al., 1989). But this in itself is not very informative. The same thing is true for many dimensions of personality - the Big Three, the Big Five, a variety of other personality traits, most of the scales of the popular personality inventories, and a surprising array of things like vocational and avocational interests - see Bouchard (this volume).

A more critical question is whether P hangs together as a single genetic dimension. There have recently been a couple of papers addressed to this issue, by Heath and Martin (1990) and Heath et al. (1989), using data from the Eysenck Personality Questionnaire, which was administered to an Australian sample of nearly three thousand twin pairs. These are studies based on the covariance structure among the items of the scales, an approach which requires large sample sizes for the dependable estimation of tetrachoric correlations among individual items. The analysis is based on the idea that the covariances among items can be analyzed into genetic and environmental components, just as the variances of the individual items can.

Consider two items from the P scale of the EPQ: "Would you feel very sorry for an animal caught in a trap" ("No" is the high-P response), and "Do people tell you a lot of lies" (to which "Yes" is the high-P response). If identical twin pairs are much more likely than fraternal twin pairs to give the same response when asked "Would you feel very sorry for an animal caught in a trap," this would provide evidence for a genetic influence on response to this particular item. Likewise for "Do people tell you a lot of lies." But to analyze the genetic coherence of the underlying dimension, one asks about agreement between the responses of twin A to one item and twin B to the other. Now, Heath and his colleagues did not separately decompose the covariances between every possible pair of individual items. Rather, they fit models which assumed genetic and environmental factors common to all the items supposed to measure P (or N or E), as well as genetic and environmental factors specific to each item.

In the first analysis (Heath et al., 1989) it was assumed that the phenotypic factor structure would apply to both the genetic and the environmental causes of P, E, and N, and the question was to estimate the genetic and environmental contributions to general and item-specific factors under this assumption. The results for E and N were quite in line with the expectations of an Eysenckian model. Both shared and specific genetic factors were found, as well as shared and specific within-family environmental factors. That is, the observed item covariances were consistent with the assumption that they resulted from underlying genetic and environmental E and N dimensions, plus genetic and environmental specifics, and that the environmental contributions were nonfamilial. The results for P were quite different. The genetic contributions to the items proved to be entirely item-specific, whereas what the P items had in common was estimated to be entirely environmental - partly shared in families, partly unshared. The analyses were done separately for male and female pairs, and the above results held for both sexes. The authors noted the difficulties for the standard Eysenckian view of P; they suspected that the root problem might be that the genetic and environmental factor structures of the Psychoticism items were different.

Heath and Martin pursued this hypothesis in their second paper (1990). In this analysis, the genetic and environmental determinants of the covariation of the P items were allowed to differ. Heath and Martin found that whereas the environmental determination of the P items was mostly consistent with the notion that the items were all measuring something in common, the genetic structure suggested that there were at least two different groups of P items, and that far from forming a single uniform dimension they were negatively correlated with each other. This was why when the P items were all forced onto a single dimension in the first study it had been estimated to have a heritability of zero.

Now, I don't want to stir up trouble in the Eysenck camp, but I did notice that two of the three authors of these papers are frequent collaborators of Eysenck's, and that these two papers are not mentioned in either Eysenck's advance paper for this conference or in his more extended discussion of the Psychoticism concept that has just appeared in *Personality and Individual Differences* (Eysenck, 1992). I look forward to hearing Professor Eysenck's views on these papers.

Does a P Continuum Fit the Psychopathology Data?

The preceding has to do with P as measured in normal populations. Genetics also lets us draw conclusions about the continuity of the dimension in the abnormal range. For example, if we look at the first-degree relatives (parents, siblings, children) of a group presumably high in P genes, such as schizophrenics, we would expect to find a distribution of P-gene levels centering around a point about half-way between the schizophrenic average and the general population mean. I.e., if Eysenck's Figure 2 is literally correct, our sample of relatives ought to be typified by hostile, aggressive and impulsive individuals, with a fair supply of neighboring groups such as criminals, psychopaths, and schizoids. More generally, it is possible to fit to the data formal threshold models that predict, given the frequencies of the various

P-related conditions in a control population, what their frequencies ought to be among the relatives of schizophrenics, if the assumption of an underlying genetic dimension holds.

I will illustrate by the study of Tsuang et al. (1983) in which they compared the first degree relatives of 200 schizophrenics and 160 control surgical patients. Among the 354 first-degree relatives of schizophrenics, there were 11 individuals who themselves were diagnosed as schizophrenic, about 3%, whereas among the 541 first-degree relatives of the controls there were 3 schizophrenics, about 0.6%. A higher rate of schizophrenia in the relatives of schizophrenics would be predicted from almost any genetic theory of schizophrenia, dimensional or not. The authors then fitted a threshold model to a range of "schizophrenic spectrum" conditions, including a number of those listed in the lesser ranges of Eysenck's scheme. Eysenck's theory of P predicts that there should be more of these conditions among the relatives of schizophrenics than among the relatives of controls. In fact, there were fewer: 12% in the relatives of the schizophrenics versus 15% in the control relatives - the deviation from the model being statistically significant by a chi-square test of goodness of fit. Tsuang and his colleagues included a group of neurotics among these "spectrum" cases; they really should not be there in a test of Eysenck's model, because they belong on the N dimension, not P. However, if we exclude the neurotics from both groups, the size and direction of the difference remains unaffected.

Not all studies have turned out this way. Some studies have found excesses of other psychopathologies in the relatives of schizophrenics. Clearly, the evidence is mixed. Eysenck's own recent review (1992, p. 767) lists 2 studies in favor of a continuum view and 3 for specificity. One hypothesis that has sometimes been proposed by advocates of specificity involves assortative mating - that men who go around impregnating schizophrenic women tend to be on the psychopathic side, leading to an excess of psychopathic genes in their offspring. Such a hypothesis would not necessarily predict the same result for the parents and siblings of schizophrenics. Thus the consistency of results across various classes of relatives would be important here.

Concluding Comment

Well, there are a couple of examples relevant to genetics. My comments are, of course, not intended to settle the matter of the reality of a P dimension, but to generate discussion. In closing, let me commend once more Hans Eysenck's insistence that our approach to personality be broad, that it be empirically grounded, and that it be based on explicit models and theory.

References

Eaves, L.J., Eysenck, H.J. and Martin, N.G. (1989) *Genes, culture and personality: An empirical approach.* London: Academic Press.

Eysenck, H.J. (1992). The definition and measurement of psychoticism. *Personality and Individual Differences, 13*, 757-785.

Heath, A.C., Eaves, L.J. and Martin, N. G. (1989). The genetic structure of personality III. Multivariate genetic item analysis of the EPQ scales. *Personality and Individual Differences, 10*, 877-888.

Heath, A.C. and Martin, N.G. (1990). Psychoticism as a dimension of personality: A multivariate genetic test of Eysenck and Eysenck's Psychoticism contruct. *Journal of Personality and Social Psychology, 58*, 111-121.
Tsuang, M.T., Bucher, K.D. and Fleming, J.A. (1983). A search for 'schizophrenia spectrum disorders': An application of a multiple threshold model to blind family study data. *British Journal of Psychiatry, 143*, 572-577.

REPLY TO J.C. LOEHLIN

Hans J. Eysenck

Loehlin's comments are well taken, but on some points I would disagree. He states that "the more finely differentiated our predictors, other things equal, the better we can hope to predict". There are two points. (1) P correlates well with A and C, but also has a good deal of specific variance; perhaps this specific variance adds considerably to the predictive variance? (2) Other things are seldom equal; Reynolds and Nichols (1977) tested the question of whether the 18 scales of the CPI would predict a variety of behaviours better than the two factors E and N extracted from the intercorrelation between the 18 scales. There were 178 predicted variables, and their conclusions are well worth quoting: "For the most part, the factor scales do seem to capture the valid variance in the CPI scales .... In many instances a common factor portion of the scale's variance was actually more predictive of relevant criteria than was the total scale variance. These findings would suggest that users of the CPI might be better off to measure and interpret the two principal factors rather than attempt to derive meaning from a complex profile of scores." (p.914)

Loehlin is correct in stating that recent studies of the heritability of P have thrown up some oddly contrasting results; it would not have been appropriate to discuss them in a paper devoted to quite a different purpose, namely the proposal of criteria for a personality taxonomy. The criteria would still seem reasonable, even if they ruled out of court my own theory - after all, they were suggested as being universally applicable, not as supporting the views of the author! There seems an odd and indeed paradoxical contradiction between the results of fitting models to covariances of P scores among various classes of relatives, twins, adoptees, etc., which showed a substantial degree of genetic determination, and the results of analysing covariance structure among the items of the scale, which suggested that the genetic contribution was item-specific, with item communality being entirely environmental. The unreliability of single item responses must make one doubtful of the results, and I would prefer to await replication studies with larger samples before trying to unravel this particular conundrum.

Finally, Loehlin cites one study looking at the proportion of relatives of schizophrenics and failing to find the expected (according to my theory) number of spectrum disorders. Other studies (Eysenck, 1992) have shown results more favourable to my theory, as Loehlin points out in fairness. Overall results support the hypothesis. He also points out that in this sample there was a substantial group of neurotics among the spectrum cases which shouldn't have been there according to my theory. This is true, but only partially; Enright and Beech (1990) have shown that obsessional states, usually classed among the neuroses, show clear evidence of lack of *negative priming*, usually found in schizophrenics and high P scorers. Psychiatric terms and diagnostic labels must always be taken with a grain of salt; things are seldom as black and white as psychiatrists would like.

References

Enright, S.J. and Beech, A.R. (1990) Obsessional states: anxiety disorders or schizotypes? An information processing and personality assessment. *Psychological Medicine, 20*, 621-627.

Reynolds, C.N. and Nichols, R.C. (1977) Factor scales for the CPI: Do they capture the valid variance? *Educational and Psychological Measurement, 37*, 907-915.

PERSONALITY FROM TOP (TRAITS) TO BOTTOM (GENETICS) WITH STOPS AT EACH LEVEL BETWEEN

Marvin Zuckerman
Department of Psychology,
University of Delaware,
Newark, DE 19716, USA

Abstract

The Alternative Five Factor model was developed from factor analyses of questionnaire scales believed to measure basic personality factors and it includes the following trait factors: Sociability, Neuroticism-Anxiety, Impulsive Unsocialized Sensation Seeking, Aggression-Hostility, and Activity. Associated with each trait are characteristic desired and avoided goals, typical generalized expectancies and affective traits. I review the various biological "levels" underlying the traits, including the psychophysiological, psychopharmacological, neurological, and genetic levels. Certain biological traits, such as the augumenting or reducing of the cortical evoked potential, gonadal hormones, and the enzyme monoamine oxidese (MAO), have shown relatively consistent relationships with personality traits like impulsive sensation seeking and sociability, although there is no simple isomorphism between single biological markers and traits. The experimental literature on the behavioral functions of brain monoamine systems in other species, and the findings of biological psychiatry suggest a model for the biological basis of personality.

Introduction

The psychobiological model for personality presented in this chapter is not an elegant one. It would be nice if nature had designed personality traits in neat isomorphic relationships to neurological structures and neurotransmitters. But biology is not physics. Neural structures and neurochemically mediated pathways evolved in response to the need for mechanisms necessary for organisms to adapt to a dangerous and unpredictable environment. Nature did not do a factor analysis to arrive at the basic biological mechanisms and one mechanism had to serve many purposes. Given a new stimulus in the environment, or a change in the context of a familiar one (like a lion arising from its nap and strolling toward the herd), there are a limited number of options: approach, withdraw, or freeze. Individual differences in the relative strengths or sensitivities of such behavioral mechanisms, and the biological mechanisms underlying them, may be the basis for personality traits.

J. Hettema and I. J. Deary (eds.), Foundations of Personality, 73–100.

Seven Turtles

In line with the old story of the guru trying to respond to his student's question about what the world rests on, I present the levels of psychobiological analyses of personality in the form of seven turtles (Figure 1). A trait may be described as a summary of aggregated behaviors within certain protypical situations including covert but habitual cognitive reactions. Social behavior is the result of classical conditioning, observational learning, and instrumental shaping of what is learned through observation. Learning and biology are not alternate explanations of behavior. What we learn from exposure and experience depends on differences in our biological makeup.

Figure 1. Levels of analysis of personality traits in the form of seven turtles.

Underlying the conditioning mechanisms are physiological ones. One cannot be conditioned if arousal is at a very low level or if physiological attention mechanisms interfere with stimulus intake. Physiological mechanisms are dependent on biochemical ones. The noradrenergic system, for instance, modulates the signal to noise ratio making it particularly important in the orienting response and attention to novel stimuli. Neurochemical systems have their origins and nodal points in neurological structures like the locus coeruleus. There is no neat phrenological map for behavioral function but within structures like the amygdala and hypothalamus there are certain nuclei where lesioning may produce dramatic changes in behavior, motivation, and emotions, suggesting they serve vital roles in these functions. Seven is a magic number and rather than "turtles all the way down" we can stop at the genetic level for the individual. While the "chicken and egg" puzzle of causation may apply to the higher levels, like biochemistry, physiology and social behavior, the genes have a certain priority, being there at the beginning. Of course, if one goes beyond the individual to the species one could evoke another turtle (the ancestral one): evolutionary selection.

Figure 2 shows a recently developed model for personality based on a levels analysis. The model is more fully described elsewhere (Zuckerman, 1991). In this chapter I will only elaborate on some of the major points. This is not a reductionist model because each turtle is a distinct creature deserving study of itself even without reference to the other turtles. But we must recognize that when a turtle at the bottom moves a turtle at the top must perforce also move. Sometimes a movement of a turtle on top may disturb the turtle below it but this is not necessarily so and, after all, turtles have thick shells. If scientists hide within the shells of their most familiar turtle they may not recognize the sources of its movement in the turtles below.

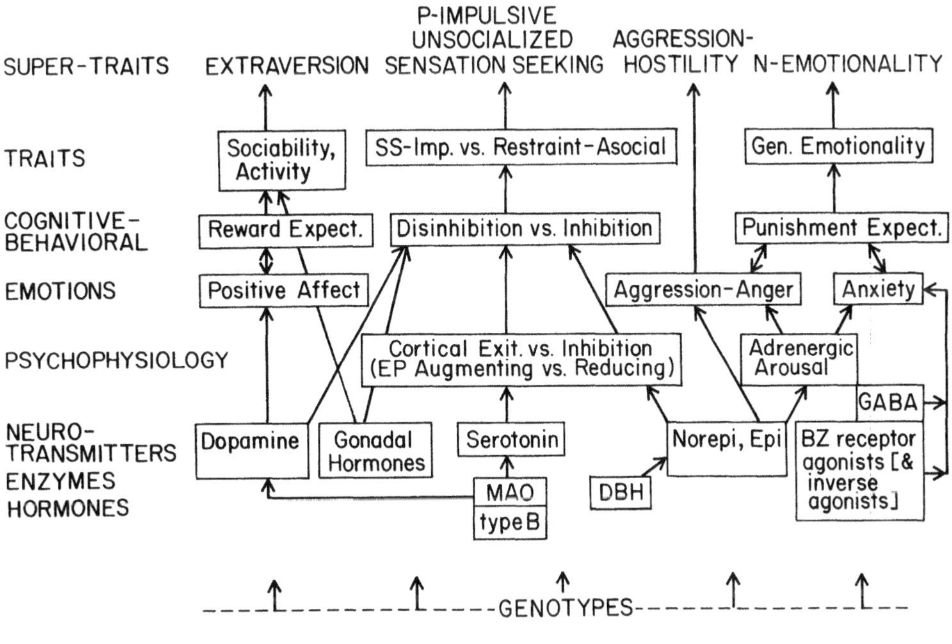

Figure 2. A psychobiological model for personality. Dopamine refers particularly to the A10 dopaminergic pathway from the ventral tegmental area to the nucleus accumbens via the

medial forebrain bundle and the A9 pathway from the substantia nigra to the caudate-putamen. Low levels of type B monoamine oxidase (MAO) may deregulate these systems. High levels of gonadal hormones, particularly testosterone, may furnish a basis for both sociability and disinhibition. High levels of serotonin in conjunction with high levels of both type A and B MAO may provide the basis for strong inhibition; low levels of serotonin together with high activity of dopaminergic systems may be involved in disinhibition, impulsivity, and aggression or hostility. Regions of the septal area are particularly involved in inhibition-disinhibition of behavior. Norepinephrine (Norepi), particularly in the dorsal ascending noradrenergic pathways from the locus coeruleus are also involved in the adrenergic arousal found in both anxiety and anger. Low levels of norepinephrine, perhaps related to low levels of the enzyme dopamine-beta-hydroxlase (DBH), may be involved in the traits of disinhibition and impulsivity. Stimulation from the central nucleus of the amygdala to the vertral tegmental areas and the locus coeruleus may increase activity in dopaminergic and noradrengergic systems. At low levels this catecholamine system activity may be rewarding and facilitating, but at high levels may be associated with anxiety, distractibility, inhibition and adrenergic arousal. When adrenergic arousal is combined with high activity of benzodiazepine receptor inverse agonists and low levels of GABA inhibition the result may be anxiety. Specific combinations of these biological traits may underlie the disposition of trait anxiety and emotionality in general.
Note. From *Psychobiology of Personality* (p. 407) by M. Zuckerman, 1991, New York: Cambridge University Press. Copright 1991 by Cambridge University Press. Reprinted by permission.

Basic Personality Traits or Temperaments

Just as seven is a magic number for turtles five seems to be one for personality traits. In the past it was 16 but now the major rivals are five and three. But the number of factors is not important because broad factors incorporate narrower factors in a hierarchal order (Eysenck, 1967, Costa and McCrae, 1985; Zuckerman et al., 1988), and a factoring of narrower factors usually reconstitutes the broader factors so that how many factors depends on the level of analysis. Disagreements on what is a basic factor often stem from differences in initial selection of variables. The importance of a factor in the analysis depends upon how many markers were chosen to represent the factor. If there is only one marker for a factor it is unlikely to emerge as an important one in the analysis.

Before beginning my book on the "Psychobiology of Personality" (Zuckerman, 1991) I decided to undertake factor analyses of questionnaire scales believed to measure basic dimensions of personality in order to provide a framework for the psychobiological analyses. Without a taxonomy of species, it is difficult to develop a theory of evolution. The theory may result in modification of the taxonomy but you have to start with something. We started with questionnaire scale markers for traits that had shown at least moderate heritability, some evidence of correlations with biological traits, and could be recognized in other species (for comparative analyses): activity, sociability, impulsivity, socialization, aggression, sensation seeking, and anxiety or general emotionality (Zuckerman et al., 1988). At the three factor level the factors that emerged were: *Sociability, Emotionality,* and one we called *Impulsive Unsocialized Sensation Seeking* (ImpUSS). Eysenck's Extraversion (E), Neuroticism (N), and Psychoticism (P) scales provided the best markers for each of three respective dimensions.

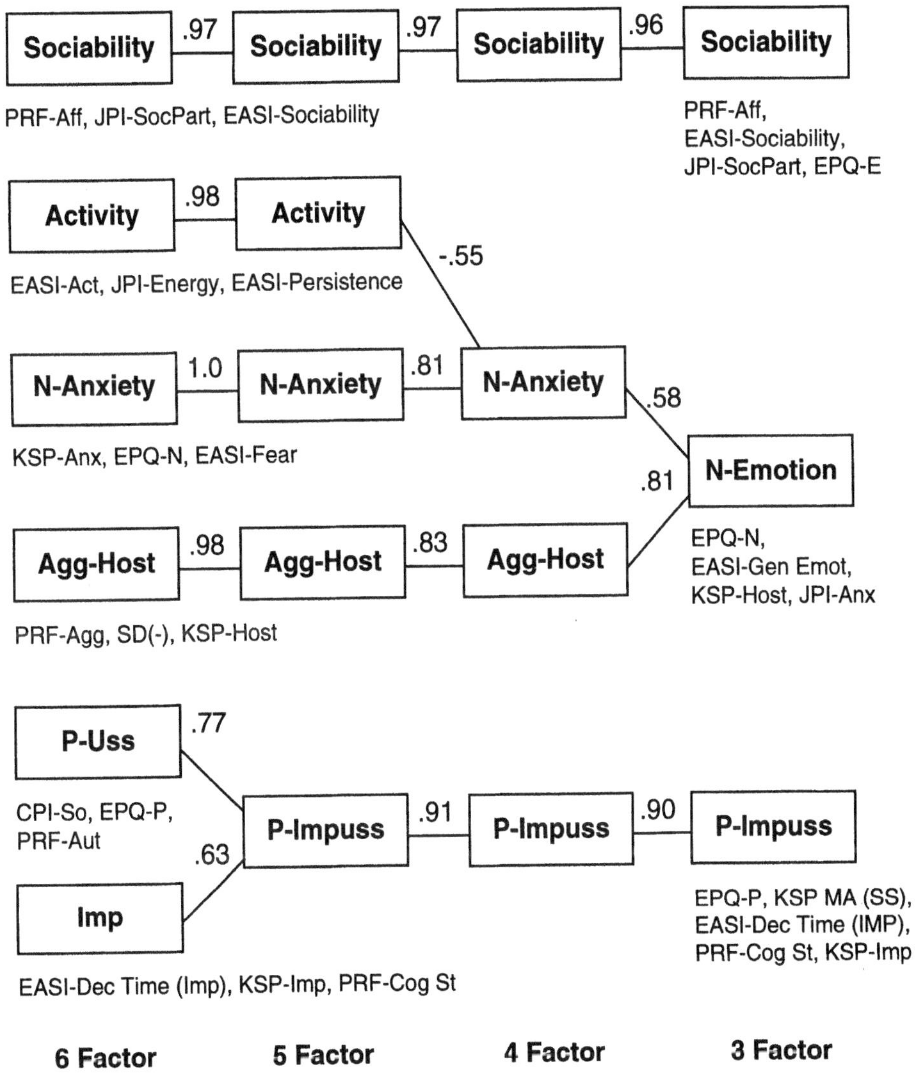

Figure 3. Factors at each level (three-, four-, five-, and six-factor analyses) and factor score correlations across levels for the total group. (N = neuroticism, Agg-Host = aggression-hostility, Emotion = emotionality, P-USS = psychopathy-unsocialized sensation seeking, Imp = impulsivity, P-ImpUSS = psychopathy-impulsive unsocialized sensation seeking. Strongest loading scales defining each factor at six and three factor levels are indicated.

Note. From "Five (or three) robust questionnaire scale factors of personality without culture" by M. Zuckerman, D. M. Kuhlman, M. Thornquist, and H. Kiers, 1991, *Personality and Individual Differences, 12*, p. 938.

In a second study, using the best markers for the narrower factors in the previous study and a much enlarged number of subjects, we found a highly replicable (across gender and four different samples of subjects) five factor solution in addition to the three factor one (Zuckerman et al., 1991). The *Sociability* and *P-ImpUSS* factors were virtually the same as in the three factor solution, but the N-Emotionality factor was split into *N-Anxiety* and *Aggression-Hostility* factors, and *Activity* was identifiable as a major factor distinct from its primary location in the E factor at the three factor level (see Figure 3).

Encouraged by the high degree of replicability of these five factors across four groups, we developed personality scales to measure the five factors (Zuckerman et al., 1992). Using these scales (Zuckerman-Kuhlman Personality Questionnaire, ZKPQ III) we compared our model with Eysenck's three factor one and Costa and McCrae's (1992) five factor one, as incorporated in their revised NEO personality inventory. Costa and McCrae's factor of *Openness to Experience* is based on the factors variously called *Culture* (Norman, 1963), or *Intellect* (Goldberg, 1990). We agreed with Eysenck that intellect is a domain apart from basic traits of personality and therefore did not attempt to include markers for it in our factor analyses.

Table 1. Factor Analysis* of EPQ, ZKPQ, and NEO Scales: Four factors

	Factor Loadings			
	Factor 1	Factor 2	Factor 3	Factor 4
NEO Extraversion	.88	-.13	-.05	.17
EPQ Extraversion	.79	-.32	.17	-.08
ZKPQ Sociability	.77	-.16	.10	-.07
ZKPQ Activity	.60	.01	-.18	.02
ZKPQ N-Anxiety	-.13	.93	-.01	.08
EPQ Neuroticism	-.16	.91	-.04	-.08
NEO Neuroticism	-.15	.90	.10	-.12
NEO Conscientiousness	.15	-.07	-.86	-.02
Eysenck Psychoticism	-.09	-.08	.80	-.28
ZKPQ Impulsive SS	.48	.08	.74	-.03
NEO Agreeableness	-.04	-.07	-.31	.81
ZKPQ Aggression-Host	.35	.34	.24	-.72
NEO Openness	.27	.14	.19	.67

Note. EPQ = Eysenck Personality Questionnaire, ZKPQ = Zuckerman Kuhlman Personality Questionnaire, NEO = Costa and McCrae NEO-PI, N = Neuroticism, Host = Hostility

* Principal Components, Varimax rotation, Four factors accounted for 74% of variance.

Table 1 shows the results of a recent factor analysis of EPQ, ZKPQ, and NEO scales. Four factors accounted for three quarters of the variance among the 14 scales. Factor 1 is E-Sociability and includes the relevant scales from the three tests plus the Activity scale from the ZKPQ. Factor 2 consists of the anxiety or neuroticism scales from the three tests. Factor 3 identifies the P and ImpSS scales at one pole of the factor and the NEO Conscientiousness scale at the other one. The fourth factor is marked by ZKPQ Aggression-Hostility at one end and NEO Agreeableness and Openness at the other. When five factors were rotated the fifth factor consisted of the Activity scale alone. When the facet (subtrait) scales of the NEO were substituted for the major trait scores, Openness did emerge as a fifth factor but it consisted only of facet scales from the NEO and was not related to the scales from the EPQ and ZKPQ. Our results show four factors which are identifiable across the three models.

Cognitive-Behavioral Traits

Cognitive traits are usually regarded as antithetical to biologically based traits of personality but they could be seen as intermediate or functional expressions of the higher order traits.

Table 2. Personality Traits, Goals, Expectancies, and Affects

TRAITS	+ GOALS	- GOALS	EXPECTANCY	AFFECTS
E-Sociability	Making and keeping friends	Avoiding social isolation	Generalized reward expectancy	Pos. Affects: joy, love
N-Anxiety	Being secure in relationships	Avoiding rejection, humiliation	Generalized punishment expectancy	Neg. Affects: anxiety, depression
Aggression-Hostility	Defeating others	Avoiding defeat	Generalized threat of punishment	Neg. Affect: anger
Impulsive Sensation Seeking	Having exciting new experiences	Avoiding boredom	Low punishment, high reward expectancy	Pos. Affects: elation, interest, Neg. affect: boredom
Activity	Achievement	Avoiding passivity	Reward expectancy Internal locus of control	Pos. Affect: interest Neg. affect: boredom

Table 2 shows a conceptualization of the typical goals and generalized expectancies associated with our five trait model. Extraversion is associated with the goal of making and keeping friends and avoiding social isolation. Activity has as its life-goal getting things done or achievement and avoiding idleness and failure. Both of these traits are related to a broad reward expectancy. Impulsive-Sensation Seeking (ImpSS) has the broad life-goal of having exciting, new experiences and avoiding restraint and boredom. This trait is related positively to reward and negatively to punishment expectancies and a behavioral mechanism controlling inhibition and disinhibition. Neuroticism-Anxiety (N-Anx) has the primary goal of being secure in human relations and avoiding rejection and humiliation. Hostility-aggression (Host-Agg) has the goal of defeating others and avoiding defeat by them, as described in the cynical aphorism: "Get them before they get you." Underlying both N-Anx and Host-Agg is the pessimistic expectation of punishment from others, but the reactions to this expectation are quite different. The N-Anx type tries to move closer and appease the potential sources of punishment whereas the Agg-Host type "moves against" them (Horney, 1939).

Underlying cognitive-behavioral traits, at a level closer to the biological substrate, are emotional tendencies. A disposition to experience positive emotional states underlies the generalized reward expectancy and a tendency to react with anger, anxiety, or depression is the basis for punishment expectancies. Elation and interest are positive emotions and boredom is the negative emotion specifically associated with Impulsive Sensation Seeking. Of course, one could argue that the cognitive tendencies are primary and the emotional responses secondary, or that both are concommitant expressions of the underlying biological traits. But from a phylogenetic viewpoint emotions have a priority over cognitions, at least those at the conscious level. Obviously, some information processing has to occur before a stimulus can elicit a conditioned emotional reaction, but this processing can occur rapidly at brain levels, like the thalamus, where only minimal analysis of the stimulus occurs (LeDoux, 1987).

Psychophysiology

EXTRAVERSION

Pavolovian theory suggested that individual differences in temperament emanated from differences in brain physiology, particularly the strengths of hypothetical excitatory and inhibitory brain centers. Eysenck's (1957) earlier theory suggested that the balance between central excitation and inhibition was the basis of the trait of introversion-extraversion. Later, following the discovery of the role of the reticulocortical activating system in arousal, he shifted to an optimal level of stimulation and arousal theory of extraversion. Essentially, the theory was that in an unstimulated state introverts tended to be close to an optimal level of arousal and therefore did not seek much additional stimulation or engage in activities which were overarousing. Extraverts, however, were generally underaroused and below their optimal levels of arousal and therefore sought more intense stimulation and exciting activities in order to reach a pleasant state of arousal. My earlier theory of sensation seeking (Zuckerman, 1969) was based on a similar premise except that novelty of stimulation was considered at least as important as intensity in producing "arousal potential."

Investigators used EEG measures of arousal to test Eysenck's theory that extraverts tend to be underaroused and introverts optimally or overaroused in an unstimulated condition. Spectrum analyses of twins show that the proportions of activity in each frequency range of the EEG are not only reliable individual characteristics but are highly (80%) heritable (Lykken, 1982). But studies comparing introverts with extraverts on EEG arousal indices have yielded mixed and indefinite results (See Gale, 1983; Gale and Edwards, 1986; and O'Gorman, 1984 for reviews). A comparison of the studies which both reviewers agreed were methodologically sound showed that the studies yielding the best results in support of the theory were those in which either mostly female or at least equally female and male (Zuckerman, 1991). Females show more arousal than males on the EEG (Deaken and Exley, 1979) perhaps accounting for the gender differences in comparisons of introverts and extraverts. There is also more recent evidence (O'Gorman and Lloyd, 1987) that the narrow trait of impulsivity (acting quickly without thinking) may be more strongly related to EEG arousal than the broader trait of extraversion (which used to include an impulsivity component in measures of E before the EPQ).

One of the many problems with the construct of "arousal" in general, and the EEG measure of cortical arousal in particular, is that arousal is quite sensitive to the conditions of the experiments and the reactions of the subjects to these conditions. Gale recognized this in a post-hoc hypothesis, devised to explain variations in results. He claimed that an intermediate level of stimulation or response demand was necessary to detect differences between introverts and extraverts because under minimally stimulating conditions (lying down with eyes closed) extraverts tend to become bored and restless and to engage in mental or physical activities to increase arousal.

New imaging techniques have been used to test the arousal hypothesis. Using the Positron Emission Tomography (PET) method, in which radioactively tagged glucose is injected and taken up in the brain and recorded by a PET scanner, Haier et al. (1987) found that extraversion correlated *positively* with the glucose use index of activity in the frontal and temporal cortex. This result is in the opposite direction to that which would be predicted from the theory. However, the subjects were anxiety patients and brain activity was measured during performance of a stressful task. No correlations with E were obtained in the resting, non-task condition. Positive correlations were found with activity in limbic areas, often associated with anxiety response. Both the subjects and the conditions do not make the test ideal for testing the extraversion-arousal hypothesis.

Another study by Mathew et al. (1984), imaging cerebral blood flow using the xenon-inhalation method with normal female subjects, found negative correlations between activity in all cortical areas in both hemispheres and extraversion. These results were in line with Eysenck's theory. The discrepancy between the two imaging studies could be due to subjects or methods.

Eysenck's (1967) optimal level of stimulation theory suggests that introverts should be more aroused by stimuli in the low range of intensity, but at higher intensities extraverts should show a stronger response because introverts are more susceptible to transmarginal inhibition. Stelmack's review (1990) proposed that auditory stimuli of low frequency and moderate intensity are most likely to show stronger reactions of introverts than extraverts. The influence of stimulus intensity on the skin conductance responses (SCR) of introverts and extraverts is illustrated in Figure 4. Introverts were more aroused by a moderate stimulus of 80 dB than extraverts and ambiverts, but in response to an intense 100 dB tone the extraverts had a stronger response than both introverts and ambiverts. These kinds of results suggest that phasic arousability rather than tonic arousal is the difference underlying the extraversion dimension of personality. Introverts react relatively more to weak stimulation while extraverts react more to strong stimulation. Introverts show greater "stimulus sensitivity" while extraverts show more tolerance for high intensities of stimulation.

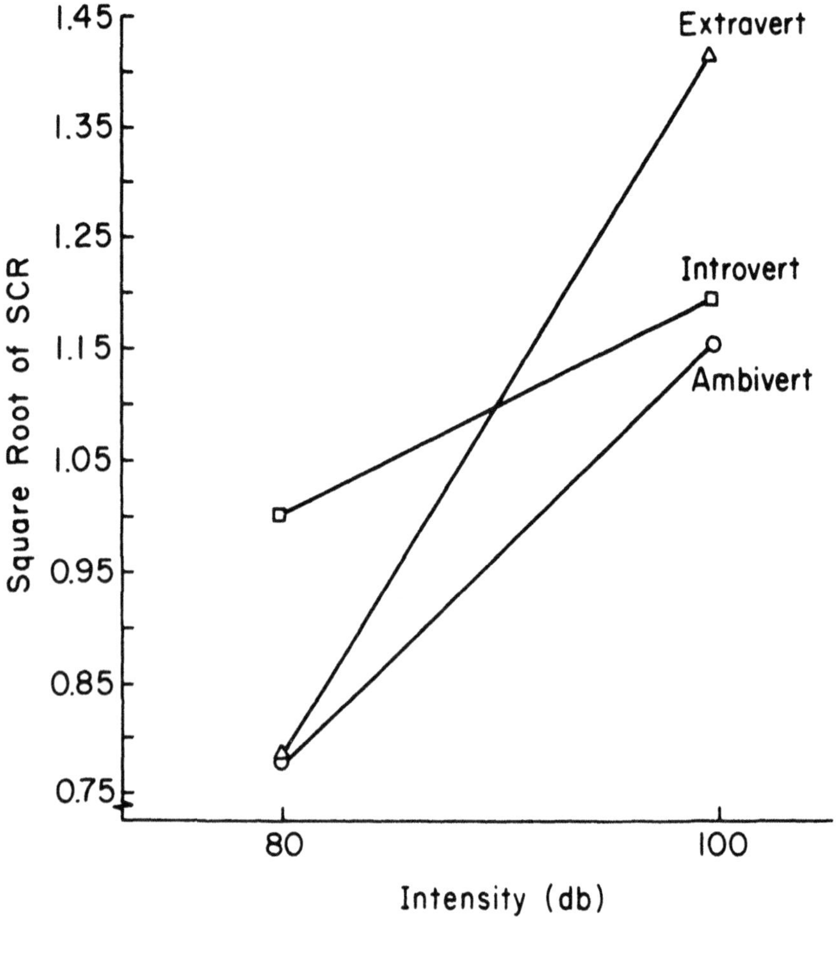

Figure 4. The effect of stimulus intensity on skin conductance response for extraverts, introverts, and ambiverts.

Note. From "Habituation and dishabituation of the electrodermal orienting response in relation to extraversion and neuroticism" by M. J. Wigglesworth and B. D. Smith, 1976, *Journal of Research in Personality, 10*, pp. 437-445. Copyright 1976 by Academic Press. Reprinted by permission.

P-IMPULSIVE UNSOCIALIZED SENSATION SEEKING

The cortical evoked potential (EP) offers a better method for testing the relation between stimulus intensity, cortical arousability and personality than the gross EEG spectrum analyses. The specific pattern of an EP evoked by a particular stimulus is highly heritable as shown by figure 5 from a study by Buchsbaum (1974).

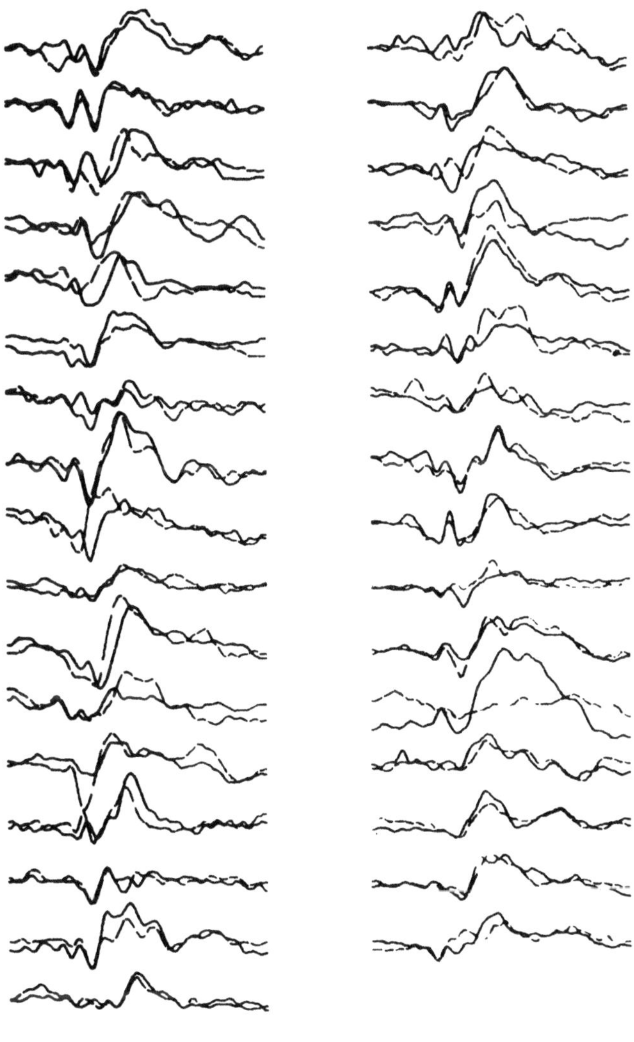

Figure 5. Avaraged visual evoked potentials of pairs of identical twins, with one twin shown as a solid line and the other twin as a dotted line.

Note. From "Average evoked response and stimulus intensity in identical and fraternal twins" by M.S. Buchsbaum, 1974, *Physiological Psychology, 2*, p. 367. Copyright 1974 by American Psychological Association. Reprinted by permission.

These look like repeat EP testing on single individuals, but in actuality the solid and dotted lines tracing the complex EP wave pattern are two identical twins. Identical twins EP patterns correlate much more highly than fraternal twins and unrelated inviduals for both visual (light flashes) and auditory (tones) EPs (Lewis et al., 1972).

Buchsbaum and Silverman (1968) developed a method for quantifying the stimulus-EP response relationship for individuals in terms of the slope of the relationship between stimulus intensity and EP amplitude. The vertex location on the scalp is generally used and the P1-N1 component of the EP is analyzed. This component, at 100 to 140 ms after the stimulus, represents the first impact of the stimulus at the cortical level. Not much information processing other than intensity can occur at this time. Slopes may vary from high positive to near zero or negative. The former is called *augmenting* and the latter is called *reducing*. Although individuals with extreme slope measures can be characterized as "augmenters" or "reducers", like most continuous biological traits this one is normally distributed.

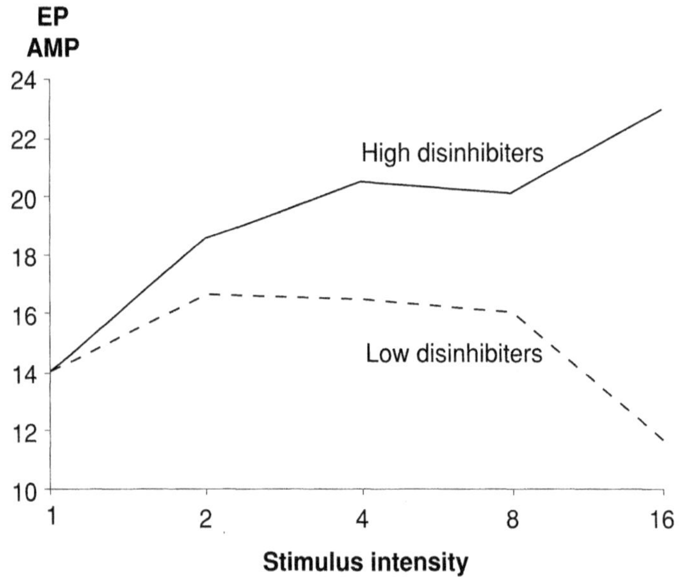

Figures 6. Mean visual evoked potential amplitudes (P1-N1) at 5 levels of light intensity for low and higd scorers on the Disinhibition subscale of the Sensation Seeking Scale.
Note. From "Sensation seeking and cortical augmenting-reducing" by M. Zuckerman, T.T. Murtaugh and J. Siegel, 1974, *Phychophysiology, 11*, p. 539. Copyright 1974 by the Society for Psychophysiological Research. Reprinted by permission.

Augmenting and reducing of visual and auditory EPs have been related many times to a subtrait of sensation seeking called *disinhibition* (Zuckerman, 1990). Disinhibition is the social form of sensation seeking in the form of parties, drinking, sexual variety seeking, and finding excitement through other people in general. It is the type most characteristic of the psychopathic or antisocial personality (Zuckerman, 1979). While correlated somewhat with extraversion it really forms the core of the ImpUSS factor in our three and five factor analyses. High disinhibiters tend to be augmenters and low disinhibiters tend to be reducers on both visual and auditory EPs as shown in figures 6 and 7. The results for the visual EP (Zuckerman et al., 1974) are particularly striking at the highest stimulus intensity where the high disinhibitors show an increasingly augmented reaction while the low disinhibiters show clear evidence of reduction.

The findings on the auditory EP (Zuckerman et al., 1988) show a weak reducing tendency for low disinhibiters contrasted with a markedly augmented cortical response of high disinhibiters. Impulsivity has been related in a similar fashion to EP augmenting-reducing (Barratt et al., 1987) as one would expect based on the close connection between disinhibition and impulsivity in the ImpUSS dimension.

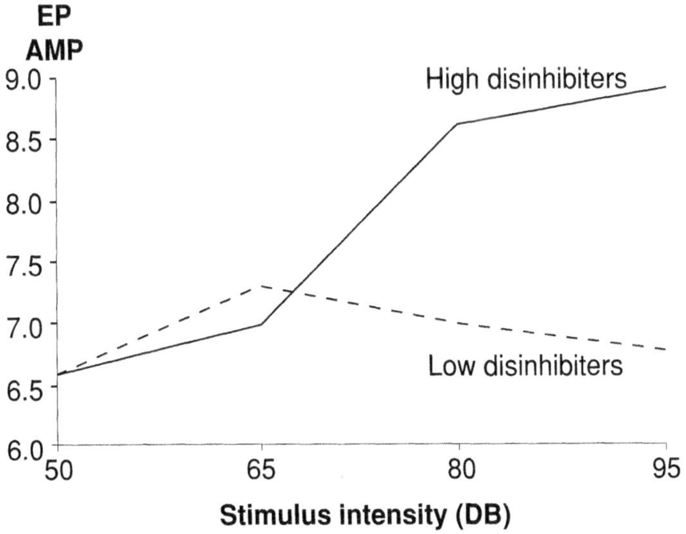

Figure 7. Mean auditory evoked potential amplitudes (P1-N1) at 4 levels of sound intensity (50-95 dB) for low and high scorers on the Disinhibition subscale of the Sensation Seeking Scale.
Note. From "Sensation seeking and stimulus intensity as modulators of cortical, cardiovascular, and electrodermal response: A cross-modality study" by M. Zuckerman, R.F. Simons and P.G. Como, 1988, *Personality and Individual Differences, 9*, p. 368. Copyright 1988 by Pergamon Press. Reprinted by permission.

Augmenting is more characteristic than reducing in human disinhibitory disorderslike delinquency and alcoholism (Zuckerman et al., 1980). Individual differences in augmenting-reducing have been found in cats where they have been related to explorativeness, activity, aggressiveness, and approach or fight-flight responses to a novel stimulus (Hall et al., 1970; Lukasand and Siegel, 1977). Augmenter cats were higher on all of these traits. Saxton et al. (1987) also compared augmenting and reducing cats on two instrumental conditioning paradigms. The augmenter cats were more responsive on a simple fixed interval (FI) schedule of positive reinforcement, but they did more poorly than reducer cats on a differential reinforcement for low rate of response (DRL) schedule.

The differences between augmenting and reducing cats reflect the relationship of the psychophysiological phenomenon to the capacity to inhibit rewarded behavior when there is the possibility of punishment (or loss of reward) for overeager responding. This is the essence of the disinhibition-inhibition mechanism postulated to underlie the P-ImpUSS dimension. Newman and Kosson (1986) showed a similar phenomena among psychopathic and non-psychopathic prisoners, classified as such using Hare's (1980) Psychopathy Check List. The psychopathic types showed more failures to inhibit responses which were punished by loss of money in a situation where there was reward for correct responding, punishment for incorrect responding, but no consequences of withholding response. The psychopathic group scored higher on Eysenck's P scale, *but not on extraversion or neuroticism scales*, higher on a sensation seeking type of scale, and lower on socialization than non-psychopaths. This pattern of test responses places them at the extreme of the P-ImpUSS dimension. Perhaps antisocial personality is not a combination of personality dimensions, as suggested by Eysenck, but represents the extreme cases on the continuous P-ImpUSS dimension (Zuckerman, 1989).

Augmenting-reducing paradigm may represent an openness to intense experience in the augmenter and a protective cortical mechanism in the reducer. Augmenting provides an operational definition of the Pavlovian construct "strength of the nervous system." Augmenters were shown to have the behavioral characteristics supposed to be associated with strong nervous systems or non-reactive types (Strelau, 1983). Lukas and Mullins (1985) found that while augmenters and reducers did not differ in performance in low work-load conditions, the EP augmenters performed better than the reducers under high work-load, stressful conditions. Also relevant are the findings that high sensation seekers have high thresholds for pain (Zuckerman, 1979) and like loud rock music (Litle and Zuckerman, 1986). The low sensation seeker or low P-ImpUSS types may have inherited a type of brain that shuts down under stressful conditions. While confering some protective advantages this defensive tendency would be maladaptive in conditions of inescapable threat where high cortical activation is required.

NEUROTICISM-ANXIETY

Tachycardia and breathing difficulties are typical symptoms of panic attacks (DSM III-R, 1987). Eysenck's earlier theories suggested the involvement of a hyperarousable autonomic nervous system in the trait of neuroticism. In his second theory, however, he proposed that the source of neuroticism was in the limbic brain where autonomic reactivity connected with emotional behavior begins. The limbic brain contains many systems, only some of which are connected with anxiety (Gray, 1982).

Studies correlating personality scales with electrodermal, respiratory and heart rate measures of autonomic activity in normals have generally failed to find significant correlations with trait neuroticism or anxiety (Fahrenberg, 1987; Hodges, 1976; Myrtek, 1984; Navateur and Baque, 1987). Results were also negative for both tonic basal levels of activity and responses to physiological, social, or performance-induced stress. However, longitudinal studies have shown that tonic heart rates have some low predictive level for shyness and inhibition in children (Kagan et al., 1988) and general adjustment in college men 15 years later (Vaillant and Schnurr, 1988).

In contrast to these results with the measured dimension of N-anxiety in normal populations certain types of anxiety disorders show significantly higher tonic heart rates and spontaneous skin conductance fluctuations than normals and other types of cases (Zuckerman, 1991, chpt. 8). Panic disorders, agoraphobics, social phobics, obsessive-compulsives, chronic anxiety, and agitated depressions typically show more electrodermal and cardiac tonic arousal than simple, situational phobics or normals. Figure 8, made from Kelly's (1980) data, shows the relation between the means of scores of various diagnostic groups on the Taylor Manifest Anxiety Scale and the mean heart rates for these groups. Groups who report high levels of subjective anxiety tend to have high heart rates ($r = .73$, $p < .01$). Notable exceptions are personality disorders, whose heart rates are relatively lower, and schizophrenics, whose heart rates are higher, than would be predicted from anxiety scale.

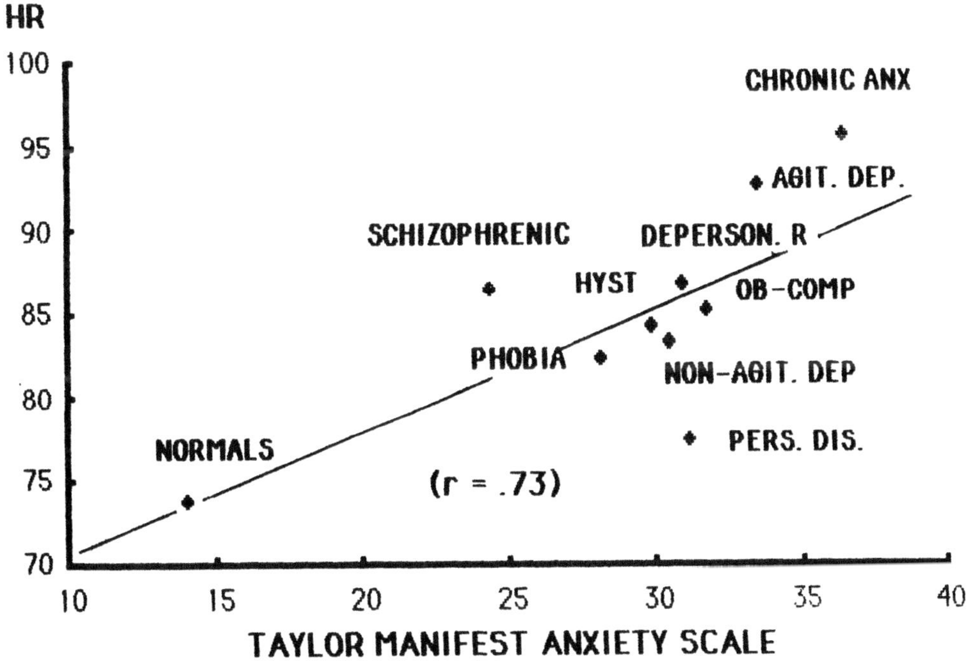

Figure 8. Relationship between mean heart rates and mean scores on the Taylor Manifest Anxiety Scale of a normal and nine diagnostic groups of patients. Data used to make chart is from Kelly (1980).

Although neurotics show higher heart rates under stress than in basal recording conditions the difference in change between their already elevated levels and reactions to stress is *weaker* than that found in normals (Kelly, 1980). What distinguishes anxiety disorders from normals is the tonic and chronic elevation of heart rate, not their reactions to stress. The normal is like a good runner who has a low tonic heart rate allowing more increase in response to the physical stress of running. Anxiety disorders reach their limits very quickly because they start so high. Situational phobics do not show elevated tonic heart rates compared to normals, but when exposed to their specific phobic stimulus or situation they show stronger responses than normals to those situations (Öst, 1987). Socially anxious persons do not show greater heart rate responses during interactions with same sex peers but do show greater heart rate reactions than controls during interactions with opposite sex peers. Since neuroticism or trait anxiety scales mostly reflect social anxiety, they may fail to correlate with heart rate because most stress tests do not employ the specific type of situation which would reveal their autonomic overactivity. However, this type of specific autonomic reactivity is likely an effect of a learned fear rather than reflecting a generalized autonomic instability. The fact that even anxiety neurotics may return to baseline when their disorder is abated suggests that if the psychophysiological effect is causal rather than reactive it does not indicate a permanent trait, but rather a temporary failure of some regulatory mechanism. Panic attacks, when they first occur, are examples of seemingly spontaneous autonomic disinhibitions. Whereas most anxiety disorders, with the exception of simple situational phobias, show evidence of a broad additive heritability (about 30%) disposition, panic disorder also has a specific heritability, possibly a dominant gene controlling the primary physiological symptoms and their regulation (Pauls et al., 1980).

Psychophysiological reactivity depends on more basic neurochemical substrates. The central regulation of both cortical and autonomic activity depends on neurotransmitters, their receptors, and the enzymes regulating their production and disposal. Therefore, let us move down to the next turtle.

Psychopharmacology

MONOAMINE OXIDASE (MAO)

In bottom-up approaches there is little interest in the enzyme monoamine oxidase (MAO) which regulates monoamine levels in the brain by catabolic breakdown. But in a top-down approach MAO is a valuable marker for some of the major personality traits, behavior patterns, and psychopathologies, and it indicates the crucial role of the monoamines (norepinephrine, epinephrine, dopamine, and serotonin) in personality. MAO type B is usually measured from blood platelets in living humans. In the brain it regulates the neurotransmitter dopamine. The relationship of platelet to brain MAO is not known but there is indirect evidence that platelet MAO varies directly with brain MAO. Ballenger et al. (1983) found a negative correlation ($-.34$, $p<.05$) between platelet MAO and homovanyllic acid (HVA), the metabolite of dopamine, in the cerebrospinal fluid. MAO inhibition in brain (assessed by PET scan) and platelets correlated .95 in a small sample (Bench et al., 1991).

MAO has been negatively related to traits of sensation seeking and extraversion in a number of studies. Athough the typical correlations between MAO and sensation seeking are weak (median $r = -.25$), and therefore not always replicable, they are nearly always in the same direction (Zuckerman, 1991). In 11 groups of normals where the full SSS was used, MAO correlated negatively with MAO in 10 of the 11 groups, and the correlations were significant in 7 of the groups (Table 3). Such replicated relationships are rare in this area. The correlations are supported in the behavioral realm where low MAO males from a normal population reported more convictions for criminal offenses, and more alcohol, tobacco, and drug use, habits associated with sensation seeking. Low MAO male monkeys observed in a natural colony environment engaged in more aggressive and sexual behavior than high MAO monkeys. MAO was negatively correlated with aggression in two Swedish studies. Low MAO newborns are more active and more motorically developed than high MAO babies. Low MAO children and adults make quicker decisions in behavioral test for impulsivity. In men there is a strong negative relationship between MAO and failure to inhibit responses in a motor task assessing passive avoidance.

Table 3. Correlations: MAO vs. Sensation seeking scale (SSS) general or total scores

Authors	Subjects	n's	r's
Murphy et al. (1977)	F students	65	.17
	M students	30	-.45**
Schooler et al. (1978)	F students	47	-.43**
	M students	46	-.52**
Ballenger et al. (1983)	M and F adults	36	-.17
Schalling et al. (1987)	M students	40	-.25
Arque et al. (1988)	M and F adult	13	-.66**
	M and F patients	44	-.25*
Ward et al. (1987)	M students	57	-.24*
	F students	30	-.15
Shekim et al. (1989)	M adults	58	-.23*

* $p < .05$; ** $p < .01$: one-tailed tests from Ballenger on.

MAO also correlated negatively with extraversion or positively with introversion in several studies done in America and Germany, but in Swedish studies the relationships were not significant. However, low MAO has been related to high levels of self-reported social activity in humans and observed social activity in monkeys.

Low MAO levels are found in marijuana users, bipolar disorders (even when not in the manic state), schizophrenics with positive symptoms (hallucination, delusions), and alcoholics. In view of the association of type B MAO with dopamine regulation, the finding of an association between substance abuse and a gene for the D_2 dopamine receptor is exciting (Uhl et al., 1992) suggests that the strong association between sensation seeking and substance abuse could be mediated by a dysregulated dopaminergic system.

GONADAL HORMONES

Gonadal hormones affect personality at several stages and in primary and secondary ways. Exposure to androgens or estrogens in the developing fetus may affect play patterns and interests in children and these hormones are involved in early gender typical patterns. Rough and tumble play, for instance, is found much more often in male than female young in both chimpanzees and humans. This type of play is also found in human females who were exposed to excessive androgens during the fetal period. Gonadal development at puberty influences the secondary physical sex changes and eventual changes in sexual-affiliative interests. Levels of testosterone in young unmarried males are related to their cumulative sexual experience and general interest in sex. Testosterone in married women is correlated with their reported sexual gratification, sexual responsivity and frequency of intercourse.

In the realm of personality there have been replicated findings of an association between testosterone and the disinhibition type of sensation seeking and extraversion and sociability (Zuckerman, 1991). Testosterone also correlates with social dominance in tests and in observed behavior. Although testosterone is related to impulsivity and lack of inhibition, its relationship with aggression is only found for groups of extremely violent males. Aggression and hostility in the normal range are not related to testosterone.

Both MAO and Testosterone play a role in at least three of the major dimensions of personality: sociability, impulsive sensation seeking, and activity. There is even some evidence of a negative relationship between testosterone and neuroticism in males. As I said in the introduction, we cannot expect to find simple one-to-one isomorphism between a single personality trait and a particular biological trait.

THE MONOAMINES

The MAO findings are what originally drew my attention to the monoamines as particularly relevant for personality. The monoamines are highly involved in motivational and emotional mechanisms discovered in comparative neuropharmacology and biological psychiatry (Zuckerman, 1984). Generalized reward, joy, fear, depression, approach and avoidance, aggression, and all the basic emotions involved in personality traits are associated with brain monoamines. Almost all of the drugs which tranquilize or stimulate emotions and are used to treat behavior and mood disturbances act directly or indirectly on the monoamine systems. This is not to deny the possible importance of the endogenous opiates (endorphins), or other transmitters like acetylcholine, but the monoamines appear to be a good place to begin the search for the "humors" of temperament.

The comparative animal research is a major source of hypotheses about the behavioral mechanisms related to the three major monoamines in the brain: the catecholamines dopamine and norepinephrine and the indoleamine serotonin. There is always a caution in comparative studies because the animal models for human traits are sometimes questionable. Table 4 summarizes and, of course, simplifies a vast and sometimes contentious literature. In the next column is the huge phylogenetic leap to the possible human equivalents in personalilty traits.

The dopamine pathway running from the ventral tegmental area to the nucleus accumbens through the medial forebrain bundle and the lateral hypothalamus and ascending to the frontal lobes is where intrinsic reward takes place. Dopamine in this pathway is essential for brain self-stimulation. Drugs which produce euphoria, including cocaine and opiates, act on this system although at different points, stimulant reward in the nucleus accumbens and opiates in the ventral tegmental end. Natural rewards such as stimulation from food or sex also activate this system.

Table 4. Some behavioral involvements of monoamine brain systems

	Animal models	**Human**
Dopamine	Reward (Brain Self Stimulation)	
	Activity-Exploration	Extraversion?
	Approach to novel stimuli	Sensation Seeking?
	Motivation	Deficit: Parkinsonism
	Feeding	
	Foraging	
	Fighting	
	F___ing	
Norepinephrine	Attention	
	Arousal (Brain and Autonomic)	Neuroticism?
	Alarm (moderate levels)	Anxiety?
	Anxiety (Gray-Redmond Hypothesis)	Deficit: Sociopathy?
		Depression?
Serotonin	Anxiety: Inhibition Function	Deficit: P Dimension
	Inhibition of Behavior in	Impulsivity?
	Approach-Avoidance Conflict	Impulsive, violent suicide and homicide
	Conflict (Passive Avoidance)	Antisocial Personality?
	Inhibition of Aggression	Depression?
	Inhibition of Sexual Behavior	

Another dopamine pathway from the substantia nigra to the neostriatum, caudate nucleus and putamen is essential for the coordination of purposive and coordinated movements. A destruction of dopamine neurons in this pathway produces Parkinsonism disease that has profound effects on general motivation and interest in the environment, apart from the difficulties in initiating or terminating movement. Animal studies also show the essential role of dopamine pathways in the classical four "F"s of basically motivated behavior: foraging, fighting, feeding and ...."frolicking." Thus dopamine is involved in the basic approach mechanism which might subsume extraversion, and impulsive sensation seeking.

Norepinephrine pathways have been associated with four "A"s: attention, arousal, alarm, and anxiety. Different theorists have put major stress on one or the other of these mechanisms but for the most part they are not incompatible. The dorsal ascending norepinephrine system originates in the locus coeruleus and extends to thalamus, amygdala, hippocampus, forebrain and the entire neocortex. It is involved in focusing of attention on any significant environmental stimuli, including threatening, novel, or potentially rewarding stimuli. As such it is essential to orienting and interest in the environment, possibly relating it to sensation seeking, a trait characterized by strong orienting responses to novel stimuli and a general openness to novel experience, even risky experience (Zuckerman, 1984, 1990). In a more general sense it innervates all parts of the cortex and seems to serve a generalized arousal function like the reticulocortical activating system. A deficit could be involved in extraversion and an excess to introversion according to the classical Eysenck hypothesis.

Both Gray (1982) and Redmond (1987) see the locus coeruleus and dorsal ascending noradrenergic system as an alarm system that directly triggers fear arousal, inhibits ongoing behavior and directs attention toward threatening or novel stimuli. Redmond concedes that at low levels of activity the system could serve as a more neutral cognitive mechanism for attention, but both investigators believe that it is the central system in the anxiety or panic response. According to this latter theory norepinephrine should be high in anxiety disorders and positively related to trait anxiety and neuroticism. A decifit in noradrenergic activity might be related to a failure of normal anxious caution, as seen in the psychopath, or clinical depression as suggested by the catecholamine hypothesis of this disorder.

As Gray points out, the typical response to sudden fear is inhibition of behavior. He sees serotonin as serving this function along with norepinephrine. But behavioral inhibition is not unique to fear, but is necessary for all kinds of adaptive behavior. Impulsivity is the opposite end of inhibition. The serotonergic system which originates in the raphe nuclei extends to many of the same limbic structures innervated by dopamine and noradrenergic systems. Serotonin is involved in the inhibition of many kinds of behavior, such as aggression, sex, and general approach behavior, which are potentiated by the catecholamine neurotransmitters. Serotonin may also inhibit emotional-impulsive behavior. A deficit in the serotonergic system could be related to the impulsive sensation seeking trait.

Evidence for the role of the locus coeruleus and dorsal ascending noradrenergic bundle (DANB) as the alarm system for anxiety is based on several types of findings including: 1. stimulation of the locus coeruleus produces fear like expressions in primates (Redmond, 1987); 2. increase in active avoidance and impairment of passive avoidance by DANB lesions (Gray, 1982); 3. benzodiazepines inhibit locus coeruleus and DANB activity; 4. noradrenergic agonists like Yohimbine increase MHPG (the norepinephrine metabolite), blood pressure, cortisol, and anxiety or panic in panic disorder patients more than in normals; noradrenergic antagonists like Clonidine produce greater decreases in MHPG, blood pressure, and cortisol in patients than in normals. There is no difference between anxiety disorders and normals in baseline levels of MHPG. Why doesn't Yohimbine produce much anxiety in normals if it does so in these patients? A more telling criticism is that a variety of pharmacological agents, including lactate and caffeine, also produce anxiety and panic in panic or generalized anxiety disorders, even though they have little or no effect on the noradrenergic system (Gorman et al., 1987). What all these agents have in common is that they produce heightened peripheral autonomic arousal. A sensitive or reactive noradrenergic arousal system may be an important component of trait anxiety, but it is neither necessary in anxiety disorders nor sufficient in normals to produce a state of anxiety.

The benzodiazepines act on receptors in the brain which seem to be designed for them. Of course the brain does not make valium but apparently there is something chemically close to valium manufactored by the brain itself. The benzodiazepine receptors are parts of a larger receptor complex for the neuroregulator gamma-aminobutyric acid (GABA). GABA is generally inhibitory on neurons in other systems. The benzodiazepines exert their antianxiety effect on structures like the locus coeruleus by augmenting the reactions of GABA. GABA itself could be the key to anxiety trait in combination with noradrenergic reactivity. Low production or nonreactivity of GABA might be associated with trait anxiety. However, the widespread distribution of GABA argues against such trait specificity. There are inverse agonists which act on GABA receptors to produce anxiety and it may be an excess of these which characterize the anxious person. A third possibility is that trait anxiety is a function of a lowered concentration of benzodiazepine receptors. An anxious strain of mice has been found to be low in these receptors. Whichever of these hypotheses is correct it is clear that the BZ receptor complex and GABA must be considered in the biological foundations of anxiety or emotionality.

Opiates also have antianxiety effects and the endogenously produced opiates (endorphins) may therefore be involved in anxiety trait as a natural extension of their primary biological function, the reduction of pain. A deficit might produce a vulnerability to arousal and anxiety or a specific vulnerability to exogenous opiate drugs.

Psychopharmacology and Personality

I have suggested some simple types of hypotheses relating the monoamines to personality traits. Dopamine systems are hypothesized to constitute the biological basis for approach tendencies and reward expectancies in E-Sociability and P-ImpUSS traits. Low MAO levels would be a deregulating factor in the systems and high gonadal hormone levels may potentiate them. Conceivably, the nigrostiatal system motoric functions may be related to the activity component of extraversion while the mesolimbic dopamine system may underlie the generalized reward expectancy and positive affect characteristic of extraverts. A serotonergic system originating in the medial raphe nucleus may mediate behavioral and cortical inhibition and deficits in the effective activity of the system may be related to P-ImpUSS. The noradrenergic dorsal tegmental bundle and the benzodiazepine-GABA receptor complex are hypothesized to interact in producing adrenergic central and peripheral arousal, negative emotionality and punishment expectancy underlying the supertrait of neuroticism-anxiety.

Interactions between neurotransmitter systems are likely since many of these pathways innervate the same limbic structures such as amygdala, septum, hippocampus and hypothalamus. Norepinephrine may potentiate activity in both dopaminergic and serotonergic systems, acting as a nonspecific arousal mechanism for either positive or negative affect. Catecholamine agonists like amphetamine and cocaine may produce euphoria, sociability and arousal at lower doses and anxiety, hostility and paranoia at higher doses (Zuckerman, 1984). Serotonin, endorphins, and endogenous benzodiazepine-type tranquilizers (not yet discovered) work in antagonism to catecholaminergic transmitters and thus may suppress approach tendencies associated with these transmitters. But endorphins and tranquilizers may also disinhibit tendencies suppressed by anxiety. The interactions between neurotransmitters, their enzymes, and hormones are bound to be complex and will ultimately yield their secrets only to a systems model. What at first appears to be so complex seems at closer inspection to be even more complex (but not insoluble).

Genetics

So far I have been discussing the six top turtles where "chicken and egg" problems of interpretation abound. For instance does acognitively or originating anxiety trait produce cardiacc and respiratory symptoms of anxiety or do instabilities in these systems result in a disposition toward anxiety? Does an active sexual life produce high levels of testosterone or is the former a function of the latter? Genes do have a priority in causal explanation so we must give some priority to the seventh of base turtle. During the past several years a number of large scale behavioral genetic studies of personality involving twins and adoptees have been published. Unless one hides within the turtle shell of social behavior, the major role of genetics in personality cannot be denied.

I reviewed the larger-scale twin studies involving a total of over 9,000 identical and 14,000 fraternal twins raised together (Zuckerman, 1991). Ninety-two % of the correlations of identical twins on the supertraits (E, N, and P-type) fall between .4 and .6; and 92% of the fraternal twin correlations on the same traits are between .01 and .28. Heritabilities, calculated either from the correlations or from the Jinks and Fulker (1970) Analysis of Variance model yield heritabilities of broad personality traits ranging between 40 and 60% with the median value around 50%.

For those who cannot accept the logic of identical-fraternal twin comparisions there are now two separated twin studies (Pedersen et al., 1988; Tellegen et al., 1988) If twins are separated soon after birth and not placed in correlated environments, the similarity of separated identicals is a pure measure of the influence of heredity. The difference in corrolation between identical twins raised together and those raised in different families is a measure of the influence of shared environment. The rest of the difference is due to non-shared environment and error of measurement. Recent results from the Minnesota separated twin study (see the chapter by Bouchard in this volume) suggest that about 40% of the variance in the E factor, 50% of the N factor, and 60% of the P or constraint factor (ImpUSS in my system) can be attributed to heredity.

There is some evidence for shared environment for the E factor but none for the N and P factors. Even in the E factor the larger part of the environmental influence is non-shared or specific to each member of the family. While the N and P factors show predominantly additive type genetic variance, the E factor shows evidence of primarily nonadditive genetic variance. The results from the Swedish separated twin study (Pedersen et al., 1988) showed about the same results for E, but lower heritability for N and some minor influence of shared environment for this trait as well as E. Two short scales were used to represent the P factor; one yielded a heritability of 23% and the other a heritability of 45%. Neither showed any influence of shared environment.

Twin studies measure broad heredity including additive, Mendelian type dominance, and epistasis (in which all relevant genes must be present to influence the trait). Only identical twins share all three types of heredity. Non-additive types of hereditary influence lower similarity in phenotype of non-twin siblings and of parents and children.This may be part of the reason why heritability estimates from non-twin studies are so much lower than those from twin studies. However, the conclusions from non-twin adoption studies do indicate one thing clearly. The shared family influence independent of a genetic influence is not a very strong one. Most of the resemblance between siblings or children and parents who are biologically related is due to genes, not to shared environment.

Any finding showing such major genetic influences in personality is counter-intuitive to most social psychologists, but even more confounding is the finding that the most important environmental aspects of personality are not those which are shared within the family or on which family units differ, but those influences that are specific to each member of the family. These findings have been misinterpreted by some to suggest that the family is unimportant in the shaping of personality. All the data can tell us is that the common influences shared by all family members are not important. But our personality theories do not suggest that all member of the family share the same influences. Dunn and Plomin (1990) give many reasons why "siblings are so different". Two siblings are not likely to be treated in the same way even though they share the same parents, if for no other reason than that they are of different ages. Differences in their own personalities may also require differential treatment by the parents. The relationships between the siblings themselves is another nonshared influence depending on their genders, age differences, and differences in personality. The complexity of family interactions is described by family therapists. Consider a normal kind of situation in which two children are raised by two parents. There are two basic dimensions found in child rearing: love vs. rejection, and controlling vs. permissive. These yield four possible patterns (controlling-love, permissive-love, controlling-rejection or punishment; permissive rejection of neglect) for any one parent depending on the particular combination of the two factors. With two parents this yields 16 possible combinations. If we add in the influence of an older sibling we now have 64 possibililties for patterns of influence on the younger child (without even considering the interaction effects). This would be a daunting challenge to any investigator who wanted to study family influences in a systematic manner. This is not to argue that the specific influences of peers outside of the family are not important. Just observe and listen to your adolescent children and ask yourself where these strange behaviors and speech come from. But basic expectancies, goals, and affects may be influenced by family constellations that are different for each member within the family.

Evolutionary Psychology

And below the seventh turtle there is still another one, a prehistoric Godzilla. To the extent that personality dispositions, or what underlies them in our biological makeup, reside in our genes these particular genotypes may be explainable in evolutionary terms of fitness and adaptation. No one knows what our human-like ancestors were like as they roamed the plains of Africa, perhaps 200.000 years ago. We can only speculate based on a few tools, evidences of ritual and forgaging habits, and observations on our evolved cousins, the great apes, who have probably changed less than we have in the years since we went our separate ways from a common ancestor. Evolutionary psychology asks an old question: "what is the natural man or woman?" "Natural" means stemming from our evolved biological selves, or our own species-specific behavior. This question suggests a common pattern of behavior but individual variation is not excluded by this approach because there has been a constant resorting of genes throughout time.

The trait of impulsive sensation seeking may represent the expression of the generalized approach-withdrawal mechanism in response to novel or ambiguous stimuli. It is likely that sensation seeking or explorativeness (to use a comparative term) conveyed advantages and disadvantages at each extreme of the trait. Our hominid ancestors lived in a dangerous environment. They were prey as well as predators. Their predators were not only saber tooth tigers but other bands of their own species as well. Goodall (1986) has observed vicious aggression and killing when intruders from another group entered the territory of the band of chimpanzess she was observing. A glance at our daily newpapers shows that nothing much has changed in this kind of reaction during the evolution of homo sapiens. But competition for resources and mates, and hunting large animals, requires a certain willingness to accept risk and an appetite for the activity itself. The appetite and willingness to accept risk is how we define sensation seeking. Sensation seeking conferred an advantage but the associated risk was a disadvantage. At the low end of the dimension the immediate risk is reduced, but in the long term there is the disadvantage of exhausting the resources in one's own area because of the unwillingness to explore new territories. The restriction in mates in one's own group is a reproductive disadvantage given the competition with other males in the group and restrictions of incest taboos. Disadvantage at both ends suggests that there would be a narrowing of the range of variation to some middle range of sensation seeking. But a number of factors may have prevented such a constriction of the range in the trait. Assortative mating is one of these factors.

If both men and women agree on what is a desirable trait in a mate then like will tend to mate with like while those low on the trait will be left to mate with each other. The trait may show a directional trend over time but the assortative mating will maintain the variability. The genetic tendency of "regression to the mean" prevents the trait from becomming bimodal in distribution. Assortative mating depends upon the perception of the trait and agreement on its desirability in a mate.

Desirable personality traits are among the most highly considered characteristics in mate selection in most cultures where the young are allowed to select their own mates (Buss et al., 1990). But surprisingly, there is little or only weak evidence of these preferences being translated into assortative mating considering the low to non-existent correlations between personality traits of married partners (Ahern et al., 1982). The one exception to this is sensation seeking which correlates substantially in unmarried couples (Thornquist et al., 1991), and married couples (Farley and Davis, 1977; Farley and Mueller, 1978; Lesnik-Oberstein and Cohen, 1984). The congruence on sensation seeking is related to couple and marital adjustment (Donaldson, 1989; Ficher et al., 1981; Ficher et al., 1988) suggesting that it is a valid trait for choices in assortative mating. The reason for the importance of sensation seeking in assortative mating may have to do with its pervasive influence in sexuality, interests, preferences, attitudes and behavior in general (Zuckerman, 1979) and suggests its importance as a major component in one of the five major traits of personality (Zuckerman et al., 1988, 1991).

Caveats

Investigators who confine their studies to the molar levels of traits and behavior often accuse us of reductionism. The goal of most psychobiological approaches is not reductionism. The complexity of the phenomena at each level of psychobiological analysis and the lack of simple isomorphism between levels precludes a reduction of one level to another.

Those two work at more molecular levels accuse us of oversimplicity. How can we speak of a relation between a particular neurotransmitter like dopamine and a trait like extraversion, or implusive sensation seeking? Which dopaminergic system are we speaking of? Is the relationship a function of production or catabolism of the neurotransmitter and the enzymes involved in these processes, or one of the several presynaptic or postsynaptic receptors, their sensitivities or qyantities? All we can reply is that the evidence is usually one of relative activity or reactivity of a system the specific source of which remains to be determined. A "top-down" approach goes from the general to the specific. One could, of course, start from a particular indicator from the molecular level and work from the "bottom-up", but this strategy would lead to many false leads. In this game of "21 questions" it is more strategic to start with the traditional "animal, vegetable or mineral" or perhaps "enzyme, neurotransmitter, or hormone".

References

Ahern, F.H., Johnson, R.C., Wilson, J.R., McClearn, G.E. and Vandenberg, S.G. (1982) Family resemblance in personality. *Behavior Genetics, 12,* 261-280.

Ballenger, J.C., Post, R.M., Jimerson, D.C., Lake, C.R., Murphy, D.L., Zuckerman, M. and Cronin, C. (1983) Biochemical correlates of personality traits in normals: An exploratory study. *Personality and Individual Differences, 4,* 615-625.

Barratt, E.S., Pritchard, W.S., Faulk, D.M. and Brandt, M.E. (1987) The relationship between impulsiveness subtraits, trait anxiety, and visual N100-augmenting-reducing: A topographic analysis. *Personality and Individual Differences, 8,* 1, 43-51.

Bench, C.J. et al. (1991) Measurement of human cerebral monoamine oxidase type B (MAO-B) activity with positron emission tomography (PET): a dose ranging study with the reversible inhibitor Ro 19-6327. *European Journal of Clinical Pharmacology, 40,* 169-173.

Buchsbaum, M.S. (1974) Average evoked response and stimulus intensity in identical and fraternal twins. *Physiological Psychology, 2,* 365-370.

Buchsbaum, M.S. and Silverman, J. (1968) Stimulus intensity control and the cortical evoked response. *Psychosomatic Medicine, 30,* 12-22.

Buss, D.M. et al. (1990) International preferences in selecting mates: A study of 37 societies. *Journal of Cross-Cultural Psychology, 21,* 5-47.

Costa, P.T. Jr. and McCrae, R.R. (1985) *The NEO Personality Inventory Manual.* Odessa, FL: Psychological Assessment Resources.

Costa, P.T. Jr. and McCrae, R.R. (1992) Four ways five factors are basic. *Personality and Individual Differences, 13,* 667-673.

Deakin, J.F.W. and Exley, K.A. (1979) Personality and male-female influences on the EEG alpha rhythm. *Biological Psychology, 8,* 285-290.

Diagnostic and statistical manual of mental disorders-revised. Washington, D. C.: American Psychiatric Association, 1987.

Donaldson, S. (1989) Similarity in sensation seeking, sexual satisfaction and contentment in heterosexual couples. *Psychological Reports, 64,* 405-406.

Dunn, J. and Plomin, R. (1990) *Separate lives: Why siblings are so different.* New York: Basic Books

Eysenck, H.J. (1957) *The dynamics of anxiety and hysteria.* New York: Praeger.

Eysenck, H.J. (1967) *The biological basis of personality.* Springfield, IL., Charles C. Thomas

Fahrenberg, J. (1987) Concepts of activation and arousal in the theory of emotionality (neuroticism): A multivariate conceptualization. In J. Strelau and H.J. Eysenck (Eds.) *Personality dimensions and arousal* (pp. 99-120). New York: Plenum.

Farley, F.H. and Davis, S.A. (1977) Arousal, personality and assortative mating in marriage. *Journal of Sex and Marital Therapy, 3,* 122-127.

Farley, F.H. and Mueller, C.B. (1978) Arousal, personality, and assortative mating in marriage: Generalizability and cross-cultural factors. *Journal of Sex and Marital Therapy, 4,* 50-53.

Ficher, I.V., Zuckerman, M. and Neeb, M. (1981) Marital compatibility in sensation seeking trait as a factor in marital adjustment. *Journal of Sex and Marital Therapy, 7,* 60-69.

Ficher, I.V., Zuckerman, M. and Steinberg, M. (1988) Sensation seeking congruence in couples as a determinant of marital adjustment: A partial replication and extension. *Journal of Clinical Psychology, 44,* 803-816.

Gale, A. (1983) Electroencephalographic studies of extraversion-introversion: A case study in the psychophysiology of individual differences. *Personality and Individual Differences, 4*, 371-380.

Gale, A. and Edwards, J.A. (1986) Individual Differences. In M.G.H. Coles, E. Donchin and S.W. Porges (Eds.) *Psychophysiology: Systems, processes and applications* (pp. 431-507). New York: Guilford.

Goldberg, L.R. (1990) An alternative "description of personality": The big-five factor structure. *Journal of Personality and Social Psychology, 59*, 1216-1229.

Goodall, J. (1986) *The chimpanzees of Gombe*. Cambridge, MA: Beklnap/Harvard Univ. Press.

Gorman, J.M., Fyer, M.R., Liebowitz, M.R. and Klein, D.F. (1987) Pharmacologic provocation of panic attacks. In H.Y. Meltzer (Ed.) *Psychopharmacology: The third generation of progress*. (pp- 985-993). New York: Raven Press.

Gray, J.A. (1982) *The neuropsychology of anxiety: An enquiry into the functions of the septo-hippocampal system*. New York: Oxford University Press.

Haier, R.J., Sokolski, K., Katz, M. and Buchsbaum, M.S. (1987) The study of personality with positron emission tomography. In J. Strelau and H.J. Eysenck (Eds.) *Personality dimensions and arousal* (pp. 251-267), New York: Plenum Press.

Hall, R.A., Rappaport, M., Hopkins, H.K., Griffin, R.B. and Silverman, J. (1970) Evoked response and behavior in cats. *Science, 170*, 998-1000.

Hare, R.D. (1980) A research scale for the assessment of psychopathy in criminal populations. *Personality and Individual Differences, 1*, 111-117.

Hodges, W.F. (1976) The psychophysiology of anxiety. In M. Zuckerman and C.D. Spielberger (Eds.) *Emotions and anxiety: New concepts, methods, and applications* (pp. 176-194) Hillsdale, NJ: Erlbaum.

Horney, K. (1939) *New ways in psychoanalysis*. New York: Norton.

Jinks, J. and Fulker, D.W. (1970) Comparison of the biometrical, genetical, MAVA, and the classical approaches to the analysis of human behavior. *Psychological Bulletin, 73*, 311-349.

Kagan, J., Reznick, J.S. and Snidman, N. (1988) Biological bases of childhood shyness. *Science, 240*, 161-171.

Kelly, D. (1980) *Anxiety and emotions*. Springfield, IL: Charles C. Thomas.

Lesnik-Oberstein, M. and Cohen, L. (1984) Cognitive style, sensation seeking, and assortative mating. *Journal of Personality and Social Psychology, 46*, 112-117.

LeDoux, J.E. (1987) Emotion. In F. Plum (Ed.) *Handbook of physiology. 1: The nervous system Vol. 5, Higher functions of the brain* (pp. 419-460). Bethesda, MD: American Physiological Society.

Lewis, E.G., Dustman, R.E. and Beck, E.C. (1972) Evoked response similarity in monozygotic, dizygotic, and unrelated individuals: A comparative study. *Electroencephalography and Clinical Neurophysiology, 23*, 309-316.

Litle, P. and Zuckerman, M. (1986) Sensation seeking and musical preferences. *Personality and Individual Differences, 4*, 575-578.

Lukas, J.H. and Siegel, J. (1977) Cortical mechanisms that augment or reduce evoked potentials in cats. *Science, 196*, 73-75.

Lukas, J.H. and Mullins, L.F. (1985) Auditory augmenters are sensation seekers and perform better under high work loads. *Psychophysiology, 22*, 580-581 (Abstract).

Lykken, D.T. (1982) Research with twins: The concept of emergenesis. *Psychophysiology, 19*, 361-373.

Mathew, R.J., Weinman, M.L. and Barr, D.L. (1984) Personality and regional cerebral blood flow. *British Journal of Psychiatry, 144*, 529-532.

Myrtek, M. (1984) *Constitutional psychophysiology*. London: Academic Press.

Naveteur, J. and Baque, E.F. (1987) Individual differences in electrodermal activity as a function of subjects' anxiety. *Personality and Individual Differences, 8*, 615-626.

Newman, J.P. and Kosson, D.S. (1986) Passive avoidance learning in psychopathic and non-psychopathic offenders. *Journal of Abnormal Psychology, 95*, 252-256.

Norman, W.T., (1963) Toward an adequate taxonomy of personality attributes: Replicated factor structure. *Journal of Abnormal and Social Psychology, 66*, 574-583.

O'Gorman, J.G. (1987) Extraversion and the EEG, I: An evaluation of Gale's hypothesis. *Biological Psychology, 5*, 257-318.

O'Gorman, J.G. and Lloyd, J.E.M. (1987) Extraversion, impulsiveness and EEG alpha activity. *Personality and Individual Differences, 8*, 169-174.

Öst, L.G. (1987) Age of onset of different phobias. *Journal of Abnormal Psychology, 96*, 223-229.

Pauls, D.L., Bucher, K.D., Crowe, R.R. and Noyes, R. (1980) A genetic study of panic disorder pedigrees. *American Journal of Human Genetics, 31*, 639-644.

Pedersen, N.L., Plomin, R., McClearn, G.E. and Friberg, L. (1988) Neuroticism, extraversion, and related traits in adult twins reared apart and reared together. *Journal of Personality and Social Psychology, 55*, 950-957.

Redmond, D.E. Jr. (1987) Studies of locus coeruleus in monkeys and hypotheses for neuropsychopharmacology. In H.Y. Meltzer (Ed.), *Psychopharmacology: The third generation of progress* (pp. 967-975). New York: Raven Press.

Saxton, P.M., Siegel, J. and Lukas, J.H. (1987) Visual evoked potential augmenting/reducing slopes in cats-2. Correlations with behavior. *Personality and Individual Differences, 8*, 511-519.

Schroth, M.L. (1991) Dyadic adjustment and sensation seeking compatibility. *Personality and Individual Differences, 12*, 467-471.

Stelmack, R.M. (1990) Biological bases of extraversion: Psychophysiological evidence. *Journal of Personality, 58*, 293-311.

Strelau, J. (1983) *Temperament, personality, and arousal*. London: Academic Press.

Tellegen, A., Lykken, D.T., Bouchard, T.J., Wilcox, K., Segal, N. and Rich, A. (1988) Personality similarity in twins reared together and apart. *Journal of Personality and Social Psychology, 54*, 1031-1039.

Thornquist, M.H., Zuckerman, M. and Exline, R.V. (1991) Loving, liking, looking, and sensation seeking in unmarried college couples. *Personality and Individual Differences, 12*, 1283-1292.

Uhl, G.R., Perisco, A.M. and Smith, S.S. (1992) Current excitement with D2 dopamine receptor gene alleles in substance abuse. *Archives of General Psychiatry, 49*, 157-160.

Vaillent, G.E. and Schnurr, P. (1988) Study of psychiatric impairment within a college population selected for mental health. *Archives of General Psychiatry, 45*, 313-319.

Zuckerman, M. (1969) Theoretical formulations: I. In J.P. Zubek (Ed.) *Sensory deprivation: Fifteen years of research* (pp. 407-432). New York: Appleton-Century-Crofts.

Zuckerman, M. (1979) *Sensation seeking: Beyond the optimal level of arousal*. Hillsdale, NJ: Erlbaum.

Zuckerman, M. (1984) Sensation seeking: A comparative approach to a human trait. *Behavioral and Brain Sciences, 7*, 413-471.

Zuckerman, M. (1989) Personality in the third dimension: A psychobiological approach. *Personality and Individual Differences, 10*, 391-418.

Zuckerman, M. (1990) The psychophysiology of sensation seeking. *Journal of Personality, 58*, 313-345.

Zuckerman, M. (1991) *Psychobiology of Personality*. New York: Cambridge University Press.

Zuckerman, M., Buchsbaum, M.S. and Murphy, D.L. (1980) Sensation seeking and its biological correlates. *Psychological Bulletin, 88*, 187-214.

Zuckerman, M., Kuhlman, D.M. and Camac, C. (1988) What lies beyond E and N? Factor analyses of scales believed to measure basic dimensions of personality. *Journal of Personality and Social Psychology, 54*, 96-107.

Zuckerman, M., Kuhlman, D.M., Teta, P., Joireman, J. and Carroccia, G. (1992) *The development of scales for a five basic personality factor questionnaire*. Paper presented at the meeting of the Eastern Psychological Association, Boston, MA, April 3-5, 1992.

Zuckerman, M., Kuhlman, D.M., Thornquist, M. and Kiers, H. (1991) Five (or three) robust questionnaire scale factors of personality without culture. *Personality and Individual Differences, 12*, 929-941.

Zuckerman, M., Murtaugh, T.T. and Siegel (1974) Sensation seeking and cortical augmenting-reducing. *Psychophysiology, 11*, 535-542.

Zuckerman, M., Simons, R.F. and Como, P.G. (1988) Sensation seeking and stimulus intensity as modulators of cortical, cardiovascular, and electrodermal response: A cross-modality study. *Personality and Individual Differences, 9*, 361-372.

IDENTIFYING BASIC DIMENSIONS OF PERSONALITY
Commentary on the chapter by M. Zuckerman

Guus L. Van Heck,
Department of Psychology
Tilburg University,
5000 LE Tilburg
The Netherlands

Moving in Circles

Magnusson (1991), recently has said that research in many fields of psychology goes in circles instead of moving forward. This seems especially true for personality psychology, judging by recent publications with inducements to go 'back to the phenomena' (Magnusson, 1991) or 'back to basics' (Hogan, 1987), and considering the constantly growing stream of articles in which the rediscovery of Tupes and Christal's (1961) and Norman's (1963, 1967) five-factor structure—comprising extraversion or surgency, agreeableness, conscientiousness, emotional stability versus neuroticism, and culture—is received with ever greater acclaim (see Wiggins and Trapnell, in press, for an overview).

The above suggests that, from time to time, personality psychologists lose the scent and, while aimlessly wandering, have to be put on the right track again. But what is the right course? According to Magnusson (1991, 1992), the appropriate use of theory, method, and statistics in personality research must be based on careful, systematic analyses and descriptions of the phenomena per se. Also Costa and McCrae (1992a, b) attempt to persuade their readers to give priority to a systematic method of description. However, in contrast to Magnusson, they strongly believe that our current knowledge of neuroscience is too primitive to guide decisions about personality description.

I do not feel that this pessimistic view is beyond doubt. Zuckerman, in exploring the biological basis of individual differences in personality, has proven in the foregoing chapter and especially in his trail-blazing book '*The Psychobiology of Personality*' (1991) that, in spite of our 'primitive' knowledge, the phenomena can be described with a tremendous richness. In my view, Zuckerman has shown convincingly that Costa and McCrae (1992a, b) are too skeptical when evaluating the level of sophistication of present-day psychobiological models of personality.

In this commentary, I first will point to the importance of biological individual factors in studying personality processes. Then, a brief sketch is given of Zuckerman's hierarchal model focussing on his selection of basic dimensions of personality. Next, this selection — the product of Zuckerman's attempt to bring some conceptual clarity into the numerous classifications of trait dimensions in psychological literature — will be contrasted with the Tupes and Christal/Norman Big-Five taxonomy of personality attributes. It will be argued that these

J. Hettema and I. J. Deary (eds.), Foundations of Personality, 101–111.

different taxonomies reflect to a certain degree different definitions of personality. Finally, I will present some recent research that fits very well within the Zuckerman model.

The Importance of a Psychobiology of Personality

To be fair, Costa and McCrae (1992a, b) do not intend to suggest that studies of personality and psychophysiology are unimportant. However, they strongly feel that such studies should begin with our current understanding of the organization of personality into five factors: They recommend to include measures of all Big Five factors in every psychophysiological study. According to Costa and McCrae (1992a, p. 659), "... empirical links between these factors and neurophysiological variables, even if they are not anticipated by theory, may give valuable clues to the biological basis of personality". So, they admit that research on personality and psychobiology is worthwhile. However, they have a very black view of the profit. In an agony of doubt, they ask themselves what should have become of Tupes and Christal's brainchild — the hope and pride of current personality psychology — if they would have 'refined' their neat structure based upon the neurobiological explanations that were possible at that time, some 30 years ago. Of course, they admit that progress has been made in this area, but they still have a low opinion of present day neuroscience. According to their view, at best it will add little, but most probably it will obscure the issue considerably. In any case, they do not expect that it will be very helpful in predicting and controlling behavior.

Magnusson (1992), in contrast, states that the primary goal for psychological research is not to predict and control behavior, but to identify the relevant factors operating in the process of individual functioning and the mechanisms by which they operate. In his words: "... observations of single aspects of individual functioning have to be interpreted and given meaning by reference to context including other psychological and *biological individual factors* and factors in the environment" (p. 12; italics added).

The latter point of view raises the question 'How to refer to the biological context?'. The literature contains many tentative attempts at answering this question. Perhaps the most elaborated and comprehensive 'solution' is provided by Zuckerman's hierarchal model.

Zuckerman's Hierarchal Model

Zuckerman (1991, 1992a, b) has presented a hierarchal model encompassing from top to bottom the following levels: Traits, social behavior, conditioning, physiology, biochemistry, neurology, and genetics. It is his belief that personality should be studied at all levels. For a complete understanding of a particular level one should scrutinize the next level down who forms the foundation of the level above. Traits are based on aggregations of instances of social behavior or covert reactions, which, in turn, are based on conditioning and observational learning, which is based on information processing systems and motivational systems in the brain, and so on.

Zuckerman's model is predominantly a *top-down* model, in contrast to, for instance, Gray's (1987) *bottom-up* approach. The main argument, according to Zuckerman (1992a) for preferring top-down above bottom-up is that a top-down approach does not require an isomorphism between the basic dimensions of personality and specific neurological or neurotransmitter systems. I am not sure that such an isomorphism is inevitable in the case of a bottom-up approach. What is wrong with starting to collect data in various types of situation using a broad range of psychobiological measures and analyzing these data looking for stable, situation specific state patterns and linking later on these patterns to personality? It has been done before; and not without success (see, e.g., Hettema's (1989) distinction between effort, arousal, and activation states). There are, however, practical limitations. Of course. For instance, such an endeavor is expensive, extremely time-consuming and, even more important, it requires more knowledge about environments than currently is available (Forgas and Van Heck, 1992). Moreover, there are the limitations due to current measurement technology. Thus, there is enough to make a bottom-up approach an up-hill struggle. So, I am not convinced that the isomorphism of dimensions and systems is the major obstacle. There are several other, more compelling reasons that make the development of a top-down model the more attractive alternative.

In a top-down model, the first thing that should be done is to decide what is on top. Zuckerman follows Eysenck in deciding that on top there should be the *basic dimensions of personality*. The selection of these basic dimensions is guided by the following criteria: (a) reliable identification of dimension factor structures across methods, genders, ages, and cultures; (b) temporal stability of measured dimensions; (c) identification of similar kinds of behavioral traits marking the factor in other species of animals, particularly mammals that live in social groups; (d) at least moderate heritability for the dimension; (e) the identification of the dimension with some significant biological markers; and (f) the ultimate identification of the biological systems comprising the neural substrate for the dimension. In his recent book *The Psychobiology of Personality*, Zuckerman (1991, p. 5) remarks that "a more socially oriented personality theorist would not see the necessity for the last group of criteria". I guess he is right about that.

Using these criteria, Zuckerman et al. (1991) have created their own Big Five: Sociability, Neuroticism-Anxiety, Aggression-Hostility, Impulsive-Unsocialized-Sensation Seeking, and Activity. This five-factor model does not bear much relation to the Tupes and Christal/Norman Big Five (cf. Eysenck, 1992). The rationale for the selection of markers was predominantly the frequency of usage in genetic and psychobiological studies (cf. Zuckerman, 1992a, p. 679). Consequently, this robust five-factor solution will meet the above six criteria better than the Tupes and Christal/Norman type of five-factor solution. I cannot abstain from thinking that this distinction is somehow related to a distinction between two definitions of personality which has major conceptual and research implications.

Two Definitions of Personality

In order to provide a framework for psychobiological analyses within a top-down approach, the basic dimensions of personality have to be identified. It should be realized that the identification process is heavily dependent upon the particular definition of personality chosen. It is an often neglected issue that there are in personality psychology at least two fundamentally different definitions of personality, both with major conceptual and research implications. I refer to the perspectives of the *observer* and the *actor*.

In a very inspiring chapter, *Personality Psychology: Back to Basics*, Hogan (1987) points to these two definitions of personality. According to Hogan (1987, p. 85), they are "overlapping but incommensurable, mutually translatable but deriving from different spheres of experience":

> From an observer's perspective, personality consists of an actor's reputation, his or her distinctive social stimulus value. This reputation is not value-neutral; reputations are explicitly evaluative and, *au fond*, they consist of an appraisal of the actor's potential or demonstrated contribution to his or her social group (Hogan, 1987, p. 85).

Hogan states that every culture has a vocabulary that it uses to make these evaluations. Furthermore, referring to White (1980), Hogan points to the fact that the structure of these vocabularies is much the same across cultures. It is this particular structure of the vocabulary that observers use to describe actors that is reflected in the Tupes and Christal/Norman Big Five model.

In contrast, from the perspective of actors, personality consists of the inner psychological structures, qualities, and characteristics that *cause* them to generate their unique reputation.

The kind of research one will do depends on which definition one chooses. In the case of studying personality from the observer's perspective research will focus on social perception and social cognition. When choosing the actor's perspective, investigations will focus on "the relationship between an actor's *temperament* and developmental history and his or her values, aspirations, self-images, and social knowledge structures, and the subsequent relationship between the foregoing and his or her social behavior" (Hogan, 1987, p. 86; italics added).

Now, we can ask the question what would be the most logical perspective for the identification of basic dimensions of personality suitable as a starting-point for a psychobiology of personality: the observer's perspective or the actor's perspective? Costa and McCrae (1992a) obviously opt for the first option. Looking once more at Zuckerman's criteria for basic traits of personality, it becomes clear that he is inclined to use the other definition of personality. As he states in his 1992 *Personality and Individual Differences* article: 'Conscientiousness' and 'agreeableness' are not traits that can be used as descriptions for animal behavior; 'impulsivity' and 'aggressiveness', however, are applicable. I agree with Zuckerman that this is not just a semantic issue. For, impulsivity and aggressiveness are not simple opposites for conscientiousness and agreeableness.

In Zuckerman's hierarchal model, illustrated with the metaphor of a 'tower' of seven turtles, the top turtle represents the dimensions that underlie personality traits measured by questionnaires that have been frequently used in research on the biological bases of personality. As said above, this selection criterion is responsible for the fact that the Zuckerman five-factor structure reflects more temperament and the actor perspective, while the Tupes and Christal/Norman Big Five reflect more social reputation and the observer perspective.

A good illustration of this point is the following. Conscientiousness (C) and agreeableness (A) are, in my opinion, the two factors that most clearly reflect an observer's perspective. In daily life, people constantly evaluate others in terms of how dependable they are and how rewarding they are to deal with. According to Eysenck (1992), C and A are primary factors which, together with many others, make up the superfactor Psychoticism (P), reflecting tough-minded, antisocial tendencies vs. socialized humaneness. In the analyses by Zuckerman et al. (1991), impulsivity was highly loading on the Psychopathy-Impulsive Unsocialized Sensation Seeking (P-ImpUSS) factor, while Aggressiveness was highly loading on the Aggression-Hostility (Agg-Host) factor. In an earlier study (Zuckerman et al., 1988), impulsivity, lack of socialization, sensation seeking, and aggression combined to form P. As said above, it is possible to apply 'impulsivity' and 'aggressiveness' to animals, from rats to chimps, while this would be more difficult for 'conscientiousness' and 'agreeableness'. P in Eysenck's PEN-model and P-ImpUSS and Agg-Host in Zuckerman's model are closer to what traditionally has been seen as temperament variables compared with C and A.

Strelau (1983) has differentiated temperament and personality along five dimensions. Two of these dimensions are: (1) biological factors are assumed to play a *relatively* stronger role in temperament whereas social determinants are more prominent in personality; (2) temperament can be seen in other species of animals whereas personality describes the phenomena that are specific to humans. Seeing this, I have no doubt that Strelau will agree with me that the Zuckerman five-factor structure is a better candidate for the top turtle than the Tupes and Christal/Norman Big Five. At least, when it should be a five-factor turtle.

In any case, the Zuckerman et al. (1991) classification guarantees that impulsivity and sensation seeking will be studied as separate traits that guide and steer future investigations of the biological bases of human personality.

In the remainder of this commentary, I will point to the fruitfulness of scrutinizing these two variables. When, following the recommendations of Costa and McCrae (1992a), surgency, agreeableness, conscientiousness, emotional stability, and culture/intellect/openness would have been the doors to the world of the biological context, then, much of the findings summarized by Zuckerman (1992b) as well as our own results still would have been undetected. Below, I will briefly discuss some of these outcomes.

Impulsivity and Sensation Seeking

Under the heading *P-impulsive unsocialized sensation seeking*, Zuckerman (1992b) summarizes the major findings, predominantly at the level of Turtle 4 (the 'Physiology'-turtle). This overview contains fascinating data on the disinhibition-inhibition mechanism postulated to underlie the P-ImpUSS dimension. It reveals that low sensation seekers or low P-ImpUSS types have a type of brain that shuts down under stressful conditions. Above all, this review shows that impulsivity and sensation seeking are among the most promising candidates for successful studies of the psychobiology of personality.

In our own recent research, especially sensation seeking was 'doing quite well'. In a first study (Krijns et al., submitted), slow brain potentials were recorded in an S1–S2 paradigm. S1 was a warning signal that indicated to respond either as quickly as possible or to delay the response until 1 sec after S2. Low sensation seekers showed larger N1 amplitudes at the right frontal electrode, as well as larger Positive Slow Wave (PSW) amplitudes at all leads. Low sensation seekers had smaller Negative Slow Wave (NSW) amplitudes than high sensation seekers. Moreover, low sensation seekers showed a laterality effect: they had larger NSW amplitudes in the right than in the left hemisphere. The differences in ERP amplitude between high and low sensation seekers were independent of the manipulations at S1. See Figs. 1 and 2 for graphical illustrations of the outcomes of this study.

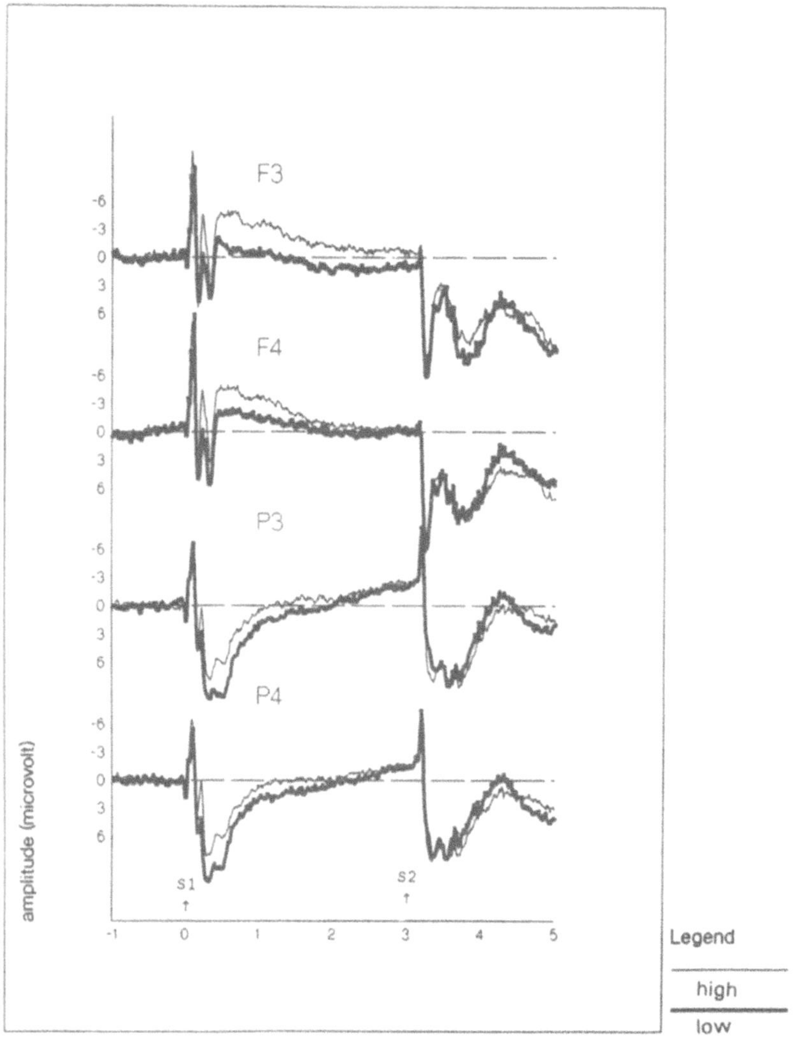

Figure 1. Grand average ERP for high and low sensation seekers.

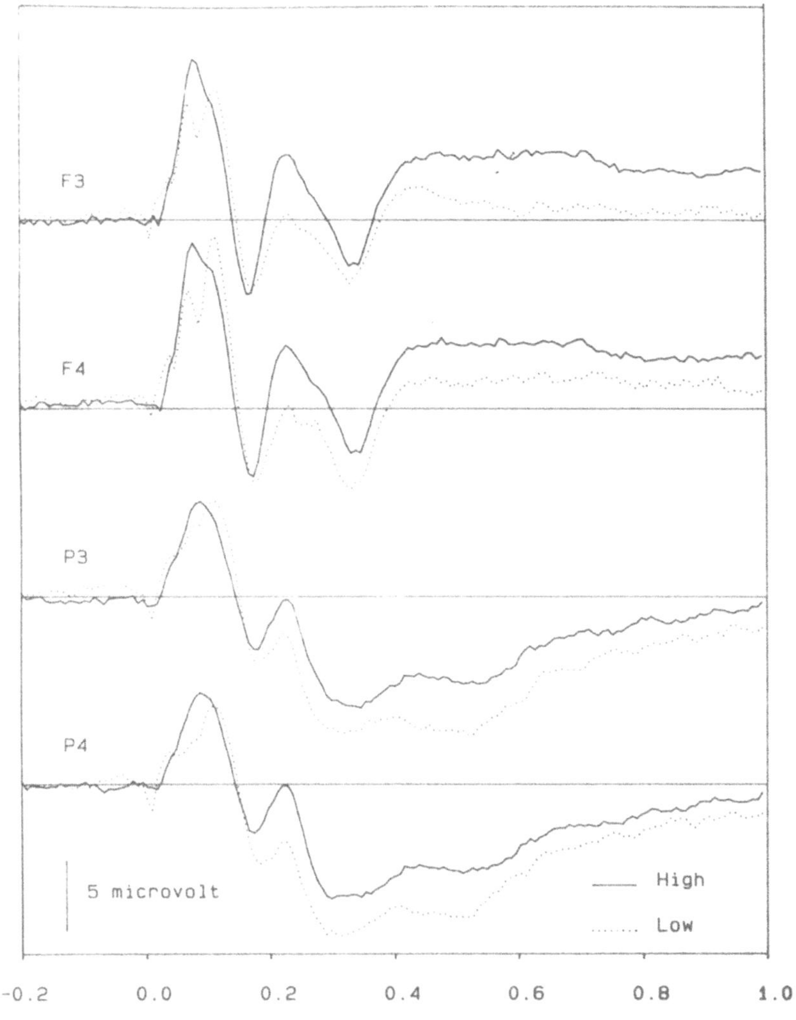

Figure 2. ERP to S1 for high and low sensation seekers.

Earlier studies (see Zuckerman, 1990; Zuckerman et al., 1988) have demonstrated that the high sensation seeker shows stronger orienting responses and augmented cortical responses (vertex recordings of the N1-P1 complex) to stimuli that are either novel or of specific interest. The low sensation seeker, on the other hand, shows defensive responses (cardiac acceleration) and reduced cortical responses.

Extending these findings to the NSW, the larger NSW of the high sensation seekers would reflect approach behavior, whereas the smaller NSW amplitudes of the low sensation seekers would reflect withdrawal behavior. The small frontal laterality effect for the NSW can be explained in terms of Davidson's (1984) distinction between approach-related emotions (mediated in the left hemisphere) and avoidance-related emotions (mediated in the right hemisphere). Low sensation seekers show a slightly larger NSW in the right hemisphere compared with the left hemisphere. This might reflect the presence of avoidance-related emotions.

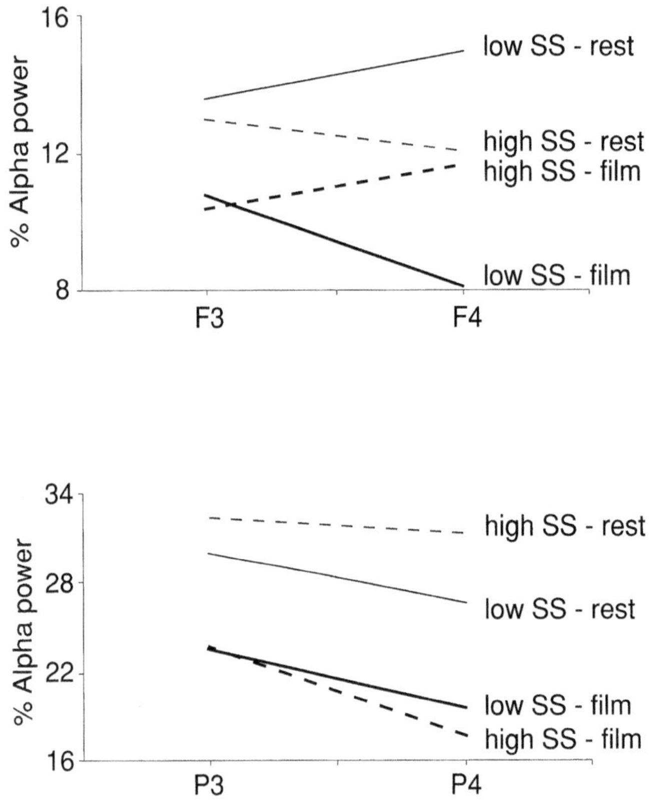

Figure 3. Alpha activity and sensation seeking

In a second study (Van Heck et al., in preparation) cortical alpha rhythm was recorded while subjects watched a film (a rape episode). Low sensation seekers showed less alpha activity over the right frontal cortex than over the left frontal cortex. For high sensation seekers the opposite was found. In the rest period preceding the film, low sensation seekers showed more alpha activity over the right frontal cortex than over the left frontal cortex. Again, for high sensation seekers the opposite pattern was recorded (see Fig. 3). Also these outcomes can be interpreted in terms of the involvement of approach-related and avoidance-related emotions (cf. Davidson, 1984).

So, applying Zuckerman's model of basic dimensions to the study of the psychobiology of personality has much appeal. However, further clarification of the basic dimensions still appears to be necessary before the living pyramid of turtles can be scrutinized in an optimal way. Several questions force on us. For instance: Is a five-factor turtle the best starting point, or is a three-factor system—comprising P, E, and N—or a four-factor solution (e.g., PEN + Openness to experience) more promising? It is hard to say; the near future hopefully will cy this issue. For now, Zuckerman has given the field of personality psychology a much-needed model of the basic dimensions of personality, a model that has considerable attraction for those who want to study personality from an actor's perspective, who not only want to forecast important behaviors but also want to explain why the forecasting models work as they do, and who realizes that the latter can only be accomplished when the cognitions, affects, actions and reactions of individuals are studied in their biopsychosocial contexts.

Moving Forward?

Several personality psychologists have recently expressed an optimism regarding the future of personality psychology (Buss and Cantor, 1989; Caprara, 1992; Caprara and Van Heck, 1992; Magnusson, 1992; Pervin, 1990). Since a long time, there is a feeling that there now is a genuine possibility of a true movement forward.

When personality psychology nevertheless relapses into the old pattern of moving in circles, then — I promise — I will write in the year 2000 an article entitled "Back to the core of personality psychology: the actor/observer distinction and the basic dimensions of personality". I will ask Marvin Zuckerman to help me.

References

Buss, D.M., and Cantor, N. (Eds.) (1989). *Personality psychology: Recent trends and emerging directions*. New York: Springer-Verlag.

Caprara, G.-V. (1992). Presidential address at the 6th European Conference on Personality Psychology, Groningen, The Netherlands, June.

Caprara, G.-V. (1992). Personality psychology. Some epistemological assertions and historical considerations. In G.-V. Caprara and G.L. van Heck (Eds.), *Modern personality psychology. Critical reviews and new directions* (pp. 3-26). New York: Harvester-Wheatsheaf.

Costa, P.T., Jr., and McCrae, R.R. (1992a). Four ways five factors are basic. *Personality and Individual Differences, 13*, 653-665.

Costa, P.T., Jr., and McCrae, R.R. (1992b). Reply to Eysenck. *Personality and Individual Differences, 13*, 861-865.

Davidson, R.J. (1984). Hemispheric asymmetry and emotion. In K.R. Scherer and P. Ekman (Eds.), *Approaches to emotion* (pp. 39-57). Hillsdale, NJ: Erlbaum.

Eysenck, H.J. (1992). Four ways five factors are *not* basic. *Personality and Individual Differences, 13*, 667-673.

Forgas, J.P., and Van Heck, G.L. (1992). The psychology of situations. In G.-V. Caprara and G.L. Van Heck (Eds.), *Modern personality psychology. Critical reviews and new directions* (pp. 418-455). New York: Harvester-Wheatsheaf.

Gray (1987). The neuropsychology of emotion and personality. In S.M. Stahl, S.D. Iverson, and E.C. Goodman (Eds.), *Cognitive neurochemistry* (pp. 171-190). Oxford: Oxford University Press.

Hettema, P.J. (Ed.) (1989). *Personality and environment: Assessment of human adaptation.* Chichester: Wiley.

Hogan, R. (1987). Personality psychology: Back to basics. In J. Aronoff, A.I. Rabin, and R.A. Zucker (Eds.), *The emergence of personality* (pp. 79-104). New York: Springer Publishing Company.

Krijns, P.W., Gaillard, A.W.K., Van Heck, G.L., and Brunia, C.H.M. (submitted). *Personality effects on brain potentials in an S1–S2 paradigm.*

Magnusson, D. (1991). Back to the phenomena. *Zeitschrift für Psychologie*, Supplement 11, 61-70.

Magnusson, D. (1992). Back to the phenomena: Theory, methods, and statistics in psychological research. *European Journal of Personality, 6*, 1-14.

Norman, W.T. (1963). Toward an adequate taxonomy of personality attributes: Replicated factor structure in peer nomination personality ratings. *Journal of Abnormal and Social Psychology, 66*, 574-583.

Norman, W.T. (1967). *2800 personality trait descriptors: Normative operating characteristics for a university population.* Ann Arbor, MI: Department of Psychology, University of Michigan.

Pervin, L.A. (Ed.) (1990). *Handbook of personality: Theory and research.* New York: The Guilford Press.

Tupes, E.C., and Christal, R.E. (1961). Recurrent personality factors based on trait ratings (*USAF ASD Technical Report* No. 61-97). Lackland Air Force Base, TX: U.S. Air Force. Reprinted in *Journal of Personality, 60*, 225-251.

Van Heck, G.L., Gaillard, A.W.K., Knipscheer, J.W., and Böcker, K. (in preparation). *Cortical and cardiovascular reactivity and psychosocial stressors.*

Wiggins, J.S., and Trapnell, P.D. (in press). Personality structure: The return of the Big Five. In S.R. Briggs, R. Hogan, and W.H. Jones (Eds.), *Handbook of personality psychology.* Orlando, FL: Academic Press.

White, G.M. (1980). Conceptual universals in interpersonal language. *American Anthropologist, 82*, 75-81.

Zuckerman, M. (1990). The psychophysiology of sensation seeking. *Journal of Personality, 58*, 313-345.

Zuckerman, M. (1991). *Psychobiology of personality.* Cambridge: Cambridge University Press.

Zuckerman, M. (1992a). What is a basic factor and which factors are basic? Turtles all the way down. *Personality and Individual Differences, 13*, 675-681.

Zuckerman, M. (1992b). Personality from top (traits) to bottom (genetics) with stops at each level between. Paper presented at the Nato Advanced Research Workshop: Biological and Social Approaches to Personality. August 17-20, Oisterwijk, The Netherlands.

Zuckerman, M., Kuhlman, D.M., and Camac, C. (1988). What lies beyond E and N? Factor analyses of scales believed to measure basic dimensions of personality. *Journal of Personality and Social Psychology, 54*, 96-107.

Zuckerman, M., Kuhlman, D.M., Thornquist, M., and Kiers, H. (1991). Five (or three) robust questionnaire scale factors of personality without culture. *Personality and Individual Differences, 12*, 929-941.

Zuckerman, M., Simons, R.F., and Como, P.G. (1988). Sensation seeking and stimulus intensity as modulators of cortical, cardiovascular, and electrodermal response: A cross-modality study. *Personality and Individual Differences, 9*, 361-372.

THE LOCATION OF THE REGULATIVE THEORY OF TEMPERAMENT (RTT) AMONG OTHER TEMPERAMENT THEORIES

Jan Strelau
Department of Psychology
University of Warsaw
Stawki 5/7
Warsaw, Poland

Abstract

Temperament theories in general and the regulative theory of temperament in particular pay special attention to the formal characteristics of behavior. The roots and some basic issues of the regulative theory of temperament are discussed against this background. During the last few years the theory has witnessed a major modification. After several theoretical and empirical stages, the present theory features seven major traits: briskness, perseverance, mobility, sensory sensitivity, endurance, activity and emotional reactivity.

These results have been derived in a study based on a sample of over 2000 subjects of both sexes, with ages ranging from 15 to 80 years. The traits are related with selected characteristics measured with popular temperament inventories like the Eysenck Personality Questionnaire-Revised, Zuckerman's Sensation Seeking Scale and the Strelau Temperament Inventory-Revised.

Introduction

When looking at conceptualizations and research on temperament from a historical perspective it may be concluded that there are some common denominators in describing the nature of temperament. Hippocrates and Galen (Hutchins, 1952) introduced the idea that individual differences in temperament may be explained by means of biological mechanisms (for a detailed description see Stelmack and Stalikas, 1991). This idea was further developed in empirical research in the first quarter of this century. For example, Heymans and Wiersma (1906-1909) studied the inheritance of temperamental traits, Kretschmer (1944) tried to explain the links between the types of temperament and the types of the physical make-up by means of the hormonal system, and Pavlov (1951-52) interpreted individual differences in temperament by referring to several properties of the central nervous system.

In contemporary studies on temperament much research has been conducted with the aim showing that given temperamental traits may be characterized by psychophysiological or neurophysiological correlates. This is especially evident in research on extraversion and neuroticism (Eysenck, 1967, 1970; Eysenck and Eysenck, 1985; Stelmack, 1990), anxiety (Gray, 1982, 1991), sensation seeking (Zuckerman, 1979, 1991), impulsivity (Barratt, 1983;

J. Hettema and I. J. Deary (eds.), Foundations of Personality, 113–132.

Schalling and Asberg, 1985) and on the Pavlovian properties of the central nervous system (Nebylitsyn, 1972; Teplov, 1964). The fact that heredity contributes importantly to individual differences in temperamental traits seems to be one of the most undoubted statements in this domain of research (Buss and Plomin, 1984; Eysenck and Eysenck, 1985; Loehlin, 1989; Zuckerman, 1991).

There are also other features of temperament which have been regarded for decades, even centuries, as being typical for temperament. Many temperament researchers consider temperamental traits to be present in man and animals (e.g., Adcock, 1957; Diamond, 1957; Garau and Garcia-Sevilla, 1985; Simonov, 1987). The presence of temperament in both man and animals has been used, among other things, as an argument that temperamental traits play an important adaptive role in the individual's functioning (Adcock, 1957; Buss and Plomin, 1984; Diamond, 1957).

The stability or resistance to change of temperamental traits, though not commonly accepted, has for many years been regarded as being typical for temperamental characteristics (e.g., Heymans and Wiersma, 1906-09; Ewald, 1924; Hagekull, 1989; Kagan, 1982; Rusalov, 1985). Another feature to which many temperament researchers refer is the presence of temperament in humans from early childhood. Evidence has been presented by Thomas and Chess (1977; Thomas et al., 1968) and their colleagues (Buss and Plomin, 1975, 1984; Kohnstamm et al., 1989; Rothbart, 1981) which shows the richness of temperamental characteristics in infants.

Some temperament researchers limit temperament to the emotional domain, following the arguments of Wundt who described the four ancient Greek temperaments by means of emotional characteristics only. The understanding of temperament by Allport (1937, p. 54) as referring "to the characteristic phenomena of an individual's emotional nature", supported this view (see also Eysenck, 1970). In contemporary research on temperament the emotion-centered understanding of temperament is represented mainly by Mehrabian (1978; 1991), Goldsmith and Campos (1982; 1986), and Gray (1991).

After many years of reviewing the literature on temperament which was published in this and the last centuries I am convinced that formal characteristics of behavior are the most frequently met common denominator used to describe the nature of temperament (Strelau, 1969, 1974a, 1983).

In the eighteenth century Kant introduced the idea that the four Hippocrates-Galen types of temperament may be characterized by means of such formal characteristics as strong versus weak emotion and activity. Since then almost all researchers who referred to the ancient Greek typology portrayed the temperamental types by referring to formal characteristics of behavior. For example, Stern (1921) described 11 typologies of temperament which were developed at the turn of the century, and which are based on the classification of the four Hippocrates-Galen temperaments. In all of them individual differences in such behavior and/or emotional characteristics as, weak versus strong, slow versus rapid, broad versus narrow, deep versus shallow, and short-lasting versus long-lasting have been used as the basis for distinguishing the four classic temperaments.

Moreover, in conceptualizations of temperament which were developed in the second half of the present century the formal characteristics of behaviour have been taken as definitional criteria for describing temperament. Some examples will illustrate this. According to Pavlovian-oriented psychologists, temperament may be understood as the dynamic characteristics of behavior expressed in individual differences in speed and intensity of reaction (behavior), and determined by the type of the nervous system (Ilin, 1978; Merlin, 1973; Rusalov, 1979;

Teplov, 1964). Diamond (1957) defined temperament: "...in terms of the ease of arousal of unlearned (underlined by J.S.) patterns of adaptive behavior" (p. 95). Again, this definition underlines that the formal characteristics are essential for temperament. In definitions of temperament given by Strelau (1974a, 1983, 1989), Eliasz (1990), Rusalov (1985, 1989), and partially also by Rothbart (1981), formal characteristics of behavior are used as definitional criteria of temperament. Additionally, in emotion-centered conceptualizations of temperament formal characteristics are the main criteria employed in describing individual differences in temperament (see Goldsmith and Campos, 1982, 1986). This tradition goes back to Allport's understanding of temperament, where characteristics of emotions such as "strength", "susceptibility", "speed", "fluctuation", "intensity" became definitional components of temperament (1937, p. 54).

The formal characteristics of behavior are also present in conceptualizations in which temperament is regarded as the style of behavior; this is expressed in the question *how* does an individual behave?, by contrast with such questions as *why*? (motivation), *what*? and *how well*? (abilities). The stylistic definition of temperament, introduced by Thomas and Chess (1977), gained high popularity among American researchers, especially in the domain of children's temperament (see e.g., Carey, 1983; Maziade, 1988; Windle et al., 1988). In this understanding of temperament, any behavior, whatever its content is, may be characterized in terms of "how" (the way it proceeds) from the beginning of a person's life.

Experimental studies conducted in Eysenck's laboratory also show that the formal characteristics, such as speed of conditioning, sensory thresholds, and speed and accuracy of performance, have been the main criteria which permitted extraverts and introverts to be distinguished (Eysenck, 1970, p. 456). The formal characteristics are the main bases of hypotheses regarding on animal's extraversion-introversion (Garau and Garcia-Sevilla, 1985; Simonov, 1987), emotionality (Broadhurst, 1975), and other temperamental traits.

The Regulative Theory of Temperament

The framework of the regulative theory of temperament (RTT) was published for the first time in the seventies (Strelau, 1974a). Taking into account the data collected in our laboratory as well as other findings, the theory has been modified and revised (see Strelau, 1983, 1989). The following sources might be regarded as most influential for the development and contemporary state of the RTT.

(1) Psychophysiological and psychological studies and experiments (see Strelau, 1969) conducted during a period of over ten years within the framework of the Pavlovian typology. From these studies I learned, among other things, that central nervous system (CNS) properties should not be understood as referring to specific neurophysiological mechanisms but as explanatory concepts. Of special significance for my conceptualization of temperament was the concept of strength of excitation. According to Pavlov strength of excitation manifests itself in the individual's ability to endure intense and/or long-lasting stimulation without passing into transmarginal inhibition. The concept of strength of excitation has been modified by Teplov and Nebylitsyn (1963). They regarded this CNS property as a dimension with two poles -one being endurance (E) and the other one sensitivity (S)- of the CNS. According to the two authors the ratio between endurance and sensitivity is relatively stable (E/S = constant). Based on this assumption Teplov and Nebylitsyn developed a variety of experimental methods aimed at assessing strength of excitation by means of measuring different aspects of sensory sen-

sitivity (see Nebylitsyn, 1972; Strelau, 1983; Teplov, 1964). The concentration on sensory sensitivity phenomena allowed: (1) the more efficient study of strength of excitation without breaking into the individual's upper threshold of capacity, and (2) the search for links between the concept of strength of excitation and arousability (Gray, 1964). Pavlov's idea, that temperament plays an important role in the individual's adaptation to the environment, supported by many studies conducted in our laboratory (Strelau, 1969, 1974b), nourished my view regarding the functional significance of temperamental traits.

(2) The theories of arousal (activation), especially the concept of arousability. Hebb's (1955) concept of arousal, which refers to the energetic (non-specific) function of sensory events, is better than other concepts for formulating hypotheses regarding the links between the formal characteristics of behavior (temperament) and their physiological correlates. The concept of optimal level of arousal (Hebb, 1955) was important for the development of the RTT. This concept states that the individual regulates by means of behavior the level of arousal in such a way as to attain or maintain an optimal level. Very important for the development of the RTT was the concept of arousability (Gray, 1964), the latter being understood as a trait that refers to relatively stable individual differences in the level of arousal. Gray's interpretation of strength of excitation in terms of arousal and arousability was helpful in moving beyond the Pavlovian tradition in temperament research (Strelau, 1983). The constructs of arousal and arousability were also very useful in interpreting the links between our concept of temperament and other biologically-oriented conceptualizations in this field of study. Among the theories which are supposed to have some elements in common with the RTT, besides the Pavlovian approach, are the following: Eysenck's theory on extraversion and neuroticism (1970, Eysenck and Eysenck, 1985), Zuckerman's sensation seeking temperament (1979, 1991), Gray's (1982) dimensions of anxiety and impulsivity, and Petrie's (1967) augmenting/reducing dimension.

(3) The theory of action as developed within the framework of the Russian tradition by Vygotsky (1962) and Leontev (1978). According to this theory, actions are treated as the core of human behavior. This conceptualization, thoroughly modified by my master, Tomaszewski (1978), facilitated the investigation of temperamental traits from the point of view of reciprocal relations between humans and their environment, where human activity plays the most important role in regulating these relations.

(4) My own knowledge based on the literature of temperament and related topics, accumulated during more than thirty years of study. This experience has allowed me to locate my own reflections on temperament in a historical context and in a broad perspective regarding conceptualizations of temperament and personality.

(5) Intensive psychometric research conducted in our laboratory (Zawadzki, 1992; Zawadzki and Strelau, 1991) aimed at studying the structure of temperament and at developing a temperament inventory which corresponds with the RTT.

THE REGULATIVE THEORY OF TEMPERAMENT: COMPLEMENTARY STATEMENTS AND FINDINGS

For a better understanding of the modifications of the regulative theory of temperament which took place under the influence of studies conducted during the last decade (see Eliasz, 1981; Klonowicz, 1992; Strelau, 1983; 1985a), information of the RTT in its original formulation is needed. In this chapter I must limit the presentation to some basic statements. The interested reader is referred to publications in which the RTT has been described in detail (see Strelau,

1983, 1985b, 1989; Strelau and Plomin, in press). In this section the postulates underlying the RTT and the way in which the structure of temperament has been revised will be presented.

Postulates Underlying the RTT

From the statement that temperament refers to formal characteristics of behavior the following postulates may be explicitly or implicitly formulated.

(1) There exist relatively stable individual differences with respect to formal characteristics of behavior, encompassed by two basic categories: intensity (the energetic aspect of behavior) and time (the temporal aspect of behavior). Thus individual differences in temperament are obvious.

(2) Every behavior, whatever its kind and content, can be characterized by means of energetic and temporal characteristics. Therefore temperament may be expressed in all kinds of behavior and reactions.

(3) From early infancy children differ in the intensity and the temporal characteristics of their basic drives, reactions and behaviors, which means that from the beginning of postnatal life they may be characterized in terms of temperamental traits.

(4) Regardless of the specific behaviors that are typical for man and other animals, all mammals (at least) may be characterized by means of properties which refer to the categories of intensity and time, thus temperament occurs in both humans and animals.

(5) Taking into account postulates (3) and (4) it is reasonable to assume that temperamental characteristics are a product of biological evolution, and there must, therefore, exist some genetic bases as well as physiological mechanisms co-determining individual differences in temperament.

Considering the postulates formulated above as well as the theoretical considerations on temperament to be met in the literature I propose to define temperament in the following way: *Temperament refers to basic, relatively stable personality traits which apply mainly to the formal aspects of reactions and behavior (energetic and temporal characteristics). These traits are present since early childhood and they occur in man and animals. Being primarily determined by inborn physiological mechanisms, temperament is subject to changes caused by maturation and by some environmental factors.*

This definition of temperament, together with postulates (1)-(5), not published until now, is the starting point for developing the regulative theory of temperament.

Revising the Structure of Temperament

After years of study I became convinced that the RTT, strongly influenced by the Pavlovian tradition, has besides its advantages also some disadvantages. The weakest points of the RTT seem to be the following: (1) The concept of activity which plays such an important role in the RTT has not been sufficiently operationalized. (2) The relationship between endurance and sensitivity as assumed by Teplov and Nebylitsyn (1963), and taken by me as a presumption for the understanding of the reactivity trait, needs empirical evidence in order to be accepted. (3) The emotional characteristics present in almost all theories of temperament did not have a proper place in the RTT. (4) There was no inventory or other diagnostic instrument available which fully corresponds with the RTT.

The modification of the RTT, which took into account the critical remarks formulated under (1)-(3), consisted mainly in revising the view on the structure of temperament. In its original form the RTT postulated three following traits: reactivity and activity representing the energetic characteristics of behavior, and mobility which refers to the temporal domain. For a better understanding of the postulated changes in the structure of temperament a short characterisation of the three traits is needed.

According to the RTT *reactivity* reveals itself in the intensity (magnitude) of reaction to acting stimuli. This temperament trait co-determines the individual's sensitivity (sensory and emotional) and endurance (capacity to work). Reactivity resembles the Pavlovian concept of strength of excitation. This was one of the main reasons why reactivity was mostly measured by means of the Strength of Excitation (SE) scale from the Strelau Temperament Inventory (STI, Strelau, 1983). It should be noted, however, that the SE scale is aimed at measuring only the endurance aspect but not the individual's sensitivity, the opposite pole of the reactivity dimension. The term reactivity has been used in our research in two different meanings. First, as an explanatory construct which allowed us to interpret the relationship between sensory sensitivity and endurance. Second, as a descriptive term, as measured by the STI-SE scale.

Activity has in the RTT a specific meaning. "*Activity* is a temperamental trait which reveals itself in the amount and range of undertaken actions (goal-directed behaviors) of a given stimulative value" (Strelau, 1989, p. 40). By means of activity the individual regulates the stimulative value of behavior and/or situations in such a way as to satisfy his/her need for stimulation (Eliasz, 1981; Strelau, 1983). The latter is also co-determined by the individual's level of reactivity. The need to maintain or attain an optimal level of arousal is regarded in the RTT as a standard for the regulation of stimulation (Eliasz, 1981). The stimulative value of activity consists in the fact that activity by itself is a source of stimulation. The more complex and difficult the activity, the higher the stimulation being generated. One of the most efficient generators of stimulation is the emotional connotation of activity (risk-taking, threatening actions, etc.). By means of activity the individual may also modify the stimulative value of the environment. Activity aimed at approaching or avoiding stimulation stemming from the surroundings exemplifies this idea (activity as the "organizer" of stimulation). It is important to emphasize that any kind of human activity, at any period of life, has a given stimulative value. The criterion for high or low activity is the energetic characteristic, i.e., the stimulative value of activity. It has to be mentioned that activity was measured in our studies only in experimental settings, mainly by assessing the so-called style of action (Strelau, 1983, 1989). In most of our research a reverse relationship between level of reactivity and the stimulative value of activity has been found, as predicted by the RTT.

Mobility, understood as the ability to switch behavior in response to changes in the surroundings, was regarded as a secondary trait. Since mobility as postulated by the RTT has much in common with the Pavlovian concept of mobility of the nervous system, the Mobility (MO) scale from the STI was often used in our laboratory for measuring RTT mobility. A factor analytic study (Gorynska and Strelau, 1979) has shown that mobility is composed of traits like speed and tempo (with positive loadings), and persistence and recurrence (with negative loadings). Our studies have shown that mobility plays an important role in human adaptation (Klonowicz, 1992; Strelau, 1983), and that this temperament trait allows us to search for links between temperament and abilities (Strelau, 1977).

PSYCHOMETRIC RESEARCH AIMED AT IDENTIFYING THE STRUCTURE OF TEMPERAMENT

As mentioned above, the fact that temperament refers first of all to the energetic and temporal characteristics of behavior, is one of the most essential assumptions of the RTT. However, previous studies on temperament within the RTT have convinced me that the three traits -reactivity, activity and mobility- which have been the center of our attention for a period of almost twenty years, do not exhaust the whole variety of the energetic and temporal characteristics on which individuals differ. Research efforts were undertaken in two directions to make the formal characteristics of behavior as complete as possible. These efforts are described as stages I and II.

Stage I. In the *temporal* domain factor analytic studies have been undertaken already in the seventies in order to describe in details the traits which refer to characteristics described by time parameters (Gorynska and Strelau, 1979). The RTT postulated six following temporal traits: persistence, recurrence, mobility, regularity, speed, and tempo. However, with few exceptions only mobility was subject to empirical studies. This was mainly due to the fact that in the STI (Strelau, 1983) only the Mobility scale referred to the temporal characteristics of behavior. Taking into account the earlier results of Gorynska and Strelau (1979), we separated the following five temporal characteristics:

(1) *Mobility*: speed of shift in behavior in response to changes in the surroundings, i.e., the ability to react adequately to quickly varying stimuli.

(2) *Speed*: speed of reactions (as measured by reaction time) to stimuli or other external demands.

(3) *Tempo*: frequency of occurrence of homogeneous reactions within a given time-unit (e.g., number of words or movements being naturally performed in a given period of time).

(4) *Recurrence*: tendency to repeat the same behavior (reaction) after the stimulus (situation) evoking the given behavior (reaction) has terminated.

(5) *Persistence*: tendency to maintain reactions (behaviors) after the stimulus (situation) evoking these reactions (behaviors) has terminated.

With respect to the *energetic* characteristics of temperament the two traits -reactivity and activity- have been divided into lower-level traits, taking into account former findings and some of our assumptions. Thus, considering that reactivity comprises two poles -sensitivity and endurance- both have been treated as separate characteristics. Endurance has been divided into two different kinds depending whether it refers to the individual's own activity or to resistance against external factors. Further, taking into account the importance of emotions for describing temperament, we separated the emotional components within the sensitivity and endurance characteristics. According to the RTT, activity by itself has a stimulative value. On the other hand, by means of activity the individual is able to regulate the stimulative value of the situation. Taking this distinction into account we separated also two kinds of activity (that which is a direct and that which is an indirect source of stimulation). As a result, the following traits referring to the energetic characteristics of behavior have been distinguished:

(1) *Sensory sensitivity*: the ability to react to stimuli of very low stimulative value within all kinds of sensory stimulation.

(2) *Emotional sensitivity*: the ease with which emotions are generated as a response to stimuli (in physical and social situations).

(3) *Endurance to fatigue*: the ability to react adequately in situations demanding long-lasting or high stimulative activity.

(4) *Endurance to distractors*: the ability to react adequately to intensive external stimulation, such as physical discomfort; and tolerance to noise, pain, temperature or other strong stimuli.

(5) *Emotional endurance*: resistance against strong stimulation, such as physical and social stress, expressed in emotional behavior.

(6) *Activity as a direct source of stimulation*: the tendency to undertake behaviors of high stimulative value.

(7) *Activity as an indirect source of stimulation*: the tendency to undertake behaviors the aim of which is to provide stimulation stemming from outside the organism.

Stage II. The 12 characteristics (5 referring to the temporal and 7 to the energetic aspect of behavior) distinguished as a result of former studies, and critical remarks and discussions regarding the RTT, served as the starting point for the refined analysis of the structure of temperament based on these formal attributes of behavior.

A thorough study by Zawadzki (1992) was aimed first at exploring the structure of temperament as postulated by the RTT, and second, at constructing a new temperament inventory.

About 600 behavior statements were generated and then classified into the twelve categories of behavior characteristics by means of prototypicality ratings as well as by item analysis as suggested by Angleitner et al. (1986). Data from over 2000 subjects aged from 15 to 80 (1166 females and 857 males) were collected. To explore whether the findings were replicable, the subjects were divided randomly into two samples: the construction sample and the test sample. Each sample comprised over 1000 subjects and they were matched for basic group characteristics. All psychometric and statistical analysis were carried out for each sample separately. Principal components were computed separately for the temporal and energetic characteristics of behavior. The structure of temperament as revealed by Varimax rotation is presented in Table 1.

As can be seen from this table within the temporal domain two factors emerged, which explained 67.6 % of the total variance (Factor I - 45.5 % and Factor II - 22.1%). Factor I, termed Briskness, has the highest loadings in speed and tempo. Both persistence and recurrence have highest loadings on Factor II which was labelled as Perseverance. Since mobility shares its loadings on both factors we decided for the time being to treat it as a separate trait. The functional significance of mobility found in our former studies (see Strelau, 1983) was an additional argument for maintaining this trait.

The energetic characteristic of behavior is represented by four factors. These factors explained 84,3 % of the total variance, scoring on the separate factors as follows: Factor I - 46.9%, Factor II - 16.0%, Factor III - 12,9%, and Factor IV - 8.5% of the total variance. Factor I comprises both forms of endurance (fatigue and distractors) and is identified as Endurance. Factor II, named as Emotional Reactivity, has the highest loadings on both emotionality characteristics (emotional sensitivity and emotional endurance). Factor III, comprising both forms of activity is identified as Activity. The only high loading found in Factor IV -Sensory Sensitivity- occurred in sensory sensitivity. The factors found in both samples are virtually the same and therefore highly robust.

Table 1. Varimax Rotated Factor Matrix of Temporal and Energetic Characteristics of Behavior

(A) Temporal characteristics

Sample:	Construction sample			Test sample		
Traits	Factor			Factor		
	I	II	h^2	I	II	h^2
Mobility	.68	-.44	.66	.74	-.31	.64
Speed	.70	-.33	.60	.74	-.29	.63
Tempo	.82	.17	.70	.80	.19	.68
Recurrence	.02	.87	.75	.03	.86	.75
Persistence	-.26	.77	.67	-.29	.77	.68

(B) Energetic characteristics

Sample:	Construction sample					Test sample				
Traits	Factor					Factor				
	I	II	III	IV	h^2	I	II	III	IV	h^2
Sens. sens.	.04	-.01	.10	.99	.99	.07	.09	-.01	.99	.99
Emot. sens.	-.23	-.87	-.23	.10	.88	-.22	-.19	-.89	.10	.89
End-fatigue	.87	.21	.20	.05	.84	.88	.20	.20	.08	.86
End-distrac	.83	.31	.11	.01	.80	.79	.14	.38	.03	.78
Emot.endur.	.38	.79	.21	.11	.83	.42	.22	.75	.13	.80
Act-direct	.28	.23	.78	.08	.74	.25	.84	.19	.07	.80
Act-indir	.05	.16	.89	.06	.83	.08	.90	.15	.06	.84

Note. All loadings >.60 are underlined

The factor analytic study forced us to reduce the number of the 12 separated formal characteristics of behavior to seven traits, representing the energetic and temporal characteristics of behavior. These traits, representing the structure of temperament, may be described as follows:

(1) *Briskness*: reveals itself in the speed of reaction to stimuli and in the tempo of performing given activities. It is expressed mainly in motor reaction time to sudden events, in the tempo of movements and speech, and the rate of performing different daily activities.

(2) *Perseverance*: reveals itself in the duration of behavior (persistence) and in the number of repetitive behaviors (recurrence) after cessation of stimuli. It is expressed in the persistence of emotional states, in the duration of cognitive associations, in the unintentional repetition of particular movements arising from emotional background, or in the existence of verbal and motor stereotypes.

(3) *Mobility*: reveals itself in the ability to react quickly and adequately to varying stimuli.

(4) *Sensory sensitivity*: reveals itself in the ability to react to stimuli of very low stimulative value. It is mainly expressed in reactions to tactile, olfactory and visual stimuli.

(5) *Endurance*: reveals itself in the ability to react adequately: (a) in situations demanding long-lasting or high stimulative activity and (b) under intensive external stimulation (physical discomfort, noise, painful stimuli, temperature, etc.).

(6) *Activity*: reveals itself in the tendency to undertake behaviors of high stimulative value or aimed at supplying strong stimulation from the surroundings (e.g., performing risky tasks, getting involved in social activity, seeking new experiences).

(7) *Emotional reactivity*: reveals itself in the intensity of emotional reactions to stimuli. It is expressed in high emotional sensitivity (ease of generating emotional reactions to weak stimuli) and in low emotional endurance (low resistance to situations generating strong emotions).

It has to be said that there are some links between the seven temperamental traits presented above. Table II illustrates the empirical relations between them. From a psychometric perspective it is expected that scales should be orthogonal to each other. However, considering temperamental traits from a functional view, it has often been shown that traits related to each other may play specific adaptive functions. This has been demonstrated, among other things, in case of strength of excitation and mobility (see Strelau, 1983; Klonowicz, 1992), in studies related to different kinds of rhythmicity (e.g., Windle, 1992; Windle, et al., 1986), in studies conducted by Thomas and Chess (1977; Chess and Thomas, 1986) as well as in research on different components of sensation seeking (Zuckerman, 1979, 1991).

Table 2. Pearson Correlations Between the RTTI traits

Scale		(1)	(2)	(3)	(4)	(5)	(6)	(7)
				CONSTRUCTION SAMPLE				
(1)	Briskness		-.08	.33	.15	.22	.11	-.14
(2)	Perseverance	-.06		-.31	.15	-.36	-.04	.56
(3)	Mobility	.38	-.25		.22	.50	.32	-.48
(4)	Sensory sensit.	.19	.12	.28		.00	.07	-.01
(5)	Endurance	.26	-.34	.53	.07		.25	-.54
(6)	Activity	.16	-.03	.31	.05	.24		-.41
(7)	Emotional react.	-.19	.55	-.48	-.03	-.59	-.39	
			TEST SAMPLE					

Note. All coefficients of correlation >.10 are statistically significant ($p < .01$).

Conclusions

The main conclusions to be drawn from the empirical study described in stages I and II are as follows:

(1) The formal characteristics of behavior may be described by means of seven temperamental traits, of which four refer to the energetic aspect (sensory sensitivity, endurance, activity and emotional reactivity) and three to the temporal characteristics (briskness, perseverance, mobility) of behavior.

(2) Sensory sensitivity does not correlate with endurance. This is in disagreement with Teplov and Nebylitsyn's (1963) statement regarding the reverse relationship between both phenomena under discussion as well as with our former assumption regarding reactivity (see p. 11 and Strelau, 1983). Since these results are replicated it means that the statement regarding the ratio between sensitivity and endurance has no support in these psychometric data.

(3) Emotional characteristics constitute a separate trait. In the emotional area the two poles -sensitivity and endurance- are inversely related to each other. This supports Teplov and Nebylitsyn's statement about the sensitivity-endurance relationship. Such a finding is also in agreement with our former assumption regarding the reactivity trait, thus the emotionality dimension has been identified by us as emotional reactivity.

(4) We were able to separate the activity trait with the specific meanings of this dimension, emphasizing its function in regulating the stimulative value of behavior and the surroundings.

THE RTT AS RELATED TO OTHER TEMPERAMENT THEORIES

Because a thorough analysis of the RTT within the context of other temperament theories is not possible here I will concentrate on those aspects in which the regulative theory of temperament differs essentially from other conceptualizations. Finally, I will present some preliminary findings showing the relations between the temperament traits as measured by the Regulative Theory of Temperament Inventory (RTTI, Zawadzki, 1992; Zawadzki and Strelau, 1991) and some other temperamental traits such as the Eysenckian PEN factors, the Pavlovian traits, and Zuckerman's sensation seeking characteristics.

A THEORETICAL PERSPECTIVE

As regards the comparison from a theoretical perspective, attention will be paid to several features typical for the RTT as compared with other temperament theories; below are some of them.

(1) The RTT emphasizes most consistently the fact that temperament comprises the formal characteristics of behavior. The RTT offers a structure of temperament in which the formal characteristics are presented in the broadest range, taking into account the energetic as well as the temporal aspects of behavior.

(2) Among the temperamental traits to be proposed there are two which differ essentially from other contemporary temperament concepts, i. e., activity and perseverance. As already mentioned activity as a temperament trait occurs in almost all temperament theories. In all of them, however, it refers to the *motor* component of behavior only, whereas in the RTT, any goal-directed behavior, including motor and mental activity, when considered not from the content but from the point of view of individual differences in stimulation regulation is considered as an expression of the activity trait. The closest conception to my understanding of activity can be found in Heymans and Wiersma's (1906-1909) typology. These authors consider activity in terms of behavior characterized by the amount of time a person performs any kind of actions.

Perseverance, which resembles the secondary function, i.e., one of the three (beside activity and emotionality) basic temperament dimensions distinguished by Heymans and Wiersma, is not present in other contemporary theories of temperament.

(3) The RTT emphasizes that temperament has a biological basis (see postulates 3-5). However, in contrast to other theories, it does not specify the physiological mechanisms

contributing to the variance of temperamental traits. Very generally, it is assumed that the energetic characteristics of behavior may be explained by means of mechanisms responsible for the regulation of the level of arousal (activation). Certain features of the endocrine system, autonomic nervous system, the brainstem, subcortical and cortical centers should be mentioned here. These systems operate as a functional unit with a fairly stable structure. There exist intra- and interindividual differences in the level of functioning of each separate system included within the unit. Taking these differences into account I prefer to use the concept of *neuro-endocrine individuality* when the physiological bases of the energetic components of temperament are considered. The intraindividual differences are probably one of the main sources of the modality- (stimulus- and/or reaction-), specific temperament characteristics (see Strelau, 1991). As regards the temporal traits, they probably may be explained by means of neurophysiological mechanisms responsible for the speed of elicitation and extinction of nervous processes and the interaction between both of these states (Strelau, 1983). It may be also expected that some neurotransmitters, such as e.g., dopamine, are involved in mediating the temporal characteristics of behavior (see e.g., Netter, 1991).

The physiological mechanisms underlying temperamental traits are probably much more complex and unknown than assumed by some authors (Eysenck 1967; Eysenck and Eysenck, 1985; Gray 1982, 1991). To illustrate the complexity I will give one example from the domain of inanimate objects. The "acceleratability" of a car (its typical acceleration as measured in time units) depends on many of its aggregates and elements, of which the most important seem to be the following: type of engine, capacity of cylinders, ignition, carburetor, shape, size and weight of the car, and type of wheels. Other conditions being equal, the interaction among these aggregates and the "make-up" elements of the car determines its acceleratability. Drivers know, however, that factors like dirty spark plugs, disordered ignition, choked carburator, leaky cylinder, etc. influence directly the acceleratability of a car by lowering its capacity. Even one single element in the systems of aggregates determining acceleratability, for example, leaky cylinder, decreases essentially this feature. This gives an erroneous impression as if acceleratability were determined by the one factor only. The temperamental metaphor of a car's acceleratability suggests, that in case of man the physiological bases must be much more complex and, I strongly believe, it will take decades until the physiological mechanism underlying any temperamental trait will be identified.

(4) The RTT, because of its strong roots in the Pavlovian tradition, is often identified as a neo-Pavlovian approach. Taking the Teplov-Nebylitysn studies as the most representative for the neo-Pavlovian approach to human temperament, I will refer to some aspects in which both concepts differ most essentially.

(a) The Teplov-Nebylitsyn research may serve as a typical example of a bottom-up approach, whereas the RTT represents a rather top-down one.

(b) The Teplov-Nebylitsyn approach to temperament was exclusively based on experimental -psychophysiological, psychophysical and neurophysiological- studies, whereas research within the RTT combined the experimental approach with the psychometric one, the latter being unique during the almost two last decades of study in the neo-Pavlovian domain.

(c) The concepts of strength of excitation, strength of inhibition, mobility, dynamism and lability of nervous processes were used by Teplov and Nebylitsyn as explanatory constructs, with scanty information regarding the behavioral level in which these CNS properties may be expressed. The temperamental concepts used in the RTT have mainly the status of descriptive terms, if the identification of biological mechanisms underlying traits is used as the main criterion for ascribing concepts explanatory functions. It has to be added, however, that most of the constructs separated within the RTT, such as e.g., sensory sensitivity, emotional reactivity, endurance or activity may be used as explanatory concepts in that they allow an explanation of the regulative function that temperament has in human behavior.

(d) According to the Teplov-Nebylitsyn school, nervous system properties are regarded not only as temperament characteristics, as has been affirmed by Pavlov, but also as explanatory constructs referring to aptitudes, (so-called partial nervous system properties) and general abilities (see Golubeva, 1980; Nebylitsyn, 1972; Rusalov, 1979; Teplov, 1964). The RTT considers the traits separated within the theory as specific temperament characteristics only. A more detailed comparative characteristics regarding the neo-Pavlovian approach as related to the regulative theory of temperament has been presented elswhere (see Strelau, 1985b).

PRELIMINARY FINDINGS

Table 3. Pearson Correlations Between the RTTI scales and the STI-R, EPQ-R and SSS-V Scales

Remaining				The RTTI scales			
scales	Br	Pe	Mb	SS	En	AC	ER
SE	.29	-.45	.57	.03	.62	.49	-.60
SI	.06	-.23	.23	.10	.30	-.01	-.31
MO	.27	-.24	.68	.08	.36	.49	-.39
BA	.22	-.22	.32	-.05	.31	.42	-.29
E	.18	-.13	.48	.01	.18	.75	-.30
N	-.29	.69	-.49	.05	-.52	-.13	.74
P	.04	-.23	.01	-.16	.08	.19	-.19
L	.06	-.16	.18	.13	.14	-.03	-.16
To	.19	-.25	.25	.01	.21	.52	-.33
TAS	.18	-.27	.33	.00	.33	.46	-.38
ES	.08	-.09	.10	.13	.09	.31	-.17
Dis	.12	-.13	.10	-.06	.09	.34	-.15
BS	.12	-.19	.16	.01	.03	.34	-.22

Note. All coefficients of correlation >.14 are statistically significant ($p < .01$).

Br - Briskness
Pe - Perseverance
Mb - Mobility
SS - Sensory Sensitivity
En - Endurance
Ac - Activity
ER - Emotional Reactivity

SE - Strength of Excitation
SI - Strength of Inhibition
MO - Mobility of Nervous Proc.

BA - Balance of Nervous Processes
E - Extraversion
N - Neuroticism
P - Psychoticism
L - Lie scale
To - Sensation Seeking (total)
TAS - Thrill and Adventure Seeking
ES - Experience Seeking
Dis - Disinhibition
BS - Boredom Susceptibility

To illustrate the relations between the temperamental traits as proposed by the RTT and some selected traits I will refer to a study conducted in our laboratory in 1991. Data from the Eysenck Personality Questionnaire-Revised (EPQ-R), the Sensation Seeking Scale-Form V (SSS), the Strelau Temperament Inventory-Revised (STI-R) and the RTTI were collected from 317 subjects of both sexes (155 females and 162 males). The subjects were university and high-school students aged from 15 to 23 years (M = 19.07; SD = 2.41).

Table 3 shows the correlations between the RTTI scales and the scales of the other temperament inventories applied in this study. The most striking findings are the following:

(1) Briskness is less than moderately related to strength of excitation (.29), mobility (.27) and neuroticism (-.29).

(2) Perseverance correlates highly with neuroticism (.69) and moderately with strength of excitation (-.45).

(3) Mobility shows high correlations with mobility of nervous processes (.68) and strength of excitation (.57), and moderate correlations with extraversion (.48), neuroticism (-.49), thrill and adventure seeking (.33), and balance of nervous processes (.32).

(4) Sensory sensitivity does not correlate remarkably with any of the traits being investigated.

(5) Endurance correlates high with strength of excitation (.62) and neuroticism (-.52), and moderately with mobility of nervous processes (.36), thrill and adventure seeking (.33), balance of nervous processes (.31) and strength of inhibition (.30).

(6) Activity shows high correlation with extraversion (.75) and sensation seeking (total, 0.52), and moderate with strength of excitation (.49), mobility of nervous processes (.49), the four scales of SSS (.31-.46) and with balance of nervous processes (.42).

(7) Emotional reactivity correlates highly with neuroticism (.74) and strength of excitation (-.60). Moderate correlations occur with mobility of nervous processes (-.39), two scales of SSS (TAS= -.38; To= -.33), strength of inhibition (-.31) and extraversion (-.30).

In general, it might be stated that in case of scales comparable to the STI and STI-R (Endurance resembles the Strength of Excitation scale and Mobility the Mobility of the Nervous Processes scale), the configurations of correlations are similar to the ones found in research in which the STI (Strelau, 1983; Strelau et al., 1990) or STI-R (Ruch et al., 1991) were applied. The high correlations of the Activity scale with Extraversion and Sensation Seeking scales as well as the close relationship between the Emotional Reactivity and Neuroticism scales are understandable when looking at the characteristics of the activity and emotional reactivity traits. Curiously enough, sensory sensitivity does not correlate with extraversion and strength of excitation as might be expected from the theories (see Eysenck, 1970; Teplov and Nebylitsyn, 1963).

To gain a better understanding of the relationships between the temperamental traits as proposed by the RTT and the other traits under study we factor analyzed the data by means of the principal components method taking into account three factor solutions: 8, 5 and 3-factors.

To keep the factor construction as specific as possible, eight factors, which explained 88.2% of the total variance, were separated. The data (from the SSS scales only the Total score is included) are presented in Table 4. As can be seen, the results are comparable to those delineated in Table 3, especially as regards the RTT traits in relation to other temperament characteristics. Factor I is represented by Perseverance, Emotional Reactivity, and Neuroticism; Factor II by Activity, Extraversion, and Sensation Seeking; Factor III by Endurance, and Strength of Excitation; Factor IV by Mobility, and Extraversion; Factor V by Psychoticism, and Sensation Seeking (total). It came out that Sensory Sensitivity (F VI), Briskness (F VII), and Strength of Inhibition (F VIII) were independent factors.

To allow for a better allocation of the RTT traits within the most popular personality/temperament concepts it is useful to present the data within a three- and five-factor solution. For both solutions the factors will be characterized by scales with loadings >.50, in decreasing order.

Table 4. Varimax Rotated Factor Matrix of Temperamental Scales (RRTI, STI-R, EPQ-R and SSS-V-Total)

Scales	Factors								
	I	II	III	IV	V	VI	VII	VIII	h^2
Br	-.17	.09	.12	.14	.03	.07	.96	.00	.99
Pe	.83	-.01	-.22	-.07	-.16	.14	-.07	-.01	.80
Mb	-.42	.34	.22	.63	-.03	.12	.23	.08	.82
SS	.10	.02	-.02	.07	-.06	.98	.07	.04	.98
En	-.42	.08	.80	.12	-.01	-.07	.14	.09	.88
Ac	-.04	.87	.21	.20	.17	.04	.07	.02	.88
ER	.78	-.26	-.28	-.17	-.11	.05	-.07	-.17	.83
SE	-.31	.31	.71	.29	.13	.06	.06	.14	.81
SI	-.20	-.08	.14	.08	-.11	.05	.00	.94	.99
MO	-.13	.30	.19	.87	.02	.03	.08	.06	.92
E	-.09	.87	.03	.29	-.05	-.04	.04	-.12	.88
N	.90	.00	-.13	-.12	-.02	-.03	-.10	-.13	.87
P	-.11	.00	.01	.04	.93	-.11	-.01	-.10	.92
Tot	-.18	.52	.14	-.08	.64	.12	.11	.00	.81

Note. For description of scales see Table 3. Loadings with >.50 are underlined.

The three factor solution, which explained 62.5% of the total variance, resulted in separating the following factors. Factor I, which contributed 36.9% of the total variance, comprised: Emotional Reactivity (-), Mobility (+), Strength of Excitation (+), Endurance (+), Mobility of Nervous Processes (+), Neuroticism (-), Activity (+), and Extraversion (+). This factor resembles a mixture of the Eysenckian Neuroticism and Extraversion dimensions. Factor II (14.2% of the total variance) comprised the following scales: Extraversion (+), Activity (+) and Strength of Inhibition (-). It resembles to a high extent the Extraversion trait. The third factor, responsible for 11.3% of the total variance, can be easily identified as being opposite to Psychoticism because of its loadings on the Psychoticism (-) and Sensory Sensitivity (+) scales. Summarizing, one may conclude, that the three factor solution resulted in separating dimensions rather similar to the Eysenckian PEN dimensions, with some reservation regarding the first factor. On the basis of our results this factor, which is a kind of a "omnibus" one, may be characterized as rather referring the the energetic characteristics of behavior covering both activity- and emotion-oriented dimensions.

The five factor solution does not reflect the expections represented by the enthusiasts of the Big Five model. Altogether the five factors explained 75.7 % of the total variance. The characteristics of the consecutive factors is as follows: Factor I - Neuroticism (-), Emotional Reactivity (-), Perseverance (-), Endurance (+), Strength of Excitation (+), Strength of Inhibition (+), and Mobility (+). Again, this factor which combines, as it was in the three factor solution, action-oriented traits with emotional characteristics of behavior, may be identified as referring to the energetic characteristics of behavior. Factor II - Extraversion (+), Activity (+), Mobility of Nervous Processes (+), Mobility (+), and Strength of Excitation (+). It might be identified as the Extraversion dimension. Factor III - Psychoticism (+) and Sensation Seeking (+), has much in common with the Impulsive-Unsocialized Sensation Seeking (P-ImpUSS) dimension as proposed by Zuckerman et al. (1988). Factor IV has its loading only on the Sensory Sensitivity scale and Factor V on the Briskness scale. The two latter factors can hardly be interpreted within the Big Five model. These results illustrate that the configuration and quality of factors depends to a high extent on the data used as a starting point for factor analysing.

One must note that the scales being used in our study represent different factor orders (e.g., the RTT are second order factors whereas the Eysenckian PEN dimensions are third order factors), which poses problems when considering the relationships between the temperament traits under study.

Before any reasonable conclusions regarding the links between the RTT traits and other temperamental constructs can be drawn, studies have to be replicated with a broader range of temperamental traits being included, what is the aim of our following research.

Author's Note

The first draft of this paper was written during my one-year-visit (1991/92) as a Humboldt Research Award Winner at the University of Bielefeld, Department of Psychology, Bielefeld, Germany; the final version was written during my fellowship (1992/93) at the Netherlands Institute for Advanced Study in the Humanities and Social Sciences, Wassenaar, The Netherlands.

I gratefully acknowledge the comments to the first draft given by Professor Alois Angleitner.

Requests for reprints should be sent to Jan Strelau, Department of Psychology, University of Warsaw, Stawki 5-7, 00-183 Warsaw, Poland.

References

Adcock, C.J. (1957). The differentiation of temperament from personality. *The Journal of General Psychology, 57*, 103-112.

Allport, G.W. (1937). *Personality: A psychological interpretation.* New York: Holt.

Angleitner, A., John, O.P. and Lohr, F.J. (1986). It's *what* you ask and *how* you ask it: An itemmetric analysis of personality questionnaires. In A. Angleitner and J.S. Wiggins (Eds.), *Personality assessment via questionnaires: Current issues in theory and measurement* (pp. 61-108). Berlin: Springer.

Barratt, E.S. (1983). The biological basis of impulsiveness: The significance of timing and rhythm disorders. *Personality and Individual Differences*, 4, 387-391.

Broadhurst, P.L. (1975). The Maudsley reactive and non-reactive strains of rats: A survey. *Behavior Genetics, 5*, 299-319.

Buss, A.H. and Plomin, R. (1975). *A temperament theory of personality development.* New York: Wiley.

Buss, A.H. and Plomin, R. (1984). *Temperament: Early developing personality traits.* Hillsdale, NJ: Erlbaum.

Carey, W.B. (1983). Some pitfalls in infant temperament research. *Infant Behavior and Development, 6*, 247-254.

Chess, S. and Thomas, A. (1986). *Temperament in clinical practice.* New York: The Guilford Press.

Diamond, S. (1957). *Personality and temperament.* New York: Harper and Brothers, Publishers.

Eliasz, A. (1981). *Temperament a system regulacji stymulacji* [Temperament and the system of stimulation control]. Warszawa: Panstwowe Wydawnictwo Naukowe.

Eliasz, A. (1990). Broadening the concept of temperament: From disposition to hypothetical construct. *European Journal of Personality, 4*, 287-302.

Ewald, G. (1924). *Temperament und Charakter* [Temperament and character]. Berlin: Verlag von Julius Springer.

Eysenck, H.J. (1967). *The biological basis of personality.* Springfield, IL: Charles C. Thomas.

Eysenck, H.J. (1970). *The structure of human personality.* London: Methuen.

Eysenck, H.J. and, Eysenck, M.W. (1985). *Personality and individual differences: A natural science approach.* New York: Plenum Press.

Garau, A. and Garcia-Sevilla, L. (1985). Drug postulate of Eysenck in the rat. *Personality and Individual Differences, 6*, 189-194.

Goldsmith, H.H. and Campos, J.J. (1982). Toward a theory of infant temperament. In R.N. Emde and R.J. Harmon (Eds.), *The development of attachment and affiliative systems* (pp. 161-193). New York: Plenum Press.

Goldsmith, H.H. and Campos, J.J. (1986). Fundamental issues in the study of early temperament: The Denver Twin Temperament Study. In M.E. Lamb, A.L. Brown and B. Rogoff (Eds.), *Advances in developmental psychology* (Vol. 4, pp. 231-283). Hilsdale, NJ: Erlbaum.

Golubeva, E.A. (1980). Some problems of the experimental study of the inborn prerequisites of general capacities. *Voprosy Psikhologii, 4*, 23-37 (in Russian).

Gorynska, E. and Strelau, J. (1979). Basic traits of the temporal characteristics of behavior and their measurement by an inventory technique. *Polish Psychological Bulletin, 10*, 199-207.

Gray, J.A. (Ed.) (1964). *Pavlov's typology.* Oxford: Pergamon.

Gray, J.A. (1982). Precis of the neuropsychology of anxiety. An enquiry into the functions of the septo-hippocampal system. *The Behavioral and Brain Sciences, 5*, 469-534.

Gray, J.A. (1991). The neuropsychology of temperament. In J. Strelau and A. Angleitner (Eds.), *Explorations in temperament: International perspectives on theory and measurement* (pp. 105-128). New York: Plenum Press.

Hagekull, B. (1989). Longitudinal stability of temperament within a behavioral style framework. In G.A. Kohnstamm, J.E. Bates and M.K. Rothbart (Eds.), *Temperament in childhood* (pp.283-297). Chichester: Wiley.

Hebb, D.O. (1955). Drives and the C.N.S. (conceptual nervous system). *Psychological Review, 62*, 243-254.

Heymans, G. and Wiersma, E. (1906-1909). Beitrage zur speziellen Psychologie auf Grund einer Massenuntersuchung [Contributions to special psychology based on large-scale investigation]. *Zeitschrift fur Psychologie, 42*, 81-127, 258-301; *43*, 321-373; *45*, 1-42; *46*, 321-333; *49*, 414-439; *51*, 1-72.

Hutchins, R.M. (Ed.). (1952). *Great books of the Western world: (1) Hippocratic writings, (2) On the natural faculties* (Vol. 10, Hippocrates, Galen). Chicago, IL: Encyclopaedia Britannica.

Ilin, E.P. (1978). *A study of nervous system properties*. Yaroslavl: Yaroslavl University Press (in Russian).

Kagan, J. (1982). The construct of difficult temperament: A reply to Thomas, Chess, and Korn. *Merrill-Palmer Quarterly, 28*, 21-24.

Klonowicz, T. (1992). *Stres w Wiezy Babel: Roznice indywidualne a wysilek inwestowany w trudna prace umyslowa* [Stress in the Babel Tower: Individual differences and allocation of effort]. Wroclaw: Ossolineum.

Kohnstamm, G.A., Bates, J.E. and Rothbart, M.K. (Eds.). (1989). *Temperament in childhood.* Chichester: Wiley.

Kretschmer, E. (1944). *Korperbau und Charakter: Untersuchungen zum Konstitutionsproblem and zur Lehre von den Temperamenten* [Physique and character: Research concerning problems of constitution and knowledge on temperaments] (17-18th ed.). Berlin: Springer.

Leontev, A.N. (1978). *Activity, consciousness and personality*. Englewood Cliffs, NJ: Prentice-Hall.

Loehlin, J.C. (1989). Partitioning environmental and genetic contributions to behavioral development. *American Psychologist, 44*, 1285-1292.

Maziade, M. (1988). Child temperament as a developmental or an epidemiological concept: A methodological point of view. *Psychiatric Developments, 3*, 195-211.

Mehrabian, A. (1978). Measures of individual differences in temperament. *Educational and Psychological Measurement, 38*, 1105-1117.

Mehrabian, A. (1991). Outline of a general emotion-based theory of temperament. In J. Strelau and A. Angleitner (Eds.), *Explorations in temperament: International perspectives on theory and measurement* (pp. 75-86). New York: Plenum Press.

Merlin, V.S. (Ed.). (1973). *Outline of the theory of temperament* (2nd. ed.). Perm: Permskoye Knizhnoye Izdatelstvo (in Russian).

Nebylitsyn, V.D. (1972). *Fundamental properties of the human nervous system.* New York: Plenum Press.

Netter, P. (1991). Biochemical variables in the study of temperament: Purposes. approaches,and selected findings. In J. Strelau and A. Angleitner (Eds.), *Explorations in temperament: International perspectives on theory and measurement* (pp. 147-161). New York: Plenum Press.

Pavlov, I.P. (1951-52). *Complete works* (2nd ed.). Moscow and Leningrad: SSSR Academy of Sciences (in Russian).

Petrie, A. (1967). *Individuality in pain and suffering*. Chicago: The University of Chicago Press.

Rothbart, M.K. (1981). Development of individual differences in temperament. In M.E. Lamb and A.L. Brown (Eds.), *Advances in developmental psychology* (Vol. 1, pp. 37-86). Hillsdale, NJ: Erlbaum.

Ruch, W., Angleitner, A. and Strelau, J. (1991). The Strelau Temperament Inventory-Revised (STI-R): Validity studies. *European Journal of Personality, 5*, 287-308.

Rusalov, V.M. (1979). *Biological bases of individual-psychological differences*. Moscow: Nauka (in Russian).

Rusalov, V.M. (1985). On the nature of temperament and its place in the structure of human individual differences. *Voprosy Psikhologii, 1*, 19-32 (in Russian).

Rusalov, V.M. (1989). The natural prerequisites of development of human individuality. *Studia Psychologica, 31*, 53-67.

Schalling, D. and Asberg, M. (1985). Biological and psychological correlates of impulsiveness and monotony avoidance. In J. Strelau, F.H. Farley and A. Gale (Eds.), *The biological bases of personality and behavior: Theories, measurement techniques, and development* (Vol. 1, pp. 181-1940). Washington: Hemisphere.

Simonov, P.V. (1987). Individual characteristics of brain limbic structures interactions as the basis of Pavlovian/Eysenckian typology. In J. Strelau and H.J. Eysenck (Eds.), *Personality dimensions and arousal* (pp. 123-132). New York: Plenum Press.

Stelmack, R.M. (1990). Biological bases of extraversion: Psychophysiological evidence. *Journal of Personality, 58*, 293-311.

Stelmack, R.M. and Stalikas, A. (1991). Galen and the humor theory of temperament. *Personality and Individual Differences, 12*, 255-263.

Stern, W. (1921). *Differentielle Psychologie* (3rd ed.). Leipzig: Johann Ambrosius Barth.

Strelau, J. (1969). *Temperament i typ ukladu nerwowego* [Temperament and type of the nervous system]. Warszawa: Panstwowe Wydawnictwo Naukowe.

Strelau, J. (1974a). Temperament as an expression of energy level and temporal features of behavior. *Polish Psychological Bulletin, 5*, 119-127.

Strelau, J. (Ed.). (1974b). *Rola cech temperamentalnych w dzialaniu* [The role of temperamental traits in action]. Wroclaw: Ossolineum.

Strelau, J. (1977). Behavioral mobility versus flexibility and fluency of thinking: An empirical test of the relationship between temperament and abilities. *Polish Psychological Bulletin, 8*, 75-82.

Strelau, J. (1983). *Temperament, personality, activity*. London: Academic Press.

Strelau, J. (Ed.). (1985a). *Temperamental bases of behavior: Warsaw studies on individual differences*. Lisse: Swets & Zeitlinger.

Strelau, J. (1985b). Pavlov's typology and the regulative theory of temperament. In J. Strelau (Ed.), *Temperamental bases of behavior: Warsaw studies on individual differences* (pp. 7-40). Lisse: Swets & Zeitlinger.

Strelau, J. (1989). The regulative theory of temperament as a result of East-West influences. In G.A. Kohnstamm, J.E. Bates and M. K. Rothbart (Eds.), *Temperament in childhood* (pp.35-48). Chichester: Wiley.

Strelau, J. (1991). Are psychophysiological/psychophysical scores good candidates for diagnosing temperament/personality traits and for a demonstration of the construct validity of psychometrically measured traits? *European Journal of Personality, 5*, 323-342.

Strelau, J., Angleitner, A. and Ruch, W. (1990). Strelau Temperament Inventory (STI): General review and studies based on German samples. In J.N. Butcher and C.D. Spielberger (Eds.), *Advances in personality assessment* (Vol. 8, pp. 187-241). Hillsdale, NJ: Erlbaum.

Strelau, J. and Plomin, R. (in press). A tale of two theories of temperament. In G.V. Caprara and G.L. Van Heck (Eds.), *Modern personality psychology: Critical reviews and new directions.* Chichester: Wiley.

Teplov, B.M. (1964). Problems in the study of general types of higher nervous activity in man and animals. In J.A. Gray (Ed.), *Pavlov's typology: Recent theoretical and experimental developments from the laboratory of B.M. Teplov* (pp. 3-153). Oxford: Pergamon Press.

Teplov, B.M. and Nebylitsyn, V.D. (1963). Experimental study of properties of the nervous system in man. *Zhurnal Vysshei Nervnoi Deyatelnosti, 13*, 789-797.

Thomas, A. and Chess, S. (1977). *Temperament and development.* New York: Brunner/Mazel Pulishers.

Thomas, A., Chess, S. and Birch, H.G. (1968). *Temperament and behavior disorders in children.* New York: New York University Press.

Tomaszewski, T. (1978). Tatigkeit und Bewusstsein: Beitrage zur Einfuhrung in die polnische Tatigkeitspsychologie [Action and consciousness: Contributions to the introduction to Polish theory of action].. Weinheim and Basel: Beltz Verlag.

Vygotsky, L.S. (1962). *Thought and language.* New York: Wiley.

Windle, M. (1992). Temperament and social support in adolescence: Interrelations with depressive symptoms and delinquent behaviors. *Journal of Youth and Adolescence, 21*, 1-21.

Windle, M., Hooker, K., Lenerz, K., East, P.L., Lerner, J.V. and Lerner, R.M. (1986). Temperament, perceived competence, and depression in early and late adolescents. *Developmental Psychology, 22*, 384-392.

Windle, M., Iwawaki, S. and Lerner, R.M. (1988). Cross-cultural comparability of temperament among Japanese and American preschool children. *International Journal of Psychology, 23*, 547-567.

Zawadzki, B. (1992). *Cechy tempermentu w ujeciu regulacyjnej teorii temperamentu i ich pomiar metoda kwestionariusza* [Temperamental traits as formulated by the regulative theory of temperament and their measurement by an inventory method]. Unpublished doctoral dissertation, University of Warsaw, Warsaw.

Zawadzki, B. and Strelau, J. (1991, September). *The Regulative Theory of Temperament Inventory (RTTI): Preliminary information.* Paper presented at the International Workshop of Cross-Cultural Studies on Temperament. Nieborow, Poland.

Zuckerman, M. (1979). *Sensation seeking: Beyond the optimal level of arousal.* Hillsdale, NJ: Erlbaum.

Zuckerman, M. (1991). *Psychobiology of personality.* Cambridge: Cambridge University Press.

Zuckerman, M., Kuhlman, D.M. and Camac, C. (1988). What lies beyond E and N? Factor analyses of scales believed to measure basic dimensions of personality. *Journal of Personality and Social Psychology, 54*, 96-107.

THE REGULATORY THEORY OF TEMPERAMENT: THE VIEW FROM THE TOP
Commentary on the chapter by J. Strelau

Robert M. Stelmack
Department of General Experimental Psychology
University of Ottawa
Monpetit Hall
Ottawa, Ontario KIN 6N5,Canada

Introduction

At the beginning of my research career, my exploration of extraversion with psychophysical and psychophysiological methods was drawn towards Pavlov's typology of the nervous system properties and Sokolov's model of the orienting reaction (OR). Western psychophysiologists, in fact, owe much to Sokolov for providing a fecund theory of perception and conditioning that used psychophysiological methods. This model of the OR, gave a big boost, and satisfying employment opportunities, to psychophysiologists, that in fact continue to this day. My own introduction to the Soviet psychology was greatly assisted by a paper that Jan Strelau published in 1969 in the first issue of the *Polish Psychological Bulletin.* I have followed the work of Strelau and his colleagues since that time and I have benefited from their efforts. I am pleased to comment on their most recent work on the Regulative Theory of Temperament (RTT). My main objective in this discussion will be to introduce a schema of descriptive and explanatory constructs that are employed in research on temperament and personality that will provide a context for illustrating the RTT research program as I see it and for discussing the merits and limitations of RTT.

In the introduction to his paper, Strelau acknowledges the influence of the Pavlovian typology of fundamental nervous system properties that was the focus of research in Soviet psychology for so many years. These nervous system properties include, particularly, strength of excitation and inhibition, and mobility. The influence from Western sources on his work is also acknowledged, especially the work of Hans Eysenck and various authors who made use of the arousal concept. It is important to recognize that the political antagonism between East and West did not sever the exchange of intellectual information in this area, although certainly this did introducc somc impediments. During the past decade, Strelau and his colleagues made good use of the information from personality inventories and experimentation that was conducted in Western laboratories and indeed he has collaborated with Western colleagues on a number of significant projects. For these reasons, the RTT work has more in common with the Western temperament research than it has differences.

In the RTT, three explanatory constructs are emphasized. These are:

* Reactivity: a concept that involves sensitivity to stimulation,
* Activity: a concept that is concerned with the regulation of stimulation, and
* Mobility: a concept that refers to response flexibility.

J. Hettema and I. J. Deary (eds.), Foundations of Personality, 133–138.

These constructs are defined directly by psychophysical or psychophysiological measures. For example, reactivity is defined, in part, by sensory thresholds. Reactivity is said to determine individual sensitivity (sensory and emotional) and endurance (capacity to work) that are manifest in personal and social experience and behaviour. A recent undertaking that is a focus of Strelau's present paper is the development of psychometric measures that are derived from these three constructs. The psychometric work that is presented yielded seven factors that now constitute the RTT Inventory. In the remarks that follow, I would like to comment on some of the general issues and considerations that stem from the general approach that Strelau advocates rather than on a psychometric assessment which, at this stage and with the information available, would be premature.

Descriptive and Explanatory Constructs

Table 1 shows a schema of descriptive and explanatory constructs in the biological approach to personality research that I find useful in focusing my own work in this research enterprise and which may be helpful for discussing the location of the RTT with respect to other research programs.

In this schema, two general types of constructs are noted: descriptive and explanatory. The descriptive constructs refer primarily to psychometric definitions of personality concepts that we are all familiar with such as extraversion, neuroticism and sensation-seeking. The methods used to define these concepts are self reports, ratings and observable behaviours and the measures that quantify the constructs are test scores. The kind of research undertaken with descriptive constructs seems to fall under six main categories. Concurrent validation studies relate one personality concept to another and attempt to identify the common meaning in those terms. In that work, we have the major typologies such as the *Eysenck Personality Questionnaire* and the big five typologies. Another important body of work has investigated the consistency of descriptive concepts across the lifespan. The construct validity studies of descriptive concepts are designed to show that persons who are defined by the concept actually behave in ways that they are expected. For example, studies are designed to show that extraverts actually talk more in social situations, initiate more social contacts and occupy positions, e.g. sales, that require more social demands than introverts. Research that also contributes to this type of descriptive construct validity are works that examine person-situation interactions or person-person interactions, although they may have an examination of personality dynamics as their primary objective e.g. Endler on anxiety. The behavioural genetic studies examine the relation of descriptive constructs between relatives and these data are used to make inferences concerning the heritability of personality concepts.

Explanatory concepts that have been proposed to account for variations in the descriptive constructs are defined in terms that are more general than the descriptive concepts. In early work by Eysenck, cortical excitation and inhibition were understood, at least initially, as hypothetical neural structures that determined individual variations in the descriptive concept of extraversion. More recently, explanatory concepts such as arousal and behavioural inhibition have been defined in terms of specific neural circuits or neurotransmitter pathways. For example, Eysenck (1967) has defined arousal in terms of the ascending reticular activating system and Gray has defined behavioural inhibition in terms of the septo-hippocampal system (Gray et al., 1983). Zuckerman has referred sensation-seeking to variation in catecholamine systems (1979).

Table 1. Descriptive and Explanatory Constructs in Biologically Based Personality Research

I. Descriptive Constructs			
Concepts	*Methods*	*Measures*	*Explorations*
Extraversion	Self-report	Testcores	Concurrent validation
Anxiety	Ratings		Construct validity
Sensation-seeking	Behaviour		Consistency
Conscientiousness	check-list		Person-situation interactions
Intelligence	performance		Person-person interactions
Reactivity			Behavioural genetics
Activity			
Mobility			
II. Explanatory Constructs			
Concepts	*Methods*	*Measures*	*Explorations*
Arousal	Conditioning Learning	Acquisition Extinction Disinhibition Response frequency	
Response excitation	Psychophysical	Sensory Thresholds	Relation to descriptive construct (Theory testing)
Behavioural inhibition		Reaction time	
Strength of nervous system Reactivity Activity Mobility	Psychophysiological	GSR HR EEG ERP	Concurrent validity between measures and between methods
ARAS SHS	Physiological	Neurophysiology Animal models	Neural circuits and their functions
	Biochemical	enzymes catecholamines neurotransmitter	Neurochemical pathways and their functions
	Genetic	RNA, DNA genetic code	

The methods and measures used in explorations of personality with explanatory concepts have been drawn from a broad range of experimental psychology, including psychophysics, learning and conditioning, psychophysiology and physiological psychology and psychopharmacology. Thus far, few direct genetic analyses have been conducted, although analysis of blood antigens for which there are known chromosomal loci is a promising avenue for that work.

The validity of an explanatory construct is established by confirming the functions that it purports to serve. For example, the validity of the arousal construct was assessed by demonstrating that the various indices of arousal (EEG desynchronization, skin conductance etc.) were systematically related to variation in attention, sensitivity and conditioning and displayed an inverted-U relation with performance. The majority of explorations of personality differences with explanatory constructs have aimed to determine the relation between descriptive and explanatory constructs as required by the personality theory i.e. to assess the extent to which descriptive constructs (e. g. extraversion) are dependent on variation in explanatory constructs (e.g. arousal). In my view, this approach can be referred to as the top down approach to personality research. Here individuals are defined in terms of the descriptive constructs and comparisons are made between subjects on the basis of explanatory construct measures such as sensory threshold or autonomic nervous system response. Less frequently, there are studies of the concurrent validity of the explanatory construct measures or methods. For example, there has been a good deal of research examining the relations between various autonomic nervous system measures of arousal - GSR, heart rate, and vasomotor responses. The relatively low correlations between arousal measures has indicated that the arousal construct is not a unitary one, at least under neutral conditions. Explorations relating measures of an explanatory construct such as arousal that are obtained by different methods, e.g. psychophysical and psychophysiological, are even less frequently conducted.

In this schema, the traditional Neo-Pavlovian work on temperament is seen primarily as explorations with explanatory constructs. There has been an enormous amount of work that has attempted to refine the concepts such as strength of excitation, strength of inhibition and mobility by intercorrelating the various measures of these constructs and by determing the conditions under which effects are observed. There has also been some fine work with animals that has been inspired by these concepts. In this work, the nervous system properties have been examined within conditioning paradigms and the effects of anatomical lesions on performance were assessed. This work is currently well represented by Pavlov Simonov (1987). His research strategy is very similar to that adopted by Jeffrey Gray in his research on anxiety and impulsivity using animal models.

In Strelau's work, reactivity is an explanatory construct that has been defined in terms of sensory thresholds. The nature of the construct and its regulatory processes (functions) has been probed by examining individual differences in psychophysiological response levels and showing, for example, that high reactivity is related to greater electrodermal fluctuations and other autonomic indices (Klonowicz, 1987). There has also been a good deal of work referring this construct of reactivity to rather specific behaviours and occupational performances. For example, high reactives have been shown to be less sensitive to the social content of stories when in an anxiety inducing examination situation (Eliasz, 1987). This work is very similar to the construct validity research with descriptive constructs that has taken place in the west, except that here individual differences are defined in terms of the explanatory construct rather than a descriptive construct.

RTT: An Explanatory or Descriptive Model?

To this point, the work of the Neo-Pavlovians, including Strelau, can be seen essentially as a bottom up approach to the understanding of personality that has focused on refining and defining explanatory constructs that are assumed to determine individual differences in the acquisition, extinction and expression of behaviour. The development of the Strelau Temperament Inventories and this more recent RTT Inventory now enables explorations that view personality differences from the top down. It can be seen from Strelau's Table 4, that the dimensions that compose the RTT Inventory relate rather unequivocally to the established dimensions Extraversion and Neuroticism. The RTTI scales do not load on Zuckerman's P-ImpUSS factor. The factor pattern of the RTT is quite clear, with the seven scales displaying high factor loadings on single factors. The nine factors that are described in this analysis also appear to be relatively independent. These considerations will allow information from experiments with any of these inventories to be understood in terms of the other.

At this time, the correspondence between the RTT descriptive constructs and the RTT explanatory constructs from which they are drawn, i.e. reactivity, activity and mobility is not determined. Further, it is not clear what measures are used to define the explanatory constructs of activity and mobility in this new proposal. Indeed, it is not certain from this text whether the explanatory constructs of reactivity, activity and mobility are retained or whether they are replaced by the arousal construct. In general, it can be said that attempts to derive psychometric measures from explanatory constructs such as arousal, reactivity, or behavioral inhibition have not met great with great success. A good example of this problem is given by Paul Kohn (1987) with his research on a psychometric reactivity scale that was derived from the Neo-Pavlovian work. In that case, the reactivity scale did not correlate significantly with any of the measures that defined the explanatory construct of reactivity. I am not clear on the reason for this situation but it may be that the range of items selected for the descriptive constructs were too narrow to be of practical utility since some of the measures of the explanatory construct (e. g. sensory thresholds) did correlate inversely with the broader extraversion trait measure. Overall, the relations between the descriptive and explanatory constructs using the bottom up approach have not been compelling. In any case, explorations using explanatory constructs within this RTT framework, from the top down, will be an interesting and worthwhile challenge and quite a coup if strong relations can be shown.

My final comment is also a challenging one. Thus far, specific neural circuits have not been proposed for the hypothetical constructs of reactivity, activity and mobility that are the bases of the RTT theory. It appears to me that defining the neural circuits and their functions for these explanatory constructs should be an important objective that will give substance to the theory and that will be helpful in focusing work in this research program.

References

Eliasz, A. (1987). A temperament contingent cognitive orientation for various aspects of reality. In J. Strelau and H.J. Eysenck (Eds.). *Personality dimensions and arousal*. New York: Plenum Press.

Eysenck, H. J. (1967). *The biological basis of personality*. Springfield, Ill.: Thomas.

Gray, J. A., Owen, S., Davis, N., and Tsaltas, E. (1983). Psychological and physiological relations between anxiety and impulsivity. In M. Zuckerman (Ed.), *Biological bases of sensation-seeking, impulsivity and anxiety*. Hillsdale, N.J.: Erlbaum.

Klonowicz, T. (1987). Reactivity and the control of arousal. In J. Strelau and H. J. Eysenck (Eds.). *Personality dimensions and arousal.* New York: Plenum Press.

Kohn, P.M. (1987). Issues in the measurement of arousability. In J. Strelau and H.J. Eysenck (Eds.). *Personality dimensions and arousal.* New York: Plenum Press.

Simonov, P.V. (1987). Individual characteristics of brain limbic structures interactions as the basis of Pavlovian/Eysenckian typology. In J. Strelau and H.J. Eysenck (Eds.). *Personality dimensions and arousal.* New York: Plenum Press.

Zuckerman, M. (1979). *Sensation* seeking: Beyond the optimal level of arousal. Hillsdale, N.J.: Erlbaum.

BIOSOCIAL ADAPTATION: A STRATEGIC-TACTICAL APPROACH TO INDIVIDUALITY

Joop Hettema
Department of Pychology,
Tilburg University,
P.O. Box 90153,
5000 LE Tilburg,
The Netherlands

Abstract

Biological and social factors are generally assumed to be important determinants of personality. Yet, attempts to integrate both aspects are scarce. This paper advocates a biosocial conception of personality, emphasizing adaptation as a core concept. Human adaptation, it is argued, is multiply determined, including biological as well as social aspects. Although biological and social adaptation differ in several respects, both are involved in regulating individual behavior in daily life situations. The tactics people employ to deal with daily life situations are conceived as the major effects of biological adaptation, whereas the strategies governing behavior are viewed primarily as the results of social learning. Strategies and tactics are conceptualized and connected in a comprehensive system regulating individual behavior. Empirical evidence is provided indicating that tactics have a physiological basis whereas strategies are affected by social learning. Personality is conceived as a strategic-tactical coalition, i.e. the product of biological and social determinants interacting during individual development.

Introduction

Traditionally, personality theorists have agreed on the importance of biological and social determinants for the study of individuality. The history of personality psychology reveals several attempts to account for biological and social factors as well as their relationships. For instance, the work of Freud emphasized instincts as the major biological forces underlying behavior. Rather than serving an adaptive purpose for the individual, instincts were conceived by Freud as factors enhancing species survival (cf. Bowlby, 1969). Instincts were seen as blind forces of the Id, that may or may not be governed by Ego processes taking into account the demands of the external world. Social factors, on the other hand, were posited mainly in the Superego, based on incorporating the values of parents and wider society. The major function of the Superego was to force the Ego to act according to moral considerations and to counteract impulses emanating from the Id.

McDougall postulated innate driving forces to represent biological determinants. In addition, he assumed a 'group mind' acting on a higher plan to govern social behavior. Although

J. Hettema and I. J. Deary (eds.), Foundations of Personality, 139–164.

representing the major social determinants of behavior, the group mind was taken to be ultimately determined through evolutionary mechanisms as well.

A different emphasis can be found in the work of Allport (1961). In Allport's view, personality includes several systems determining his typical way of adapting to the external world. At the basis Allport postulated temperament, governing a person's susceptibility to emotional stimulation, his strength and speed of response, and the quality of his prevailing mood. Socialisation was assumed to be obtained through socially maturing and learning, yielding properties like altruism, sympathy, chivalry, and the like.

Another classical personality theorist, Murray (1938), emphasized needs as the basic concept of personality. According to Murray, needs may be conveniently divided into primary or viscerogenic and secondary or psychogenic needs. Needs were assumed to interact with external press influencing motives to engage in particular kinds of action.

Freud, McDougall, Allport and Murray all acknowledged biological and social determinants of individuality. They proposed different conceptualizations, layers or structures to reflect biological and social determinants. However, rather than testable hypotheses, biological and social determinants obtained the status of assumptions underlying broader conceptions of personality. Accordingly, no direct evidence sustaining their functioning was provided and no generally accepted framework exists to date.

SOCIAL APPROACHES

In more recent years, this picture has changed. During the mid-sixties, severe criticisms were raised against the then dominant organismic approach to personality (e.g. Mischel, 1968). The critics argued that many aspects of behavior in daily life cannot be controlled by the individual. Those aspects were primarily seen as a function of the situation and the broad sociocultural factors governing behavior in the situation. Accordingly, a social behaviorist framework was proposed to replace classical personality theory. The attention became focused upon direct situational factors (reinforcement contingencies) as major determinants of behavior. During the seventies this approach has left its footmarks in many studies that were published under headings like *social learning* and *social cognition.*

Clearly, the social behaviorist approach neither attaches much weight to the inner self of the behaving individual, nor to his biological make-up. Instead, it emphasizes social psychological and sociological concepts like roles, norms, and expectancies, that are based on a conception of behavior as the function of a constantly changing audience instead of inner dispositions. The social behaviorist approach has led to several reactions. First of all, the criticisms raised against personality have overrated the case in some ways (Kenrick and Funder, 1988). But even granted the core of the comments, personality psychologists felt uneasy with the alternative offered by social behaviorists. Without denying the impact of social forces residing in the situation, they viewed this conception as one-sided. Accordingly, situationism was rejected as the ultimate paradigm for personality psychology. While taking a firm stand against the 'social psychological invasion of personality' (Kenrick and Dantchik, 1983), alternative explanations of individual behavior were looked for. Those were derived from biology.

BIOLOGICAL APPROACHES

Biological approaches to personality emphasize survival and successful reproduction, behavior genetics, the heritability of social behavior (sociobiology), physiological processes and temperament as the major basis of individuality. Recently, the biological approach has been designated *evolutionary personality psychology* by Buss (1991). Applying evolution theory to social behavior, evolutionary personality psychology has connected social traits with ecological pressures that existed during evolution. Natural selection is held largely responsible for divergent traits like aggression, altruism, revenge, conformity, love, territory drive and xenophobia (Buss, 1991; Kenrick, et al., 1983; Rushton, et al., 1986; Wilson, 1978).

The modern biological approach provides an alternative for social behaviorism to explain individuality. Rather than relying on situational contingencies and learning, it emphasizes genetic determinants and ensuing biological processes. At first sight, explanations could not be more divergent. The social approach stresses acquired aspects, the biological approach emphasizes innate aspects of behavior. The social approach is particularly concerned with proximate determinants, the biological approach with ultimate determinants of behavior (Kenrick, 1987; 1989). In view of the differences existing, a point could be made for the social and biological approaches to be conceived as different paradigms, without really affecting each other.

However, as we have pointed out elsewhere (Hettema and Kenrick, 1989; 1992), the biological and social approaches can only offer partial answers to the questions asked. Since personality is concerned with the whole organism, solutions that focus only on social learning and social cognition or on physiological processes and temperament are not acceptable. This is especially true to the extent that biological and social factors interact in non-additive ways to obtain a coherent individual mode in any one person. Thus, rather than studying individuality from any of the two perspectives separately, we proposed to achieve a synthesis in one biosocial approach.

Adaptation

The present paper elaborates that proposition. It starts from the assumption that biological as well as social mechanisms are important in determining individuality. Obviously, a common framework is required to connect biological and social mechanisms. Adaptation is proposed here as the key concept to provide a first broad outline. Based on a general conception of human adaptation, the paper argues that behavior in specific situations is governed by several regulatory systems. Presumably, some of those systems are primarily biologically determined, whereas others are based on social learning. The identification of those systems may provide an avenue towards clarifying the biological-social issue. A major aim of this paper is to conceptualize the systems involved and to indicate how they cooperate in determining behavior in concrete situations. Empirical evidence, supporting this conception is provided. Finally, based on the understanding gained, individuality is conceived as the product of the systems interacting in personality development.

BIOLOGICAL ADAPTATION

Adaptation is a major characteristic of life in general. The concept of adaptation has been forwarded by Darwin (1859) to explain the obvious correspondence between plant and animal species on the one hand and the ecological niches they inhabit on the other. The major mechanism involved is natural selection. Lewontin (1978) has defined adaptation as the process of evolutionary change by which the organism provides a better and better 'solution' to the 'problem', and the end result is the state of being adapted. Based on morphological, physiological and behavioral characteristics, species obtain the capacity to survive in specific ecological niches: their environment of evolutionary adaptedness EEA (Bowlby, 1969). This environment may be conceived as a sub-domain of all possible environments (cf Fig. 1)

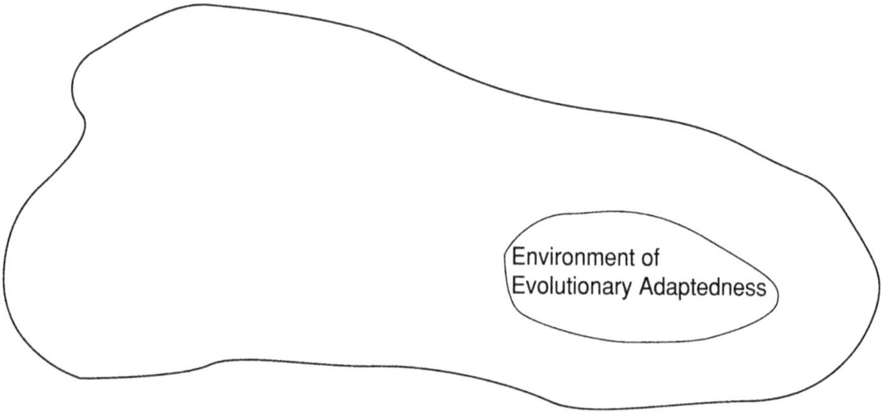

Figure 1. Environment of evolutionary adaptedness

A major adaptive modality is behavior. Animal species exhibit *species specific behavior*, i.e. collections of behaviors that nearly all organisms in a species develop relatively independently of the particular environment in which they are raised. As an example, in primates species specific behavior includes aspects like curiosity and aggression, and e.g. the use of tools in chimpanzees. The human species shares the general primate heritage as social, aggressive, sensual, curious and intelligent creatures. In addition, we possess our own uniquely evolved linguistic and cultural characteristics. Recent research in evolutionary personality psychology has provided evidence on several other aspects of 'human nature', like characteristic emotions and emotional expressions, love of kin, aspects of sexual relationships, and the like (cf Buss, this volume).

In human adaptation, *emotions* occupy a special position. In general, more complex animals are less mature at birth. Their species specific behavior includes a wider array of action patterns and a wider array of releasing stimuli interacting with each other over longer periods of time. Humans are assumed to be the least fixed and the most able to adjust and readjust as environmental conditions change. At the same time there is the danger that species members will lose sight of what is necessary for survival. What is specially needed then is a system which confers value on the crucial survival-related areas, while at the same time leaving the way open for a good deal of learning. This building-in of value is accomplished by the system of emotions. Emotions serve the survival-related purpose of sensitizing the animal to certain classes of events and of calling forth, or making more likely, certain classes of action (Breger, 1974). Most classical personality theorists (e.g. Freud and McDougall) would agree that in humans species specific behavior becomes manifest whenever one observes emotions. Also, modern emotion theorists like Plutchik (1980) and Öhman (1986) conceive of emotions as firmly rooted in evolution.

Within species, individuals exhibit differences based on *genotype*. As Plomin, et al. (1980) have pointed out, genes have the capacity to control behavior through the development and functioning of physiological mechanisms. Genes control the production of proteins, which in turn exert profound influences on behavioral structures and processes via the nervous system and the production of behaviorally relevant hormones and neurotransmitters. Thus, a basis is provided for individual differences in personality and temperament.

One of the first modern attempts to link personality with biological functions is Eysenck's (1967) theory on introversion-extraversion. In Eysenck's view, biological processes are directly involved in the regulation of behavior. Temperamental dimensions like activity, sociability, and emotionality (Buss and Plomin, 1984), activity/reactivity (Strelau, 1983), introversion-extraversion, emotional stability (Eysenck, 1967), and sensation seeking (Zuckerman, 1979) emphasize different energetic, emotional, and stylistic aspects of behavior.

Recent research in behavior genetics has demonstrated that genetic influence on individual differences is usually significant and often substantial (Plomin, 1989). Current estimates of the variance due to genetic factors in personality measures range from 30% - 50% (Pedersen, 1992; Bouchard, this volume). Another conclusion drawn from research on behavior genetics is that nongenetic factors are responsible for sizable parts of the variance. Interestingly however, shared family environments do not make children similar to each other, as might be expected. There appears to be a big environmental impact upon personality, but it is an impact that takes place uniquely, on an individual level. As an explanation, Plomin (1989) proposed that measured environmental influences in families are not solely environmental but confound heredity with environment and are causally ambiguous with respect to direction of effects (cf Willerman, 1979). Heredity can affect measures of family environments, it can also mediate associations between measures of family environments and developmental outcomes of children. Major mechanisms involved are: genotype-environment correlation (the extent to which children receive or create environments correlated with their genetic propensities) and genotype-environment interaction (differential effects of environments on children with different genotypes) (Plomin, 1989). Genotype-environment correlations and interactions emphasize biological and social determinants. Accordingly, from the point of view of behavior genetics, social adaptation deserves our attention as well as biological adaptation.

SOCIAL ADAPTATION

The human species has managed to successfully master all possible ecological niches on earth. As a consequence, natural selection in humans is limited. Or, in the words of Mayr (1950): "If the single species man occupies successfully all the niches that are open to a *Homo*-like creature, it is obvious that he cannot speciate". We should keep in mind, however, that although natural selection is limited, genetic mechanisms continue to be fully operative. This implies, that at the level of the individual, *not* all niches are open. In order to master environments that are beyond the reach of an individual's genetic endowment, other mechanisms have to be activated to achieve control. Those mechanisms are summarized here under the heading of *social adaptation*. Social adaptation refers to our capacity to absorb the cultural inheritage left by our ancestors and to apply it to environments we cannot cope with on mere biological grounds. Culture involves the knowledge and information on environments, collected, compiled, and transmitted through education and learning. Some of this information is passed on through formal education, but the vast majority is transmitted in daily life by watching other people's behavior. Social learning processes affect cognitive development and establish elements like categories to recognize situations and rules to effectively deal with them. On the basis of social learning we are able to expand our environment of evolutionary adaptedness and behave effectively in environments that are well beyond the reach of our biological endowment. As a result, our environment of evolutionary adaptedness (EEA) is supplemented with an environment of social adaptedness (ESA) (Fig. 2).

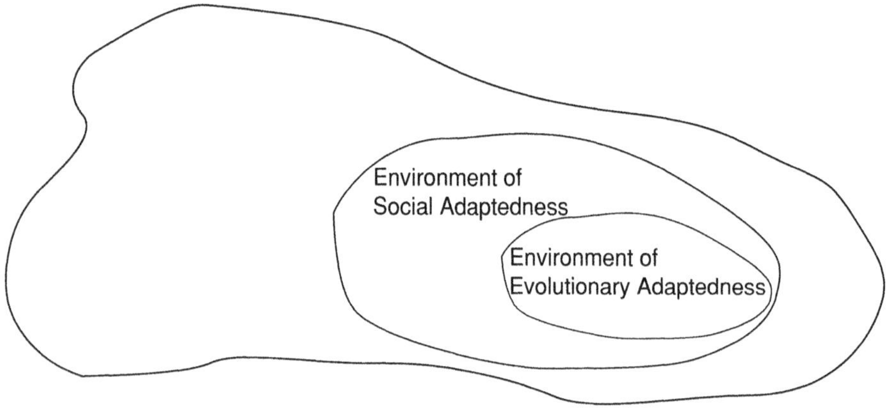

Figure 2. Environment of social adaptedness extending the environment of evolutionary adaptedness

Social adaptation differs from biological adaptation in several respects. First of all, biological adaptation is defined at the level of the species. Stable and consistent individual differences are adaptive first and foremost against that background. Social adaptation is defined at the level of individuals, as a function of ontogenetic experiences. There are also differences with regard to the environments emphasized by each. Biological adaptation determines an environment of evolutionary adaptedness accomodating the species. Social adaptation, on the other hand, is primarily concerned with the ways individuals deal with the situations met during their lifes. If successful, the main function of social adaptation is to enlarge the environment of evolutionary adaptedness and extend it with an environment of social adaptedness. Furthermore, biological adaptation is based on behavior that is environmentally stable, i.e. developing in most members of the species irrespective of upbringing (Hinde, 1970). Social adaptation, on the other hand, is based on environmentally labile behavior, i.e. behavior depending on the particular environment in which individuals are raised.

Social adaptation has the function literally to broaden a person's horizon. It can be seen as a beneficial supplement to biological adaptation. Generally speaking, for an indiviual, the two types of adaptation will support one another. However, they may also become at odds. As Barash (1986) has suggested, many of the difficulties in human life stem from the fact that two kinds of evolution have an influence on people: biological evolution and cultural evolution. The first usually takes many milennia to be completed whereas the second occurs much more quickly. As a consequence, biological evolution has prepared the human species for life in a world that was very different from the one we now inhabit. Problems are due as soon as the behavioral tendencies that are built in as part of human nature are countermanded by pressures existing in our contemporal culture.

Major differences between biological and social adaptation are due with respect to the nature of person-environment relationships emphasized by each. In a recent contribution, Hettema and Kenrick (1992) have provided a heuristic framework, summarizing the different ways in which persons interact with their environments. We introduced a distinction among six basic types of interaction: static person-environment mesh, choice of enviroments by persons, choice of persons by environments, transformation of environments by persons, transformation of persons by environments, and person-environment transaction. According to Hettema and Kenrick (1992), the first three types have obtained special attention from biological models. A common feature of these types is that they are primarily concerned with the natural correspondence between an individual's genetic endowment and his environment. The latter three types of person-environment interaction have obtained special emphasis in social models of personality. A common feature of these types is that they stress person-environment modification, rather than assuming a natural fit between the two. Modification can take different forms, implying either learning, situation transformation or both. Clearly then, the biological and social approaches study interactions from different points of view: correspondence and modification.

Buss (1985) defined three mechanisms for person-environment *correspondence*: selection, evocation and manipulation. Selection refers to nonrandom choices of interpersonal and physical milieus. Evocation involves unintentional elicitation of reactions from the environment. And, finally, manipulation is defined by different tactics used to influence and exploit the environment. Conceptually, manipulation is the broadest mechanism of person-environment correspondence, because, in principle, persons can use countless actions to influence the nature of the environment. It is also a more fundamental mechanism, because it provides a basis for the other two to be effective. Selection and evocation are based on individual characteristics with definite advantages in specific environments. Manipulation mechanisms may represent such individual characteristics and explain their functioning.

Modification is a basic interaction type emphasized by the social approach to individuality. From a learning perspective, the environment is conceptualized in terms of discriminative and reinforcing stimuli guiding behavior. Learning processes contribute to person modification in a gradual and systematic way. Social learning theory has specially emphasized observation as a powerful instrument to acquire as well as to perform new types of complex behavior in daily life situations (Bandura, 1986). The processes described by social learning theorists leave behind products in the individual that are the basis for subsequent behavior. Thus, Mischel (1973) argued that learning experiences lead to individual differences in cognition, that, in turn, guide behavior. While confronted with new situations, those cognitions are used to select, perceive, process and interpret stimuli, before reacting in one way or another. Thus, behavior is seen as an end product of a sequence of cognitive transformations. The social learning approach puts special emphasis on *cognitions* governing behavior. Examples are locus of control (Rotter, 1966), self-efficacy (Bandura, 1977), construction competencies, encoding strategies, outcome expectancies, values, and self-regulatory plans (Mischel, 1973).

Regulatory Systems

What are the major personality systems involved in biological and social adaptation? The emotion system and the cognitive system appear to provide a promising basis: *emotions* may be taken to be a core issue in biological adaptation, whereas *cognitions* are frequently associated with social adaptation.

For a long time, emotions have been predominantly considered to be maladaptive aspects of human behavior, inconveniently interfering with rationality and realism (Hettema and Van Heck, 1987; Hettema, 1991). However, during the last decades, the evaluation of emotions has changed. Current theorists agree that emotions are fundamentally adaptive in nature and play a major role in establishing and maintaining fruitful connections with the environment (Izard, 1971; Leventhal, 1984; Plutchik, 1980; Scherer, 1984; Tomkins, 1980). Several functions have been connected with emotions like, for instance, exerting powerful effects upon other behavioral systems by directing attention to the surrounding events and initiating an evaluation of the environment (Mandler, 1980), providing information about the state of the person-environment relationship (Lazarus, et al., 1982), interrupting a behavioral attempt and assessing outcome likelihood (Scheier and Carver, 1982), signalling the world's state with respect to the satisfaction of the system's driving forces, and indicating what should be done (Frijda, 1985), organizing and integrating the behavioral machinery, so that both automatic and deliberative processes aim at common goals (Leventhal, 1984), magnifying the signals activating them, thus contributing to effective need fulfillment (Tomkins, 1980).

To summarize the adaptive functions of emotions we have proposed an interactive systems conception, according to which emotions are conceived as part of a relay system (cf Hettema and Van Heck, 1987; Hettema, 1991). That relay system connects cognitive elements with sensorimotor elements (cf. Fig. 3).

According to this conception, the emotion system is activated by disruptive or interruptive events, causing a discrepancy to occur between the states of the elements of the cognitive-symbolic and the sensorimotor systems. Emotions initiate transitions in the state of any of the two systems, aiming at dissolving the discrepancy and re-establish convergence. Emotions are assumed to have the capacity to affect cognitions as well as events in the situation prevailing.

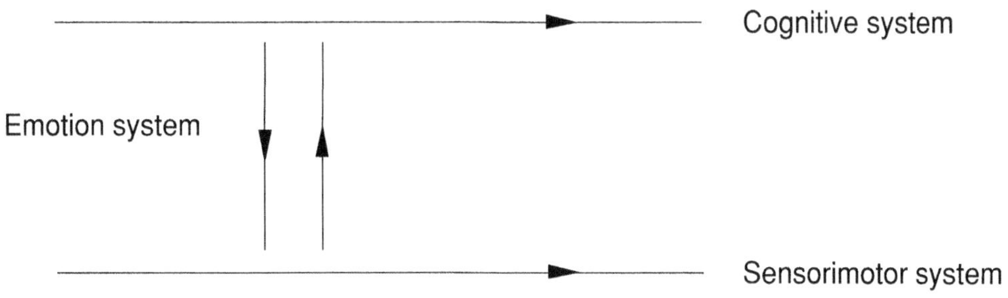

Figure 3. The emotion system connecting the cognitive and sensori motor systems

This conception provides a basis for conceptualizing the relationship with cognitions in the framework of adaptation (Hettema, 1979). The major function of emotions may be summarized as bridging the gap or reducing the distance between one's subjective appraisals, expectancies, needs, desires, and concerns on the one hand, and the state of affairs in the outside world on the other. To that effect, emotions have the capacity to introduce tactical operations like exerting influence, manipulating, and exploiting the environment. Accordingly, I have proposed to summarize the major adaptive function of emotions as *tactical* (Hettema, 1979, 1989a, 1991).

The major adaptive function of the cognitive system is *strategic* rather than tactical. Recent cognitive approaches to personality have forwarded strategies as major structures for self-regulation (Carver and Scheier, 1981; Mischel, 1973). However, in the context of adaptation, the emphasis differs. Instead of governing internal regulation, strategies are assumed to be core elements to achieve control over the environment (Hettema, 1979). This interpretation is based on the original definition forwarded by Von Clausewitz in the context of warfare. In Clausewitz' view, warfare is characterized by several peculiarities, like a basic uncertainty regarding all data available, a diversity of means leading to an end, and reciprocal action including diverse possible consequences of one's actions (cf Rapoport, 1968). As I pointed out earlier (Hettema, 1979), the peculiarities mentioned by Clausewitz not only characteristize warfare, but are typical for most real life situations in general. Accordingly, I adopted the original conception forwarded by Clausewitz as a basis to study strategies and tactics in the context of daily life. A similar emphasis may be found in the work of Cantor and Kihlstrom (1987) who defined strategies as 'sets of cognitive processes that link a person's goals to his or her subsequent behavior in a life task situation..... The strategy involves the ways in which the person interprets the "problem" and plans a "solution" so as to be consistent with his or her prevalent goals in that "task"'(p.175).

In the present conception, strategies are cognitive structures representing situations, specifying goals, and designing behaviors to reach goals. Thus, strategies are future-oriented and pro-active, using resources that are simultaneously available. The processes governed by strategies are deliberate processes, transforming situations into directions that are more favorable to the individual.

Summarizing then, strategies and tactics are conceived here as major weapons to achieve *environmental control.* Control is concerned with the agreement or disagreement between an individual's cognitions (expectancies, intentions, goals) on the one hand, and the actual state of affairs in the environment on the other (cf. Hettema, 1979, 1989a). If agreement prevails, control is present. In case of disagreement, tactical and/or strategic mechanisms may be activated to establish control.

Tactics and strategies both have the capacity to establish and maintain control, although their ways differ. Tactics are spontaneous reactions to incongruities between one's ideas, expectancies, cognitions, and the actual state of affairs in the environment. Their major function is to directly establish congruity. Strategies, on the other hand, attempt to obtain agreement indirectly through modifying situations. If effective, strategies produce new situations, offering better opportunities for control.

TACTICAL CONTROL

Tactics become visible from the moment of birth and develop in an orderly way. From the onset of life, control is achieved through emotional expression. If babies need food they start crying immediately. If their needs are satisfied, they smile. Those reactions affect parents to behave accordingly. Later on, infants exhibit curiosity and *exploration.* While playing, they discover that their limbs can be used in different ways and that there are multiple manners to manipulate objects. By trial and error they learn to replace behaviors and apply *substitution* if an original attempt fails. During childhood, they acquire a sense of autonomy through *persistence* in the face of obstacles and through exerting their will against others. Exploration, substitution and persistence are major tactical mechanisms affecting the sensorimotor system and having a direct impact upon the situation. They are summarized as *sensorimotor-operational tactics.*

Later, other tactical mechanisms become visible in close connection with cognitive development. Rather than affecting the environment, those tactics affect the cognitive structure governing behavior. Accordingly, they are called *cognitive-symbolic tactics.* Evidence for cognitive-symbolic tactics may be derived from research on cognitive development, especially the work of Piaget (Piaget and Inhelder, 1969). During cognitive development, children acquire the capacity for decentering, i.e. to go beyond egocentrism and adopt different perspectives while dealing with the same objects or situations. The tactic aiming at that effect is *reflection.* Development is also characterized by increasing complexity, including the notion that different operations may lead to the same result. Thus, children may develop the ability to replace transformation rules proven ineffective. Several new rules may be considered to replace the original one. The relevant tactic here is *uncoupling.* And, finally, children develop a broader time perspective, including future orientation and the ability to postpone rewards. Thus, goal situations may be replaced by others when unattainable. This can be done through *redirection.* All tactical mechanisms may be connected in a comprehensive system, represented in Figure 4.

Although differing with respect to the moment of development, the (sensorimotor or cognitive) level of effectivity and the aspect of behavior involved, tactics share several features. Tactics are temporary states. They are activated by frictions between cognitions and environmental events and they disappear when frictions are resolved. Tactics are local states. The nature of the tactic used is determined by the meaning of the prevailing situation for the person acting. And, finally, tactics are autonomous states. They have a definite neurological basis. They are assumed here to be closely connected with the physiological basis of individuality.

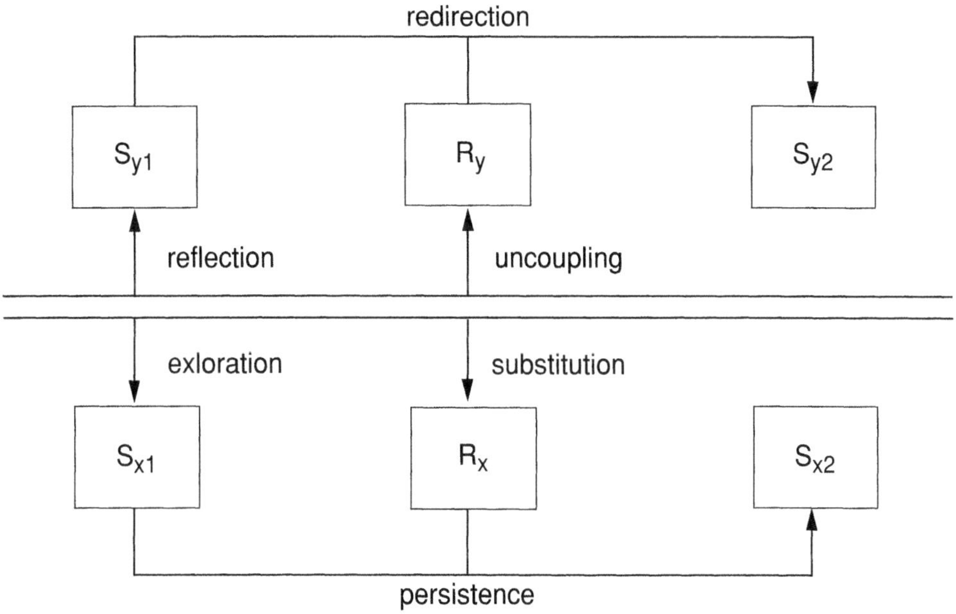

Figure 4. Tactical control system (after Hettema, 1979)

The psychophysiology of tactics

Most attempts to study the physiological basis of personality have postulated specific internal systems to regulate individual behavior. For instance, Eysenck (1967) postulated stable differences between introverts and extraverts in the amount of cortical arousal as a function of emotional stimulation. The core structure regulating cortical arousal is the ascending reticular activation system, connecting the hypothalamus and the cerebral cortex. In addition to base levels, ARAS functioning exhibits differences in reactivity, connected by Eysenck with individual differences in emotional stability.

Gray (1987) emphasized differences in reactivity between introverts/extraverts as well as between emotionally stable and labile persons. Instead of one system Gray proposed two separate systems for the regulation of cortical arousal, the behavioral approach system (BAS) and the behavioral inhibition system (BIS). Together, the two systems account for individual differences in introversion/extraversion and emotional stability/lability.

The physiology of tactics differs from the two approaches mentioned. Rather than the internal wiring of individuals, physiological reactions reflect different ways of controlling the environment. Recent developments in psychophysiology provide a basis for this approach. In the past, one general type of physiological arousal or activation has been proposed (e.g. Duffy, 1972), but current investigators have questioned the unitary nature of such a construct. Arousal has been found to be multidimensional at several levels of analysis, including the cortical (Hobson and Scheibel, 1980), autonomic (Lacey, 1967), and endocrine levels (Mason, 1975). Therefore it appears useful to consider different patterns or forms of autonomic arousal. For instance, in the area of emotions, different states are assumed to be associated with different central, autonomic, and motor patterns. According to this view, individuals may differ not only in the intensity of emotions, but also in the dominant zone of expression.

In psychophysiology, emotions are currently treated as multidimensional states, and so are cognitive processes, like vigilance, sustained attention and effort during problem solving. Major models, used to explain the energetics of information processing, are multiple resource models (cf. Hockey, et al., 1986). Multiple resource models assume information processing resources to be limited. While performing a task, individuals have the capacity of changing priorities, emphasizing e.g. now perceptual discrimination, now response selection. Multiple resource models assume individuals to be in one of many possible states at any moment in time (cf. Hockey and Hamilton, 1983). The dominant state an individual exhibits will be a function of both the individual and the situation prevailing.In this connection, cognitive processes can be considered as states, when we are able to describe or conceive of the current pattern of information processing activity in the system. Thus, for instance, Kahneman (1973) has shown that processing states can be functionally identified with autonomic and somatic states.

To study tactical states, the work of Pribram and McGuiness (1975) provides a general framework. Pribram and McGuiness proposed different processes of arousal, effort and activation to be involved in the control of attention. Arousal is conceived as a phasic neural system, concerned with the regulation of input, activation is a tonic system for output regulaton, whereas effort coordinates input/output regulation. Conceptually there are some clear links with the tactical mechanisms proposed here (cf. Hettema, 1979, 1989a). Exploration and reflection are connected with input elaboration and are conceived as states of arousal. Persistence and redirection are connected with output thus requiring activation. And, finally, substitution and uncoupling are connected with input/output regulation through effort.

Adaptation includes emotional reactivity as well as task-oriented information processing. However, as Öhman (1986) has pointed out, tasks are not the most appropriate types of situations for the study of emotions. In Öhman's view, emotions are meaning-centered rather than task-centered, so they should preferably be studied in non-task situations. Since emotions are considered here to be major tactical elements, we used meaningful non-task situations to study them. To enhance ecological validity, tactics are studied in our laboratory with films representing daily life situations as stimuli (Hettema, et al., 1989a; Van Heck, 1989). During films, a broad spectrum of (autonomous) physiological reactions from the subjects are continuously monitored.

Our studies have provided evidence on the physiological nature of tactics. For instance, persons exhibit consistent autonomous reaction patterns in specific situations. When confronted with a specific situation, a person tends to develop a consistent pattern that is maintained as long as the situation lasts. However, a new situation may elicit a completely different pattern, that is equally consistent as the former (Hettema, et al., 1989b; Geenen, 1991; Leidelmeijer, 1991). These results reflect *motivational response specificity*, implying that specific patterns are determined by the person's interpretation of the situation, rather than by either the situation or the person alone (Averill and Opton, 1968; Fahrenberg, 1986). An experimental study by

Leidelmeijer (1991) has yielded information on the specific aspects involved. In Leidelmeijer's study, instructions were employed to induce different meaning patterns to the same situations. Through different instructions specific expectancies, values and self-regulatory strategies were emphasized. The results showed that instructions had been quite effective in eliciting specific patterns of physiological reactivity. For instance, subjects who were only partly informed on the films' content had higher blood pressures and higher termoregulatory reactions than those who had complete information. With respect to stimulus values a difference became manifest between positive and negative values. Positive cognitive sets elicited more thermoregulatory reactivity, whereas negative sets gave a rise of blood pressure, cardiovascular and respiratory reactions. Also, instructions emphasizing different self-regulatory strategies gave different effects. In a condition enhancing involvement, subjects showed increased cardiovascular reactions, whereas in a detachment condition blood pressure was considerably increased. Leidelmeijers's study favors the conclusion, that physiological reactions during films refer to specific states, connected with regulation and control. On the one hand, those states are subjectively experienced as emotions and feelings. On the other hand, they represent major ways actively and directedly to deal with the situation prevailing. The results also suggest that different individuals use different mechanisms to deal with the same situations, and that the same individuals use different mechanisms to deal with different situations.

Tactics are conceived as temporary states that last as long as frictions are present. Studies by Willems (1984) and Geenen (1991) provide evidence on this aspect. In both studies physiological reactions were monitored in a series of situation films, while the whole experiment was replicated several months later. The results of these studies showed extinction of physiological reactions on the second occasion. A possible explanation is habituation of tactical reactivity in situations that match cognitive patterns established before. In Geenen's (1991) study it was interesting to note that part of the reactivity showed stability across occasions. Stability was obtained especially with variables reflecting parasympathetic reactivity. The variables concerned exhibited considerable correlations with the dimensions forwarded by Eysenck: introversion/extraversion and emotional stability. Both dimensions were positively correlated with parasympathetic reactivity. Geenen concluded that while sympathetic reactivity shows all the features of transient states, parasympathetic reactivity shows dispositional characteristics. Summarizing then, these studies have demonstrated that physiological reactions observed during situation films exhibit general features of tactics, like locality, temporarity, and elicitation by meaning patterns associated with friction.

Another line of research has been devoted to the identification of physiological patterns representing separate tactics. Using the film technique, Hettema, et al. (1989b) analyzed a number of variables with a view to identify coherent patterns of reactivity. Using generalizability analyses, signal detection methods and cluster analyses, we identified six consistent patterns, tentatively interpreted as reflection, exploration, uncoupling, substitution, redirection, and persistence. Each pattern was defined with scores on four different variables: heart rate, T-wave amplitude, galvanic skin response level, and finger temperature. To provide evidence on the interpretations forwarded, we argued that tactics can be categorized into two groups on the basis of the level at which they operate: the cognitive-symbolic or the sensorimotor-operational level. Both aspects were tested in separate experiments. A study by Hettema, et al. (1989c) was concerned with cognitive-symbolic tactics. At the cognitive-symbolic level, tactics are primarily concerned with cognitive representations and transformations. The main functon of cognitive-symbolic tactics is to affect the cognitive structure governing behavior. In effect, they may establish a new structure that is more effective in achieving the environmental transformations intended. To test this interpretation, we used cognitive tests as stimuli (cf. Hettema et al., 1989c). The stimulus material consisted of different tasks, selected to evoke specific tactics. Hypotheses were tested concerning two types of relations. State elicitation hypotheses were concerned with the capacity of specific tasks to elicit specific states. State efficacy hypotheses referred to the capacity of specific states to yield better results in specific tasks. The results of the experiment showed that five out of six elicitation hypotheses were confirmed by the data. The efficacy hypotheses obtained support in two out of six cases.

A second validation study was concerned with sensorimotor-operational tactics. Instead of affecting cognitions, those tactics are assumed to directly affect the situation prevailing. For instance, expression/exploration has the effect of emphasizing different situation elements than before. Substitution alters the behavior used earlier while preserving the general direction. And, finally, persistence re-emphasizes the goal selected earlier. All sensorimotor-operational tactics share one feature: they affect existing elements of the situation, without bringing in new elements. Accordingly it can be predicted that, compared with cognitive-symbolic tactics, sensorimotor-operational tactics will produce more behavior that is prototypical for the prevailing situation. This hypothesis was tested in a study in which the behavior of subjects was observed during role playing in six different situations (Hettema, 1989b). Before role playing, all subjects were observed with respect to physiological reactivity during films representing the same six situations. Patterns of physiological reactivity were used as a basis to predict the prototypicallity of behavior during role playing. The tests revealed that in each of the six situations the average prototypicallity score for sensorimotor-operational states exceeded the score for cognitive-symbolic states.

Concluding this section, we have provided evidence generally sustaining our definition of tactical mechanisms as patterns of physiological reactivity. The evidence stems from different studies, including reacting to situation films, solving problems in cognitive tasks and playing roles in simulated situations. As a whole, then, physiological reactions offer a promising tool for the assesment of tactical control mechanisms.

STRATEGIC CONTROL

Tactics are effective means in some environments, but may be ineffective in others. A new element enters the discussion when individuals are compelled to operate in situations outside their EEA. While no adequate tactical mechanisms are available, persons have to resort to (social) learning. When confronted with a situation, a person will take notice of environmental contingencies and take them into account in subsequent behavior. Direct reinforcement and social learning are major processes involved here. According to social learning theory, those processes establish cognitive elements like constructs, competencies, expectations, evaluations, and plans (Mischel, 1973). For environmental control, concepts for structuring and interpreting the environment as well as rules for transforming the environment seem particularly useful.

Accordingly, situation concepts and transformation rules were selected as the major elements of a person's *strategy* (Hettema, 1979, 1989a). Through the specification of goals and means, strategies have the capacity to govern goal directed behavior in any situation met. At a more complex level, strategies involve long-term sequences of contingency-behavior relationships that are effective in pursuing long time life goals and accomplishing personal projects (Little, 1983). Strategies may be represented as formal structures containing situation concepts and transformation rules as in Fig. 5. This Figure reflects strategies as transition matrices, representing initial situations and final situations connected by transformation rules.

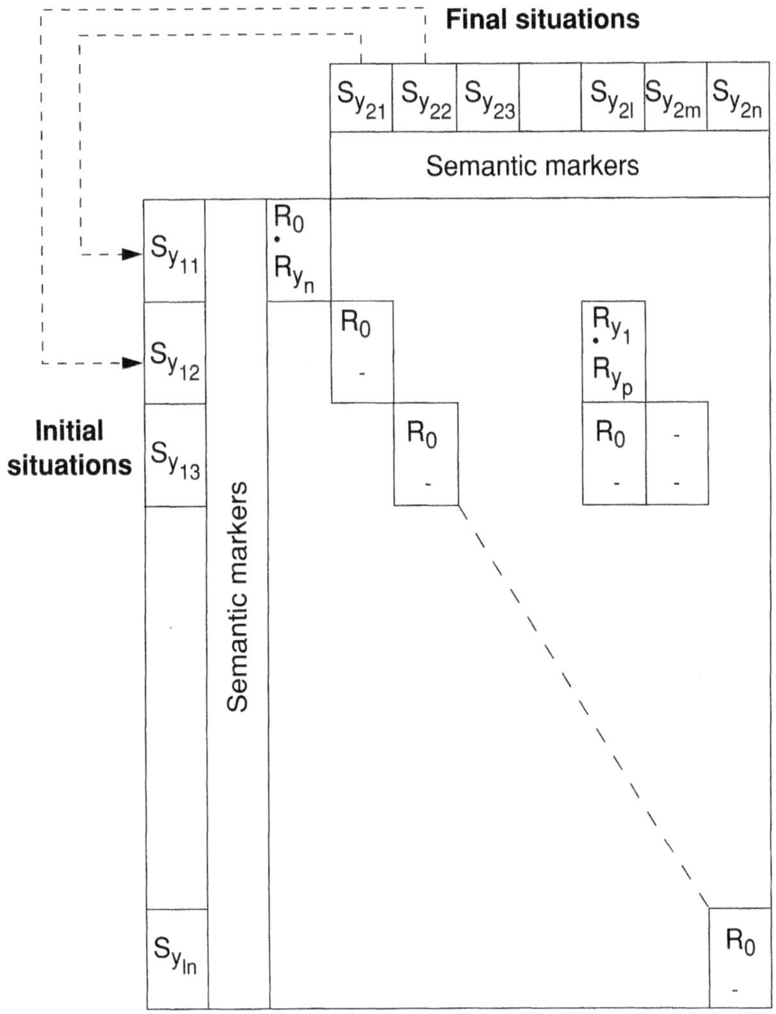

Figure 5. Behavioral strategy represented as a transition matrix (after Hettema, 1979). Situations (Sy1), represented with semantic markers, can be transformed into new situations (Sy2) through applying transformation rules (Ry).

Strategies have several features that, by nature, differ from the properties of tactics. First of all, strategies are acquired cognitive structures. They reflect competences resulting from experience. Once established, strategies are relatively stable. Strategies are goal-directed, i.e. they are applied intentionally to alter situations into a definite direction. As cognitive structures, strategies are not restricted to the situation actually prevailing. They go beyond the here and now and may extend over large situational domains and into the far future. Thus, they may be effective in governing one's development as a person and one's large scale efforts to establish achievements.

For the assessment of strategies we first developed a self-report questionnaire according to the SR format (cf. Endler et al., 1962). The questionnaire contained short descriptions of daily life situations accompanied by response alternatives to be rated with respect to probability of occurrence. With this instrument, the predictive validity of strategies in interpersonal situations was studied (Hettema, 1984)

As a criterion I used data collected through observation in role playing situations. The results were positive and indicated that behavior could be predicted quite well. However, a closer analysis of the data revealed that the positive findings could to a considerable extent be explained by general, super-individual factors, rather than individual strategies. To explain these results, I argued that if social learning processes are major determinants of strategies, individuals with the same background will tend to exhibit the same type of strategic *competence*. Thus, I forwarded the hypothesis that shared competence had been the decisive factor. In addition to competence, strategies reflect *intentionality*, an aspect that is assumed to add an individual component to strategies. Strategies are concerned with goals to be attained as well as means to materialize goals. I suggested that both aspects should be separated in strategic assessment (Hettema, 1984), to allow for conclusions to be drawn on each aspect separately. This reasoning was the basis for developing a new technique, the SRS-questionnaire (Hettema and Hol, 1989). With this instrument, subjects obtain short descriptions of concrete situations (S) and are asked to indicate what actions (R) they would take when confronted with those situations. As a result of each action, a new situation (S) is assumed to be established that may or may not be idential with the initial situation. The SRS-questionnaire has been developed on the basis of separate studies, defining situation concepts (Van Heck, 1989) and transformation rules (Hettema, 1989c). The SRS-technique answers basic requirements of internal consistency. Several aspects of strategies have been studied with this questionnaire. Major issues include: the acquired nature of strategies, the effects of learning and experience, the equifinality of strategies, the stability of goal preferences, and connections with classical personality variables, particularly temperament.

First of all, we directed attention at the acquired nature of strategies. Begeer (1984) studied strategies of managers in an organization faced with leadership problems. First he identified a number of situations typically causing leadership problems to occur. Then, Begeer developed an SRS-questionnaire and applied it to a sample of managers. He found that, on the average, the managers tended to overemphasize control as an aspect of intentionality. As a remedy, part of the sample followed a management course with the specific aim to alter their strategies. After training, the SRS-questionnaire was administered again to detect strategic change. The managers appeared to have shifted their intentionality, in that they had become more inclined to create good relationships with their associates and less inclined to demonstrate authority.

A second investigation was concerned with strategies applied in social case work. This study was done by Dooremalen (1984), who asked experts to describe problematic situations occurring frequently during counselling interviews. Then, he developed an SRS-questionnaire for the assessment of strategies in that specific domain. To evaluate the effects of training and experience, the questionnaire was administered to a sample of experienced social case workers and to a sample of students. The results showed, that while students were primarily concerned with collecting information, the experts tended to establish viable relationships with their clients. Both studies reveal a sizable influence of learning and experience upon intentionality structures revealed in strategies.

The SRS-questionnaire not only provides information on a person's actions in concrete situations. Through applying a special set of rules (Hettema, 1989c), it is possible to extrapolate the results and obtain information on the goal situations the person prefers as well as the goal situations he/she dislikes. Thus, the questionnaire offers the opportunity to test the hypothesis of *equifinality*, i.e. the convergence of preferred goal situations starting from different initial situations. This hypothesis has been confirmed in several studies (Hettema and Hol, 1989; Hol, in preparation). In these studies, subjects exhibited definite preferences for situations, relatively irrespective of the situation actually prevailing.

The stability of situation preferences has been tested in a recent study by Timmermans (in preparation). With the SRS-technique, Timmermans compared preferences of company executives over a period of more than a year. Using generalizability analysis, he obtained values in the high .80's, reflecting a high degree of stability. It may be concluded then, that individual differences in strategies are manifested first and foremost in individual preferences for situations.

To explain these results, the hypothesis was forwarded that situation preferences are determined by innate biological determinants reflected in temperament. This proposition has been tested in a study by Van Heck, et al. (1990). The results revealed several moderate but significant connections between individual situation preferences on the one hand, and temperament dimensions like emotionality, introversion/extraversion, impulsivity, thrill/adventure seeking, and disinhibition/experience seeking on the other. These results reflect the influence of biological factors upon a specific aspect of strategies: situation preferences.

Strategic-Tactical Cooperation

Strategies and tactics are conceived here as basic systems regulating person-environment transactions. Together, their functioning provides the basis for an individual's behavior in any situation met. More clarity on strategic-tactical cooperation may be obtained from a formal conceptualization of the person-environment interface. Such a conceptualization is provided in Hettema (1979) and Hettema and Kenrick (1989). The major features are represented in Figure 6.

Figure 6 summarizes the structures postulated at the person-environment interface to govern behavior. The structure allows for top-down as well as bottom-up processing to occur. Three process levels account for the major relationships involved: the cognitive-symbolic level, the control level and the sensorimotor-operational level. All three levels are assumed to be involved in any type of behavior occurring, but the sequence of events may exhibit considerable variation. For instance, the onset of behavior may be located either in the person or in the situation prevailing. The person may have the intention to attain some result in the environment. He may develop a plan and act according to that plan to materialize his goal. In this sequence, the onset of behavior is located at the cognitive-symbolic level and the (cognitive-symbolic) strategy will be activated first. The strategy determines actions to be executed at the sensorimotor-operational level. This sequence provides an example of top-down processing. On the other hand, the environment may provide the first impetus. The prevailing situation may offer a challenge or threat to the person, who will mobilize his resources to cope with the situation. In that case, the onset of behavior is located at the sensorimotor-operational level. Subsequently, at the cognitive-symbolic level, the situation will be represented and a strategy may be designed to counteract situational challenges. This sequence of events reflects bottom-up processing.

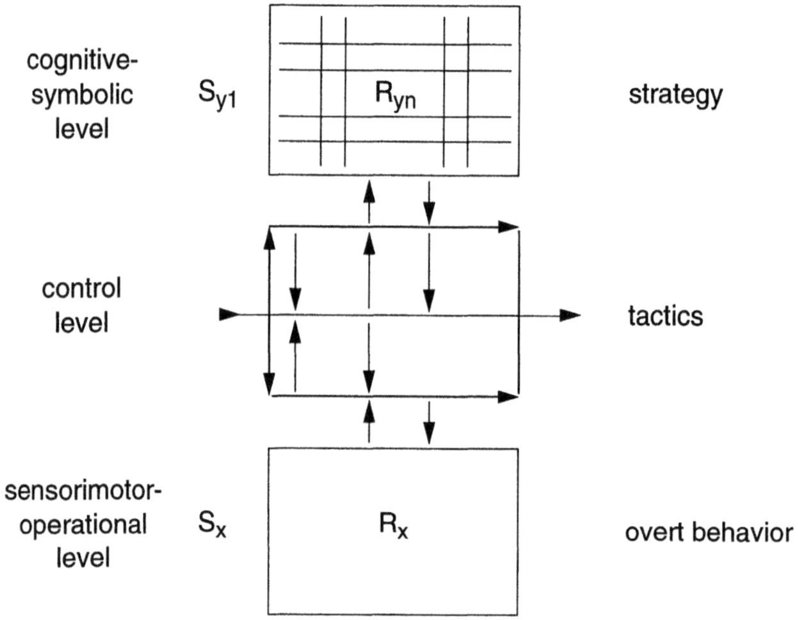

Figure 6. Strategies and tactics governing overt behavior

The two levels of the system outlined thus far are conceived as parallel levels, i.e. events at the sensorimotor-operational level have their counterparts in elements at the cognitive-symbolic level and vice versa. They will function appropriately as long as no discrepancies between the two types of elements occur. However, in real life, the levels may not be assumed to be completely compatible. Discrepancies will occur frequently, causing unbalance to occur between the two levels. The third level, control, is postulated to deal with discrepancies. Conceived as an intermediate relay level, it has the capacity of affecting the other two. At this level, discrepancies between cognitions and events activate tactical mechanisms. Tactics may affect either situational events or strategic elements, with a view to achieve a balance between the two. As a result, behavior is seen as an ongoing stream of person-situation transactions, affected by characteristics of of the person and the situation. On the person side of this process, the major structures involved are strategies and tactics. Their cooperation is primarily conceived as complementary.

Needless to say, to test this conception of person-situation transactions is far from simple. For instance, while a person is acting in a situation it is hard to indicate precisely, what structure will be involved at any moment in time. To solve this problem we argued, that, in the course of behavior, both strategies and tactics will generally play some part. Thus, we decided to predict individuals' overt behavior in concrete situations on the basis of information on strategies as well as tactics (Hettema, 1989b). In that study, strategies for a number of situations were assessed for 118 subjects completing an SRS questionnaire. Each strategy yielded information on preferred classes of behavior and non-preferred classes of behavior. Subsequently, 56 subjects drawn at random from the original sample were confronted with situation films for the assessment of tactical reactions. The tactics obtained yielded new sets of behavior preferred and non-preferred. Thus, on the basis of strategic as well as tactical information, predictions of overt behavior were derived for each subject separately in each situation. To test those predictions, the behavior of a sample of 33 subjects was observed during role playing in simulated real life situations. Although the sample was rather homogenous, clear relationships were found between behaviors predicted and behaviors observed. We concluded that the data had provided empirical support for the conception of strategic-tactical cooperation forwarded.

Individuality

Thus far, I have presented two general structures, presumably reflecting biological and social determinants of personality. A major question to be asked here refers to the conception of individuality within this framework. In general, it can be stated that phenotypic individuality involves genotypical structures as well as environmental effects. Earlier I referred to the work of Plomin, et al. (1977) and Plomin (1986), suggesting two major relationships between genotypes and environments: genotype-environment correspondence and genotype-environment interaction. However, thus far, those concepts are mainly based on macro-analyses in the context of behavior genetics. They do not provide an answer to the question, how they operate in a single person to establish his or her individuality. For defining individuality at the micro-level, we need a further conceptualization of how genotypic and environmental factors interact. A first notion may be based on the processes involved in personality development.

Initially, personality development will be governed mainly by the genotypic structures of the individual. Genes control behavior and continue to do so during ontogenetic development (Rowe, 1989). Usually, in early childhood, relationships with situations will be governed predominantly by genotype-environment correlation, i.e. correspondence between genotype and environment. Genotype-environment correlation can take different forms, like passive, active and reactive correlations (Plomin, et al., 1977). As a result, some environments are well suited to particular individuals but poorly suited to others. In the present conception, the major aspects of behavior determined genetically are tactics. Accordingly, tactics are assumed to be influential determinants of genotype-environment correlation. Tactics are effective means to establish and maintain control. However, in addition, the nature of the environment is a major issue. Some situations will be controlled more effectively by some tactics than others. For instance, in some sports persistent effort may be most effective, whereas in doing exams reflection may provide the best tactic. A major thesis of the present approach is that individuals will develop preferences for situations that can be controlled on the basis of their tactical endowment. Accordingly, individuals with different genotypes are sorted into different environments (Scarr and McCartney, 1983).

This picture may be maintained as long as persons are allowed to occupy their favored environments. However, obviously, society requires persons to go well beyond the borders of their environments of evolutionary adaptedness. For instance, in our culture, children are supposed to explore new environments, to get friends, to attend schools, to look for jobs, etc. Some of the new environments may suit the individual from the beginning but others are hard to master. New environments require specific concepts to be recognized and specific rules to be effectively dealt with. Person modification and learning are the major processes for acquiring the necessary elements. Strategies governing behavior in those environments provide a basis for explaining genotype-environment interactions. Strategies offer the opportunity to directedly transform situations, i.e. to transform unwanted situations into more desirable situations. However, strategies are far from neutral with respect to genotype. As our studies have demonstrated, strategies exhibit equifinality and directional bias. Thus, for instance, some persons may tend to transform many situations into an enjoyable get-together, others into an intellectual dispute, and still others into a quarrel. This directional bias is explained here as an outflow of situation preferences based on genotypes.

Summarizing then, a person's individuality is multiply determined, reflecting biological as well as social determinants. Biological factors are largely responsible for the more fluid aspects of personality, whereas social factors are the main basis for crystallization to occur (cf Fig. 7).

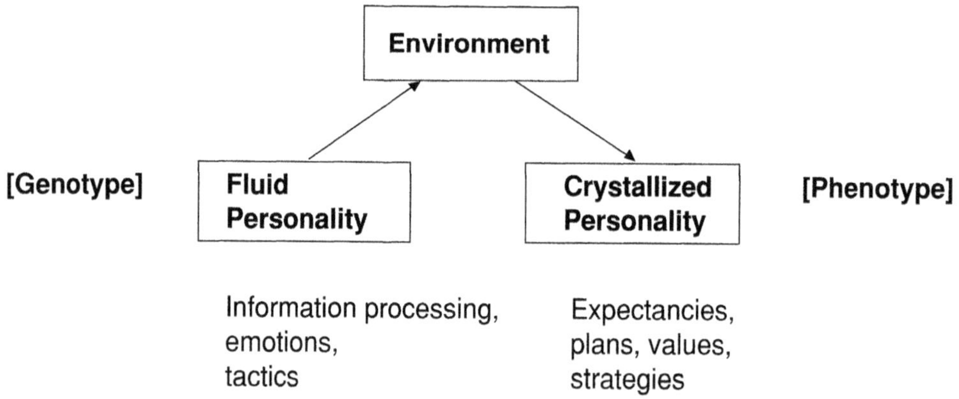

Figure 7. Fluid-crystallized conception of personality

Prospects

In view of the different adaptive functions postulated for biological and social determinants, separate assessment seems mandatory to obtain a true sense of a person's individuality. First of all, the attention should be focussed on *genotype assessment*. For the assessment of genotypes, our studies provide a basis to use multiple indices. Genotypes may be characterized by preferences for using specific tactics as well as preferences for obtaining specific situations in strategies. There is evidence sustaining both claims. A study by Buss et al. (1987) has provided evidence on tactical preferences. Studying the tactics people employ to elicit and terminate actions of others, they found consistent differences between individuals. In their view, individuals exhibit 'preferences' to use some tactics rather than others, depending on their genotypes. For instance, while trying to get others to do something, some individuals will use charm, others will coerce and still others will attempt reasoning. The differences obtained were clearly related with standard dimensions of temperament. Accordingly, Buss et al. (1987) postulated a relationship with natural selection as emphasized by evolution theory. As regards situation preferences, evidence has been obtained from the study by Van Heck et al. (1990). As indicated before, that study revealed several links with standard dimensions of temperament, suggests that situation preferences are likewise based on hereditary factors. More definite proof of the relationships postulated here may be obtained by studying tactical preferences as well as situation preferences with identical twins as subjects.

As a next step, the attention may be focussed on *genotype-environment correspondence*. A major hypothesis to be tested here is that individuals will prefer situations for which control is offered by their preferred tactics. Evidence for this assertion may be obtained by demonstrating the effectivity of specific tactics for dealing with specific situations. The hypothesis is currently studied in our laboratory, using computer simulation of tactics operating in specific situations as a major tool.

The third logical step to be taken involves the study of *genotype-environment interaction*, referring to the differential effects of the same environments upon persons with different genotypes. The study of genotype-environment interaction may be based on genotype assessment as discussed earlier. Subjects assessed with respect to genotypes may be confronted with specific situations, e.g. through role playing, to test the effects wanted. Major effects are expected with regard to the strategies they use in the domains to which those situations belong. Strategies from genotypically different subjects may be compared to obtain information on genotype-environment interaction.

A final issue of individuality refers to the relationship with phenotypical personality as conceived by classical personality theory. Clearly, the conception of individuality developed in this paper differs from classical approaches emphasizing structural dispositions of personality. Instead of traits, I have emphasized systems involved in the regulation of person-environment connections. However, I have also stressed several structural properties that may provide a basis for comparing both approaches. For instance, the goals persons strive for are a major aspect of individuality in the present view. The notion that goals are salient characteristics of their personalities has been forwarded in recent publications (cf. e.g. Read et al., 1990). This notion may eventually offer an opportunity for exploring relationships with phenotypes as exemplified in current personality factors.

References

Allport, G.W. (1961) *Pattern and growth in personality.* New York: Holt.

Averill, J.R. and Opton, E.M. Jr (1968) Psychophysiological assessment: Rationale and problems. In: P. McReynolds (Ed) *Advances in pscychological assessment.* Palo Alto: Science and behavior books, Vol I, pp.265-645.

Bandura, A. (1977) Self-efficacy mechanism in human agency. *American Psychologist,* 37, 122-147.

Bandura, A. (1986) *Social foundations of thought and action: A social cognitive theory.* Englewood Cliffs: Prentice-Hall.

Barash, D.P. (1986) *The hare and the tortoise: Culture, biology, and human nature.* New York: Penguin.

Begeer, W. (1984) *De meting en beinvloeding van de strategieen van leiders in een onderneming* [The measurement and modification of the strategies of leaders in a company]. Unpublished MA Thesis, Tilburg University, The Netherlands.

Bowlby, J. (1969) *Attachment and loss: Vol I Attachment.* New York: Basic books.

Breger, L. (1974) *From instinct to identity: The development of personality.* Englewood Cliffs: Prentice-Hall.

Buss, A.H. and Plomin, R. (1984) *Temperament: Early developing personality traits.* Hillsdale, NJ: Erlbaum.

Buss, D.M. (1985) Human mate selection. *American Scientist,* 73, 47-51.

Buss, D.M. (1991) Evolutionary personality psychology. *Annual Review of Psychology,* 42, 459-491.

Buss, D.M., Gomes, M., Higgins, D.J. and Lauterbach, K. (1987) Tactics of manipulation. *Journal of Personality and Social Psychology,* 52, 1219-1229.

Cantor, N. and Kihlstrom, J.F. (1987) *Personality and social intelligence.* Englewood Cliffs, NJ: Prentice-Hall.

Carver, C.S. and Scheier, M.F. (1981) *Attention and self-regulation: A control-theory approach to human behavior.* New York: Springer.

Darwin, C. (1859) *On the origin of the species by means of natural selection, or, preservation of favoured races in the struggle for life.* London: Murray

Dooremalen, J.C.J. (1984) *Hulpverleningsstrategieen bij maatschappelijke werkers: Op weg naar een taxonomie* [Counselling strategies in social workers]. Unpublished MA thesis, Tilburg University, The Netherlands.

Duffy, E. (1972) Activation. In: N.S. Greenfield and R.A. Sternbach (Eds) *Handbook of psychophysiology.* New York: Holt, pp. 577-622.

Endler, N.S., Hunt, J.McV, and Rosenstein, A.J. (1962) An SR-inventory of anxiousness. *Psychological Monographs,* 76, 17.

Eysenck, H.J. (1967) *The biological basis of personality.* Springfield, Ill; Charles C. Thomas.

Fahrenberg, J. (1986) Psychophysiological individuality: A pattern analytic approach to personality research and psychosomatic medicine. *Advances in Behavioral Research and Therapy,* 8, 43-100.

Frijda, N.H. (1985) Towards a model of emotion. In: C.D. Spielberger, I.G. Sarason, and P.B. Defares (Eds) *Stress and Anxiety,* 9, pp. 3-16.

Geenen, R. (1991) *Psychophysiological consistency and personality.* Tilburg, The Netherlands: Tilburg University Press.

Gray, J.A. (1987) Perspectives on anxiety and impulsivity: A commentary. *Journal of Research in Personality*, 21, 493-509.

Hettema, P.J. (1979) *Personality and adaptation.* Amsterdam: North-Holland.

Hettema, P.J. (1984) The prediction of interpersonal behavior in specific situations. In: H. Bonarius, G.L. Van Heck and N. Smid (Eds) *Personality psychology in Europe: Theoretical and empirical developments.* Lisse, The Netherlands: Swets & Zeitlinger, pp. 165-177.

Hettema, P.J. (1989a, Ed) *Personality and environment: Assessment of human adaptation.* Chichester, UK: Wiley.

Hettema, P.J. (1989b) Predictive validity. In: P.J. Hettema (Ed) *Personality and environment: Assessment of human adaptation.* Chichester, UK: Wiley, pp. 189-202.

Hettema, P.J. (1989c) Transformation rules: Towards a taxonomy of everyday behavior. In: P.J. Hettema (Ed) *Personality and environment: Assessment of human adaptation.* Chichester, UK: Wiley, pp. 71-85.

Hettema, P.J. (1991) Emotions and adaptation: An open-systems perspective. In: C.D. Spielberger and I.G. Sarason (Eds) *Stress and emotion: Anxiety, anger, and curiosity*, Vol. 14. New York: Hemisphere, pp. 47-63.

Hettema, P.J. and Hol, D.P. (1989) The assessment of behavioral strategies. In: P.J. Hettema (Ed) *Personality and environment: Assessment of human adaptation.* Chichester, UK: Wiley, pp. 87-103.

Hettema, P.J. and Kenrick, D.T. (1989) Biosocial interaction and human adaptation. In: P.J. Hettema (Ed) *Personality and environment: Assessment of human adaptation.* Chichester, UK: Wiley, pp. 3-19.

Hettema, P.J. and Kenrick, D.T. (1992) Models of person-situation interactions. In: G.V. Caprara and G.L. Van Heck (Eds) *Modern personality psychology: Critical reviews and new directions.* Hemel Hampstead, UK: Harvester-Wheatsheaf, pp. 393-417.

Hettema, P.J. and Van Heck, G.L. (1987) Emotion-cognition interactions in personality. *Communication and Cognition*, 20, 141-170.

Hettema, P.J., Van Heck, G.L and Brand, C. (1989a) The representation of situations through films. In: P.J. Hettema (Ed) *Personality and environment: Assessment of human adaptation.* Chichester, UK: Wiley, pp. 113-127.

Hettema, P.J., Vingerhoets, A.J. and Van Heck, G.L. (1989b) Patterns of physiological and biochemical reaction during films. In: P.J. Hettema (Ed) *Personality and environment: Assessment of human adaptation.* Chichester, UK: Wiley, pp. 129-146.

Hettema, P.J., Vingerhoets, A.J., Van der Molen, G.M. and Van de Vijver, F.J. (1989c) Construct validation of psychophysiological state patterns. In: P.J. Hettema (Ed) *Personality and environment: Assessment of human adaptation.* Chichester, UK: Wiley, pp. 147-161.

Hinde, R.A. (1970) *Animal behavior. A synthesis of ethology and comparative psychology.* New York: McGraw-Hill.

Hobson, J.A. and Scheibel, A. (1980) The midbrain reticular core. *Neuroscience Research Program Bulletin*, 18, 27-43.

Hockey, G.R., Coles, M.G. and Gaillard, A.w. (1986) Energetical issues in reserach on human information processing. In: G.R. Hockey, A.W. Gaillard and M.G. Coles (Eds) *Energetics and Human Information Processing.* Dordrecht, The Netherlands: Martinus Nijhoff, pp. 3-21.

Hockey, G.R. and Hamilton, P. (1983) The cognitive patterning of stress states. In: G.R. Hockey (Ed) *Stress and fatigue in human performance.* Chichester, UK: Wiley, pp. 331-362.

Hol, D.P. (in preparation) *The prediction of person-situation interactions: A model comparison.*

Izard, C.E. (1971) *The face of emotion.* New York: Appleton-Century-Crofts.

Kahneman, D. (1973) *Attention and effort.* Englewood Cliffs, NJ: Erlbaum.

Kenrick, D.T. (1987) Gender, genes, and the social environment: a biosocial perspective. In: P.C. Shaver and C. Hendrick (Eds) *Review of personality and social psychology,* Vol. 7. Newbury Park, CA: Sage, pp. 14-43.

Kenrick, D.T. (1989) A biosocial perspective on mates and traits: Reuniting personality and social psychology. In: D.M. Buss and N. Cantor (Eds) *Personality psychology: Recent trends and emerging directions.* New York: Springer, pp. 308-319.

Kenrick, D.T. and Dantchik, A. (1983) Interactionism, idiographics, and the social psychological invasion of personality. *Journal of Personality,* 51, 286-307.

Kenrick, D.T., Dantchik, A., and McFarlane, S. (1983) Personality, environment, and criminal behavior: An evolutionary perspective. In: W.S. Laufer and J.M. Day (Eds) *Personality, moral development, and criminal behavior.* Lexington, MA: D.C. Heath, pp. 217-241.

Kenrick, D.T. and Funder, D.C. (1988) Profiting from controversy: Lessons from the person-situation debate. *American Psychologist,* 43, 23-34.

Lacey, J.I. (1967) Somatic response patterning and stress: Some revisions of activation theory. In: M.J. Appley and R. Trumball (Eds) *Psychological stress: Issues in research.* New York: Appleton-Century-Crofts, pp. 14-37.

Lazarus, R.S., Coyne, J.C. and Folkman, S. (1982) Cognition, emotion, and motivation: The doctoring of Humpty-Dumpty. In: R.W.J. Neufeld (Ed) *Psychological stress and psychopathology.* New York: McGraw-Hill, pp. 218-239.

Leidelmeijer, K.C.M. (1991) *Emotions: An experimental approach.* Tilburg, The Netherlands: Tilburg University Press.

Leventhal, H. (1984) A perceptual-motor theory of emotions. In: K.R. Scherer and P. Ekman (Eds) *Approaches to emotion.* Hillsdale, NJ: Erlbaum, pp. 271-291.

Lewontin, R.C. (1978) Adaptation. *Scientific American,* 114-125.

Little, B.R. (1983) Personal projects. A rational and method for investigation. *Environment and Behavior,* 15, 272-309.

Mandler, G. (1980) The generation of emotion: A psychological theory. In: R. Plutchik and H. Kellerman (Eds) *Emotion: Theory, research, and experience. Vol. I: Theories of emotion.* New York: Academic Press, pp. 219-243.

Mason, J.W. (1975) Emotions as reflected in patterns of endocrine integration. In: L. Levi (Ed) *Emotions: Thier parameters and measurement.* New York: Raven, pp. 143-181.

Mayr, E. (1950) Taxonomic categories in fossil hominids. In: *Cold Spring Harbor symposia on quantitative biology, Vol. 15: Origin and evolution of man.*

Mischel, W. (1968) *Personality and assessment.* New York: Wiley.

Mischel, W. (1973) Toward a cognitive social learning reconceptualization of personality. *Psychological Review,* 80, 252-283.

Murray, H.A. (1938) *Explorations in personality.* New York: Oxford Press.

Öhman, A. (1986) Integrating energetic and information processing concepts: Emotion from a functional-evolutionary perspective. In: G.R. Hockey, A.W. Gaillard, and M.G. Coles (Eds) *Energetics and human information processing*. Dordrecht, The Netherlands: Martinus Nijhoff, pp. 337-352.

Pedersen, N. (1992) Genetics of personality. *Paper presented at the VIth European conference on personality in Groningen, The Netherlands.*

Piaget, J. and Inhelder, B. (1969) *The psycholgy of the child*. New York: Basic books.

Plomin, R. (1986) Behavior genetic methods. *Journal of Personality*, 54, 226-261.

Plomin, R. (1989) Environment and genes: determinants of behavior. *American Psychologist*, 44, 2, 105-111.

Plomin, R., DeFries, J.C. and Loehlin, J.C. (1977) Genotype-environment interaction and correlation in the analysis of human behavior. *Psychological Bulletin*, 84, 309-322.

Plomin, R., DeFries, J.C. and McLearn, G.E. (1980) *Behavior genetics: A primer*. San Francisco, CA: Freeman.

Plutchik, R. (1980) *Emotion, a psychoevolutionary synthesis*. New York: Harper.

Pribram, K.H. and McGuiness, D. (1975) Arousal, activation, and effort in the control of attention. *Psychological Review*, 82, 116-149.

Rapoport, A. (1968) *Clausewitz on war*. Penguin books: Pelican classics.

Read, S.J., Jones, D.K. and Miller, L.C. (1990) Traits as goal-based categories: The importance of goals in the coherence of dispositional categories. *Journal of Personality and Social Psychology*, 58, 1048-1061.

Rotter, J.B. (1966) Generalized expectancies for internal versus external control of reinforcement. *Psychological Monographs*, 80, (1, whole No. 609).

Rowe, D.C. (1989) Personality theory and behavior genetics: Contributions and issues. In: D.M. Buss and N. Cantor (Eds) *Personality psychology: Recent trends and emerging directions*. New York: Springer, pp. 294-307.

Rushton, J.P., Fulker, D.W, Neale, M.C., Nias, D.K.B. and Eysenck, H.J. (1986) Altruism and aggression: The heritability of individual differences. *Journal of Personality and Social Psychology*, 50, 1192-1198.

Scarr, S. and McCartney, K. (1983) How people make their own environments: A theory of genotype-environment effects. *Child Development*, 54, 424-435.

Scheier, M.F. and Carver, C.S. (1982) Cognition, affect, and self-regulation. In: M.S. Clark, and S.T. Fiske (Eds) *Affect and cognition*. Hillsdale, NJ: Erlbaum, pp. 157-183.

Scherer, K.R. (1984) On the nature and function of emotion: A component process approach. In: K.R. Scherer and P. Ekman (Eds) *Approaches to emotion*. Hillsdale, NJ: Erlbaum, pp. 293-317.

Strelau, J. (1983) *Temperament - Personality - Activity*. London: Academic Press.

Timmermans, F. (in preparation) *The stability of behavioral strategies*.

Tomkins, S.S. (1980) Affect as amplification: Some modifications in hteory. In: R. Plutchik and H. Kellerman (Eds) *Emotion. Theory, research, and experience. Vol I: Theories of emotions*. New York: Academic Press, pp. 141-164.

Van Heck, G.L. (1989) Situation concepts: definitions and classification. In: P.J. Hettema (Ed) *Personality and environment: Assessment of human adaptation*. Chichester, UK: Wiley, pp. 53-69.

Van Heck, G.L., Hettema, P.J. and Leidelmeijer, K.C.M. (1990) Temperament, act preferences, and situation preferences. *Nederlands Tijdschrift voor de Psychologie*, 45, 1-16.

Willems, T.R. (1984) *Een empirisch onderzoek naar de stabiliteit van persoon-situatie interacties als gereflecteerd in psychofysiologische responsvariabelen op het medium film* [The stability of person-situation interactions observed with physiological variables during films]. Unpublished MA thesis, Tilburg University, The Netherlands.

Willerman, L. (1979) Effects of families on intellectual development. *American Psychologist*, 34, 923-929.

Wilson, E. (1978) The nature of human nature. *New Scientist*, 5, 20-22.

Zajonc, R.B. (1980) Feeling and thinking: Preferences need no inferences. *American Psychologist*, 35, 151-175.

Zuckerman, M. (1979) *Sensation seeking: Beyond the optimal level of arousal*. Hillsdale, NJ: Erlbaum.

STRATEGIES AND TACTICS FOR PERSON-SITUATION INTERACTION

Commentary on the chapter by P.J. Hettema

Randy J. Larsen
Department of Psychology
580 Union Drive
University of Michigan
Ann Arbor, MI 48109-1346, USA

Throughout the history of personality psychology, enthusiasm for the social and biological determinants of individuality has waxed and waned in opposition to each other. When emphasis on social factors was popular, interest in biological factors was minimal, and vice versa. Currently, however, there appears to be a 'coming together' of the social and biological views. As evidenced by this international workshop, there is now the beginning of dialog on how to best integrate the social and biological perspectives on human nature. I believe this attests to the maturation of our science, in that formerly divergent lines of explanation and understanding are now converging and integrating. Moreover, this also represents the next great challenge for the science of personality. That challenge concerns how to best integrate the social and biological approaches to personality. What will an integrated perspective look like? How will a common framework be built out of such seemingly opposed perspectives? What will be the key themes and constructs that bring together and bridge the social and biological foundations?

The paper by Hettema presents his vision of a common framework for integrating the social and biological views of individuality. Hettema's integrating themes are adaptation and regulation. The theoretical constructs that operationalize these themes are strategies and tactics. My commentary will be organized around these key themes and constructs. I will first elaborate on Hettema's ideas, emphasizing the originality and usefulness of these themes and constructs in conceptualizing personality. I will also attempt to expand on the topics of adaptation, regulation, strategies and tactics. And finally, drawing on what I find implicit in Hettema's views, I will suggest how the metaphor of ecology might be used as one guide to our thinking about the integration of social and biological perspectives on personality.

Overview

As an overview, let me mention that the starting point for Hettema is very different from that of more traditional personality theorists. Biologically-minded theorists have traditionally emphasized individual differences in reactivity to the environment. For example, individuals with strong nervous systems (Pavlov), or Extraverts (Eysenck), or Reducers (Petrie, Vando), or low reactivity persons (Strelau) generally show diminished responses to environmental stimulation. For Hettema, however, the emphasis is not on diminished reactivity per se, but rather on how

J. Hettema and I. J. Deary (eds.), Foundations of Personality, 165–173.

that diminished reactivity becomes reflected in attempts to control the environment. At first this distinction appears subtle. But it is an extremely important shift of emphasis, one that opens up a new perspective on individuality. For example, the diminished reactivity of the extravert is, for Hettema, part of a whole system of regulating forces that function together to achieve social and biological adaptation. Because of this difference in emphasis, Hettema does not present us with a static trait model of personality. Instead, Hettema provides us with a more systems-oriented view of individuality, proposing that personality is the product of interacting regulatory systems that exist to achieve adaptive control over the environment.

As a consequence of Hettema's emphasis on adaptation and regulation, his perspective does not emphasize content-bound personality constructs like like extraversion or neuroticism. Nor does his perspective deal with the traditional issues of content-bound personality psychology, such as consistency and stability. In fact, Hettema's perspective shows little emphasis on the concept of individual differences. Instead, his point of view emphasizes more process-bound constructs like strategies and tactics for adaptation. Because of this, there is an emphasis on function rather than on content. That is, Hettema is more concerned with how persons in general function in their environments than with how they differ from one another. Clearly, people DO differ from one another in how they function, but the emphasis for Hettema is on the functional units and the mechanisms of adaptation. After we first understand these functional units (e.g., tactics and strategies) then we might seek to understand individual differences in how people function. We might also explore the connection between individual differences in strategies and tactics and the traditional trait or motivational dimensions of personality. But for Hettema, the guiding theme is that personality is first and foremost adaptation, not individual differences.

Adaptation

The environment consistently challenges humans to adjust to changing demands. This theme of adaptation may be examined using different time frames. Over an extremely long time period, such adaptation to environmental demand is called evolution. Through the process of natural selection, evolution has shaped human morphology, physiology, and behavior to reflect adaptation to environmental pressures. Regardless of whether such pressures are operative today, it is obvious that certain acquired adaptive characteristics ARE present. Natural selection has, in some ways, prepared us for a world that is different from the one in which we currently live. Social learning and culture have altered or erased some previous environmental demands while introducing new demands for adaptation. Nevertheless, understanding the social and biological factors involved in long-term adaptations -- the domain of evolutionary psychology -- is an important part of Hettema's overall perspective. Social and biological factors are thus integrated even in understanding long-term adaptation.

Employing a more moderate time frame -- the time frame of a single human life -- we see in Hettema's work an emphasis on integrating the social and biological determinants of adaptation. During development people acquire values, they learn social roles, they develop scripts and plans for achieving goals. These and other social factors influence how people adapt to environmental demands. What one values, what social role one occupies, what culture one grew up in, etc.... all influence how one thinks about, classifies, and reacts to environmental events. For example, if a person grew up in a culture that permitted or encouraged polygamy, then perhaps jealousy would be a less likely adaptation to sexual infidelity. Hettema points out how, due to such cultural and social factors, humans are extraordinarily flexible in their abilities to adjust and adapt to environmental demands and life events. Nevertheless, he points out how biology may put limits on the influence of social factors in adapting to life events over the life span.

Employing an even shorter time frame -- the time frame of a single stimulus-response event -- we again find in Hettema's perspective an emphasis on the integration of social and biological determinants of adaptation. For example, there are certain classes of releasing stimuli that trigger a specific response or action tendency. Nevertheless, even here Hettema holds that social factors play a role. Social learning may influence how people come to recognize and classify stimuli or situations. Cultural knowledge may be acquired and applied to new or unfamiliar situations, determining in part even one's immediate physiological response. The growing field of social psychophysiology (e.g., Cacioppo and Petty, 1983) contains numerous examples of physiological perspectives on social phenomena.

At all these differing time frames of adaptation, we see in Hettema an emphasis on the transaction between social and biological influences. Over long-term adaptation we see a transaction between what evolution has prepared us for, and what contemporary culture presents us with. Over the individual life span, we see the important role of social learning in adapting to life events, but such adaptations also interact with biologically-based sensitivities to specific classes of stimuli. During short-term reactions to events, we see a transaction between biological responsivity and the acquired social or personal meaning assigned to specific events, situations, and stimuli. Thus, regarding the key theme of adaptation, Hettema's theory emphasizes the transaction between social and biological processes that combine to influence or shape adaptation. The next major theme in Hettema's work concerns how adaptation is achieved by regulatory systems.

Regulation

For Hettema, personality is the product of regulatory systems. These systems exist for the regulation of everyday life. In its broadest sense, regulation occurs at the interface of the person (P) and the environment (E), and refers to different ways that persons interact with their environments. There are two broad categories of regulated P by E interactions; one kind of interaction concerns the correspondence between characteristics of persons and characteristics of the environment, with the emphasis on selection or choice. The other category of P by E interaction refers to modifying or altering the environment, with emphasis on how persons can evoke change or transformations in their environments.

The first type of P by E interaction - the selection of situations - has both a long and a short history in personality psychology. For decades we have assumed that people's traits and temperament were fixed or highly stable over time. Consequently, in such applied personality fields as personnel selection or vocational counseling, the effort was to match the right person to the right environment. For example, in employee selection, the effort is to analyze the demands of the work environment then select the person with the appropriate combination of attributes to match that situation. The assumption is that, since the person will not change, we better get them into the right environment, the environment that matches their personal attributes.

More recently, personality psychologists have begun examining situational selection in everyday life. In some of my own research (Diener et al., 1984) we have used the daily experience sampling technique to explore how people spend their time in everyday life and the kinds of situations they choose to enter. We were interested in whether people spend time in situations that are congruent with their personalities. We had people carry beepers throughout their daily life, and these beepers went off randomly twice a day for 6 weeks. When the beeper went off the person completed a brief report indicating the type of situation they were in. We asked, for example, whether they were alone or with other people, whether they were working or having fun, whether they were doing something unusual or something typical and ordinary. We then calculated how often, over the 6 week period of twice-daily reporting, that we 'caught' each person in each kind of situation. We then used standard personality questionnaires to see if the subjects' personalities were related to the situations they most frequently choose to enter. We found, for example, that extraverted subjects were more often in social situations and, when recreating, chose social forms of recreation (e.g., team sports) more often than solitary forms of recreation (e.g., swimming, running). Persons with high need for achievement were in work situations more frequently than persons low in need for achievement. More recently, in a data set just gathered using similar methods, we found that persons high on sensation seeking spent more time in unusual or novel situations, and persons high on a measure of depression spent more time alone.

In another study of ours (Emmons et al., 1986), we found that freedom of choice is an important issue in understanding situational selection. That is, people oftentimes HAVE to be in certain situations. Based on previous decisions and commitments, people often find themselves having to enter situations where they feel they have little choice. For example, the student who is working late at night in the library studying for an exam the next day might say that he has little choice in the matter. On the weekend, when he is playing football or reading a novel at home he might say he has a lot of choice. We repeated our daily sampling study, but this time added a question about the degree to which the subject felt they freely chose to be in that situation at each daily report. We found that personality showed much stronger effects on situational choice when examining those times where the person indicated feeling some choice for being in that situation. Personality exerts more of an effect on situational selection when the person has some freedom of choice.

But freedom of choice is always a matter of degree, and much of our daily lives are spent in situations in which we do not directly and explicitly choose to enter. We have commitments, duties, responsibilities to others, and these often remove the element of regulation through situational selection. This does not, however, mean that regulation stops. As Hettema points out, people interact with their environments by altering or manipulating the situation, even the ones they are forced to enter. Persons may directly evoke responses from their environments, thereby altering the situations in which they find themselves. In fact, personality maybe stable, in part, because people often re-create similar situations over time. For example, in a paper by Wachtel (1976), he discusses the case of a man who has difficulty maintaining relationships with women. The man explains that every woman he becomes involved with turns out, after awhile, to be a nagging, emotional, complainer. But Wachtel goes on to ask: "What does this man do to evoke the nasty side of every woman he encounters?" Wachtel draws on the psychoanalytic notion of the repetition compulsion to discuss the concept of situational manipulation. That is, some people repeatedly create or evoke the same kinds of problematic interpersonal situations with different persons in their lives, evoking in different people the same kinds of conflicts and interpersonal tensions. In a sense, even the psychoanalytic concept of transference refers to situational manipulation, where the patient re-creates past interpersonal problems in the current interpersonal situation.

These psychoanalytic concepts are useful to the extent that they provide examples of regulated (or disregulated) P by E interactions. As discussed by Hettema, there are many other forms of situational manipulation. For example, people cognitively transform situations. Differences in how people interpret situations may influence how they respond. Even when confronted with identical stimuli, people will often 'see' different things. For example, in some of my research (Larsen et al., 1987) we have demonstrated meaningful individual differences in how people interpret or cognitively manipulate stimuli. We had subjects look at photographic slide images of emotional scenes, such as unpleasant war scenes and pleasant scenes of children playing. We were interested in differences between high and low emotionally reactive persons, in terms of how they cognitively construe emotional stimuli. We asked subjects to write down what they were thinking about when they first looked at the slides. These thought listings were then coded along a number of dimensions by independent raters. We found that emotionally reactive persons (as determined by a questionnaire) tended to personalize the stimuli, whereas emotionally non-reactive persons remained more objective in their construals of the stimuli. For example, in response to a scene of a child with a severe head wound, one emotionally reactive person went on to describe how he once had such a wound, how painful it was for him, how frightened it made his mother, and so on. In response to the same slide image, a low emotional person went on to describe the vascular anatomy of the head and listed several facts to explain why head wounds bleed more than peripheral wounds. Clearly, these two persons did not 'see' the same stimuli; they cognitively transformed the objective stimuli such that it represented different subjective situations.

In a more recent study (Larsen, 1992), I examined how high neuroticism subjects transform information about their health situation. Neuroticism is consistently associated with inflated health complaints. In this study I found that high neuroticism subjects selectively attend to and encode minor aches and pains, and, when asked to recall their health status, they also selectively recall things as being worse than when they were actually experienced. So when you ask a high-N person how they have been feeling lately, you are likely to get information about their health situation that has been twice transformed; once by selective perception, and once by selective recall.

The regulation of situations is a key element in Hettema's perspective. He differs from other regulation-based theories precisely because his emphasis is on how people regulate and control their environments, NOT on how they control themselves or their internal states. Clearly, one's behavior, environment, and internal states are interrelated, such that if one is manipulated changes in the others are likely. Nevertheless, Hettema emphasizes the tactics and strategies for environmental control. Let me turn now to a discussion of tactics and strategies.

Tactics and Strategies of Person-environment Interaction

Regulation of the environment is achieved, according to Hettema, by tactics and strategies. Tactics are respondant acts evoked by a mis- match between the person's goals and the state of affairs. As respondant acts, tactics represent phasic reactions to events, much like emotions. Also like emotions, tactics serve the functions of interrupting behavior, directing attention, and initiating new behavioral action tendencies. Also like emotions, tactics are always evoked in response to the meaning attributed to events. The meaning of an event comes from how one represents that event vis-a-vis one's goals. Tactics are distinguished from emotions, however, in that tactics represent attempts to control the environment. Although tactics are autonomous and respondant, Hettema nevertheless stresses their functional role in terms of regulating the environment. So, for example, the tactic of exploration refers to seeking different modes of interacting with the environment. The tactic of redirection refers to changing goals when one goal is unattainable. Some tactics directly affect the situation, others affect how one thinks about the situation. In either case, the function of tactics is to achieve control over the situation.

One of the more intruiging aspects of Hettema's work is the incorporation of psychophysiological techniques in the study of tactics. One would not normally hope to differentiate such states as exploration and redirection using physiological measures. Nevertheless, using carefully designed films to elicit specific tactics, Hettema and colleagues have been able to discriminate tactical reactions using a multivariate battery of psychophysiological measures. Such measures are sensitive not only to emotional reactions, but also to cognitive activities such as whether one is shutting out or taking in information from the environment, whether one is processing information, and whether one is sustaining attention. By carefully considering cognitive and behavioral elements of each tactic, Hettema suggests that patterns of physiological response can be identified for each. Tactics thus represent the biologically-grounded phasic adaptations to disturbances between what the person wants and what the environment offers. Although biologically-based, tactics are responses to the MEANING of situations, which may be socially determined.

Strategies refer more to operant than respondant acts. Strategies are acquired competencies that function to transform situations in accordance with one's goals or projects. There is much recent work in personality psychology that emphasizes personal projects (Little, 1989) or life tasks (Cantor,1990) and the strategies that people use to reach their goals. Fulfilling one's personal projects involves controlling the environment. Hettema presents several studies showing that one strategic mechanism is the preference for selecting specific environments. Other studies will undoubtably explore other formal mechanisms whereby people pro-actively alter their environment such that outcomes become consistent with goals.

One assumption implicit in Hettema's view is that things always go well, that regulation is homeostatically achieved. We might wonder, though, about the possibility of dis-regulation and dis-order in the personality system. Clearly, not everyone is healthy, happy, social, and productive. Things sometimes go wrong in personality functioning, and it would be interesting for Hettema to speculate about the possible sources of personality disorders. Nevertheless, by providing us with a detailed characterization of the elements of regulation and adaptation, it would not be too difficult to identify the routes to dis-regulation and mal- adaptation.

Persons in Environments

Hettema's perspective on personality is, in the broadest sense, about relations between persons and environments. His concerns are with how persons adapt to and regulate their behavior in the face of demands from the environment. This stimulated me to think about the metaphor of ecology. As a science, "Ecology is the study of the interrelationships between organisms and their environment." (Emmel, 1973, p. 2). Isn't this what Hettema is talking about, how persons interrelate with, respond to, and modify their environments? Isn't he concerned with what individuals require from their environments, what they can and cannot tolerate from their environments, how they function within their environments? If so, might not some of the concepts, models, and methods of ecology prove useful in expanding and elaborating on Hettema's foundation?

Personality psychology borrows methods, concepts, and analytic strategies from many different disciplines. For example, evolution, genetics, and physiology are all disciplines that have been fruitfully drawn upon by personality psychologists. The science of ecology, however, has been relatively neglected by personality psychologists. Yet, if we are truly interested in the interrelationships between persons and their psychological environments, then perhaps ecology can provide us with a metaphor for conceptualizing these relationships. Let me turn to a few of the concepts and ideas of scientific ecology, and speculate about how they might be applied to the perspective on personality that Hettema provides.

Although there is a branch of evolutionary biology known as behavioral ecology, this discipline tends to ignore the psychological dimensions of persons and environments in mapping behaviors onto environmental characteristics. What I want to suggest is that we look at the psychological aspects of both persons and environments in conceptualizing the notion of psychological ecosystems. A psychological ecosystem would consist of the psychological resources, opportunities, as well as stressors present in the environment, combined with the psychological needs, motives, traits, abilities, and sensitivities of the persons present in that environment. The emphasis must be on the psychological aspects of both persons and environments. Then some of the ideas and principles of ecology might be employed to search for orderliness in the relation between components of the psychological ecosystem. For example, people do not distribute themselves randomly in environments. Situational orderliness exists in that there are different psychological habitats, each requiring different adaptive strategies, different psychological abilities, and satisfying different psychological needs. We might borrow the ecological concept of stratification to think about how different persons disperse within the situations available in an environment. The idea of situational selection discussed above might be thought of as a form of psychological stratification, and approached the way an ecologist approaches biological stratification. Instead of thinking about the stratification of species, we would think about the stratification of persons with different personalities, different abilities, different needs, etc. within the psychological ecosystem.

There are other analogies to be found within the ecological model. For example, there may be some psychological analogy to the food chain concept. This concept refers to orderliness in the metabolic resource exchange within an ecosystem. We might think about the transformation and exchange of psychological resources that occurs in a psychological ecosystem. Persons not only have a need for food, but they have psychological needs and motives as well. Environments not only provide food, but they offer opportunities for need satisfaction and other psychological resources as well. In a sense, psychological energy may be transformed and exchanged much like physical energy is transformed and exchanged in food chains. Additionally, we might think of strategies for extracting psychological resources from an environment as operating much like strategies developed for extracting food from the environment. Clearly, different animals have evolved different strategies and mechanisms for extracting nourishment from the environment. It seems likely that persons have also evolved different mechanisms and strategies for extracting psychological resources from the environment. Also, in the sense that some species are more efficient than others at extracting food from particular environments, it may be that some persons are more efficient than others at extracting psychological resources from particular environments. Like the ecologist studies specialized adaptive strategies, the personality psychologist, as Hettema suggests, might study different strategies for psychological transaction with the environment. What is it about some strategies that make them work better than others? In the sense that different feeding patterns lead to dispersal and distribution of species, it might be that different abilities for need satisfaction from the psychological environment lead to dispersal and distribution of persons.

Ecology has developed ways of describing environments as communities which, although differentiated, form complete systems. Ecological communities may be characterized in a variety of ways, such as species dominance, productivity, carrying capacity, and succession. Perhaps some of these concepts might be 'psychologized' and used to view psychological ecosystems. For example, do certain psychological environments show succession in the sense that changing conditions provide for replacements of some psychological resources with others.

The science of ecology has also developed ways for conceptualizing and assessing alterations of environments, whether intended or unintended. Clearly, people manipulate their physical environments and, as discussed above, they manipulate their psychological environments. Perhaps some concept like environmental impact might be useful in thinking about how persons manipulate or evoke psychological changes in various situations. For example, pollution of the physical environment may have its analogy in the destruction of a situation's ability to provide psychological resources. People may modify a situation to make it MORE stressful, as might occur in certain self- defeating neurotic behavior patterns.

Summary

In summary, Hettema links biological and social processes together in his theory of person-environment regulation. This regulation is achieved through tactics, which Hettema conceives as primarily biological, and strategies, which he conceives as primarily socially determined. Implicit in his view, though, is the notion that social and biological influences are present in both tactics and strategies and at every level and every time frame of adaptation. It seems likely that we will not achieve a psychological understanding of biological processes without considering social influences. For example, we will not understand a person's physiological response to some stimuli without knowing the meaning attributed to that stimuli. Similarly, we will not understand social learning processes without knowing the constraints imposed by biology. For example, some situations may foster particular forms of social learning, yet if a person's biological constitution provides them with an aversion to such situations, then social learning will be constrained for that person. By focusing on the interface of the person and the environment Hettema provides us with the beginings of a framework for integrating biological and social perspectives on individuality.

Author's note

Preparation of this manuscript was supported, in part, by Research Scientist Development Award KO1-MH00704 and grant RO1- MH42057 from the National Institute of Mental Health to Randy J. Larsen. Correspondence concerning this article should be addressed to Randy J. Larsen, Department of Psychology, 580 Union Drive, University of Michigan, Ann Arbor, MI 48109-1346 USA.

References

Cacioppo, J.T. and Petty, R. (1983). *Social psychophysiology: A sourcebook*. New York: Guilford.

Cantor, N. (1990). From thought to behavior: "Having" and "doing" in the study of personality and cognition. *American Psychologist, 45*, 735-750.

Diener, E., Larsen, R.J. and Emmons, R.A. (1984). Person x situation interactions: Choice of situations and congruence response models. *Journal of Personality and Social Psychology, 47*, 580-592.

Emmel, T.C. (1973). *An introduction to ecology and population biology*. New York: Norton.

Emmons, R.A., Diener, E. and Larsen, R.J. (1986). Choice and avoidance of everyday situations and affect congruence: Two models of reciprocal interactionism. *Journal of Personality and Social Psychology, 51*, 815-826.

Larsen, R.J. (1992). Neuroticism and selective encoding and recall of symptoms: Evidence from a combined concurrent-retrospective study. *Journal of Personality and Social Psychology, 62*, 480-488.

Larsen, R.J., Diener, E. and Cropanzano, R. (1987). Cognitive operations associated with individual differences in affect intensity. *Journal of Personality and Social Psychology, 53*, 767-774.

Little, B. (1989). Personal projects analysis: Trivial pursuits, magnificent obsessions, and the search for coherence. In D.M. Buss and N. Cantor (Eds.), *Personality psychology: Recent trends and emerging directions* (pp. 14-31). New York: Springer Verlag.

Wachtel, P. (1976). Psychodynamics, behavior therapy, and the implacable experimenter: An inquiry into the consistency of personality. In N.S. Endler and D. Magnusson (Eds.), *Interactional psychology and personality* (pp. 615-630). Washington, DC: Hemisphere Publishing Corp.

STRATEGIC INDIVIDUAL DIFFERENCES: THE ROLE OF PERSONALITY IN CREATING AND SOLVING ADAPTIVE PROBLEMS

David M. Buss,
Department of Psychology,
University of Michigan,
Ann Arbor, MI 48109-1346,
USA

Abstract

The science of individual differences and the science of human nature have remained separated since the days of Galton and Darwin. This paper proposes a model of strategic individual differences that integrates the two. The core of the model contains two essential propositions: (1) different individuals confront recurrently different adaptive problems, determined in part by their own personalities and by the personalities of significant others inhabiting their social environment; (2) individuals differ in the behavioral strategies they deploy to solve species-typical adaptive problems. I illustrate this model of strategic individual differences with an analysis of the problem of mate retention, and examine the role of personality in creating and solving facets of this adaptive problem. Discussion focuses on an evolutionary personality psychology that proposes new strategic units of analysis, integrates human nature and individual differences, and synthesizes "biological" and "social" determinants of personality functioning.

Introduction

Historical partitioning in psychology has separated the study of individual differences from the study of species-typical psychological functioning (cf. Cronbach, 1957). These separate lines may be traced back within "biological" approaches to Francis Galton (1865) who focused on individual differences and their possible heritable basis, and to Charles Darwin (1859) who concentrated on species-typical adaptive characteristics. The modern forms of this split include the fields of behavioral genetics and personality psychology, which concentrate heavily on individual variation, and evolutionary psychology, social psychology, and cognitive psychology, which tend to focus on species-typical mechanisms. These divisions have had two unfortunate consequences: (1) conceptually isolating the study of individual differences from theories of basic psychological functioning, and (2) isolating theories of psychological functioning from understanding the important role played by individual differences.

I believe that evolutionary personality psychology (Buss, 1991) provides a coherent theoretical framework for accomplishing the central goals of this conference - integrating human nature and individual differences and explicating "social" and "biological" determinants of

J. Hettema and I. J. Deary (eds.), Foundations of Personality, 175–189.

personality. This framework requires several key components, including: (1) *new units of analysis* for personality psychology, namely *evolved psychological mechanisms* and *behavioral strategies*; (2) placing personality in functional context; (3) understanding the role of personality in both creating and solving adaptive problems; (4) jettisoning misleading dichotomies such as "biological v. social" or "nature v. nurture" or "innate v. learned." I start with some fundamental assumptions about human nature and models of mind.

Why Our Basic Psychological Mechanisms Are Likely to be Species-Typical

Tooby and Cosmides (1990) articulate compelling arguments for why our basic psychological mechanisms are likely to be species-typical, shared by most or all humans. Essentially, all complex mechanisms require dozens, hundreds, or thousands of genes for their development. Sexual recombination, by shuffling genes with each new generation, makes it exceedingly unlikely that complex mechanisms could be maintained if genes coding for complex adaptations varied substantially between individuals. Selection and sexual recombination tend to impose relative uniformity in complex adaptive designs. This is readily apparent at the level of physiology and anatomy - all people have two eyes, a heart, a larynx, and a liver. Individuals can vary *quantitatively* in the strength of their heart or in the efficiency of their liver, but do not vary in their possession of the basic physiological mechanisms themselves (except by unusual genetic or environmental accident). This suggests that individual differences, including heritable individual differences, are unlikely to represent differences in the presence or absence of complex adaptive mechanisms. Individual differences cannot be understood apart from human nature mechanisms, any more than differences in the turning radius and stopping ability of cars can be understood apart from the basic car-nature mechanisms such as steering wheels and brakes.

The Nature of Human Psychological Mechanisms

A long-standing dogma in this century's social science has been that the nature of humans is that they have no nature (except perhaps a few basic drives and a few domain-general learning mechanisms). Evidence has been accumulating over the past decade that this view is untenable empirically (Brown, 1991; Buss, 1991). Conceptual analyses by scientists in artificial intelligence, psycholinguistics, cognitive psychology, and evolutionary psychology are showing why such a view is untenable theoretically even in principle (Tooby and Cosmides, 1992). Humans could not possibly perform the numerous, complex, situationally-contingent tasks they do routinely without considerable intricate and domain-dedicated psychological machinery. These psychological mechanisms, coupled with the adaptive problems they were "designed" to solve, coupled with the social, cultural, ecological, and internal inputs that reliably activate them, provide a starting point for a description of human nature.

Although determining exactly which couplings are part of human nature must be determined empirically, possible candidates that have emerged empirically over the past several decades of research include: *childhood fears of loud noises, darkness, snakes, spiders, and strangers; characteristic emotions such as anger, envy, passion, and love; characteristic facial expressions such as happiness and disgust; competition for limited resources; competition for desirable mates; specific mate preferences; classification of kin; love of kin; preferential altru-*

ism directed towards kin; play; deceit; concepts of property; enduring reciprocal alliances or friendships; enduring mateships; temporary sexual relationships; retaliation and revenge for perceived personal violations; sanctions for crimes against the group or its members; rites of passage; concepts of self; concepts of intentions, beliefs, and desires as part of a theory of mind; status differentiation; status striving; prestige criteria; psychological pain upon loss of status or reputation; humor; gender terminology; division of labor by gender; sexual attraction; standards of sexual attractiveness; sexual jealousy; sexual modesty; tool making; tool use; tools for making tools; weapon making; weapon use; coalitions that use weapons for warfare; collective identities; cooking and fire use; and probably hundreds more (see Brown, 1991, for an extended list of possibilities).

Since the cognitive revolution, psychologists have become increasingly aware of the necessity for understanding decision-making rules and other information processing devices inside people's heads. Although psychologists have largely jettisoned behaviorism's unworkable black box anti-mentalism, many have retained (perhaps inadvertently) the behavioristic assumption of equipotentiality: They assume that cognitive mechanisms are general-purpose and free of content-specialized procedures (Tooby and Cosmides, 1992).

Evolutionary psychologists, in contrast to the dominant social science dogma, argue that evolved psychological mechanisms cannot be solely general-purpose, must be saturated with content, and must operate differently in response to contextual input signalling different adaptive problems. Just as the body contains a large number of specific and dedicated physiological mechanisms (taste buds, sweat glands, lungs, heart, kidneys, larynx, pituitary gland), so according to evolutionary psychologists the mind must contain a large number of specialized psychological mechanisms, each "designed" to solve a different adaptive problem. Because what constitutes a "successful solution" to adaptive problems differs across domains -criteria for successful food selection, for example, differ from criteria for successful mate selection- the requisite psychological solution mechanisms are likely to be special-purpose and domain-dedicated.

Detecting Evolved Psychological Mechanisms: Identifying The Key for the Lock

Psychological mechanisms are usefully regarded as evolved solutions to adaptive problems. Analogy to the human body is useful. We have sweat glands and shivering mechanisms that solve problems of thermal regulation; callous producing mechanisms that solve the problem of repeated friction to the skin; taste preferences that solve the problem of what substances to ingest. Standards for inferring that these mechanisms are solutions to adaptive problems include *economy, efficiency, complexity, precision, specialization*, and *reliability* (Tooby and Cosmides, 1992; Williams, 1966). Mechanisms that solve adaptive problems are like keys that fit particular locks. The efficiency, detail, and complex structure of the key must mesh precisely with the inner "problem" posed by the lock.

Evolutionary analysis of psychological mechanisms proceeds in two directions - form-to-function and function-to form (Tooby and Cosmides, 1992). Imagine that one found a key, but did not know which of the thousands of possible locks it might fit. Its size, its shape, its details might suggest tentative hypotheses and rule out others. It may be too large to fit some locks, yet too small to fit others. The shape of its tines must have a corresponding mirror-image shape in the internal workings of the lock. Eventually, through an iterated process of hypothesis generation and empirical testing, we might eventually discover the exact lock that the key was designed to fit. The precision, reliability, and specialization with which a particular key fits a particular lock provide the researcher with reasonable standards for inferring that a particular key was *designed* to fit a particular lock.

Alternatively, one might identify a lock (adaptive problem), and then search for a key that might fit it. Here, the same standards would apply - precision, efficiency, complexity of design. The "bottom line" is whether the key one discovers (adaptive mechanism proposed) actually fits the lock (solves the adaptive problem with reasonable precision, efficiency, and reliability), and that alternative hypotheses about its origin (e.g., incidental by-product of some other adaptation) and function (other adaptive problems the mechanism might solve, or other mechanisms capable of solving the adaptive problem) can be reasonably ruled out.

Evolutionary psychologists proceed in both directions, form-to-function and function-to-form. Sometimes a phenomenon or form is discovered -fever, fear of snakes, male sexual jealousy, mate preferences for "kindness"- and researchers generate and test hypotheses about its function. Often, there are competing functional theories about the same phenomenon, and these may be pitted against each other in critical empirical tests (see Buss, 1990, for an example of alternative evolutionary hypotheses for the female orgasm). This method is sometimes derided as telling "just-so stories," but it is an essential process of science. The discovery of three-degree black body radiation sent astronomers scrambling for cosmological theories or "stories" to explain it. The discovery of continental drift send geologists scrambling for a theory such as plate tectonics that could explain it. The power of a theory rests with its ability to explain known facts and to generate new predictions which are then subjected to empirical test. Specific evolutionary psychological theories should be evaluated by these rigorous scientific standards. Some will pan out. Others will be jettisoned on conceptual or empirical grounds.

Evolutionary analysis provides psychologists with a powerful heuristic, guiding them to important domains of adaptive problems and guiding the development of hypotheses about adaptive mechanisms heretofore unobserved. Because fertilization and gestation occur internally within women, for example, an adaptive problem for ancestral men would have been ensuring confidence in their paternity. Men who were indifferent to this adaptive problem were less likely to become our ancestors. Identifying this adaptive problem has led evolutionary psychologists to search for adaptive solutions in psychological mechanisms such as mate preferences for chastity, fidelity, and faithfulness (Buss, 1989; Buss and Schmitt, in press) and mechanisms involved in male sexual proprietariness such as sexual jealousy (Symons, 1979; Daly et al., 1982; Buss et al., 1992). Function-to-form and form-to-function are both viable methods for discovering our evolved psychological mechanisms. These evolved psychological mechanisms form the foundation for the analysis of individual differences.

Sex Differences as One Class of Strategic Individual Differences

To get from human nature psychological mechanisms to the analysis of individual differences, it is useful to go through an intermediate step - the analysis of sex differences, which may be regarded as one class of individual differences. Evolutionary psychology provides a unique meta-theory for predicting when we should and should not expect sex differences. Men and women are expected to differ only in the delimited domains where they have faced recurrently different adaptive problems (1) over human evolutionary history, (2) during their development, or (3) over different current environments inhabited. In domains where the sexes have faced the same adaptive problems, no sex differences are expected.

Men and women historically have faced many adaptive problems that are highly similar. Both sexes needed to maintain body temperature, so both sexes have sweat glands and shivering mechanisms. Repeated friction to certain areas of the skin was a "hostile force of nature" to both sexes in ancestral environments, so men and women have evolved callous-producing mechanisms. Both sexes needed to solve the adaptive problem of identifying a good cooperator for strategic confluence when seeking a long-term mate, and this may be one reason why both sexes value "kindness" in a partner so highly across all cultures whose partner preferences have been studied (Buss, 1989).

In several domains, however, the sexes have faced different adaptive problems. For 99% of human evolutionary history men faced the adaptive problem of hunting and women of gathering, possible selective reasons for men's greater upper body strength and spatial rotation ability and for women's greater spatial location memory ability (Silverman and Eals, 1992). Internal female fertilization and gestation created the adaptive problem of uncertainty of paternity for men, but not uncertainty of maternity for women. Cryptic ovulation created the adaptive problem for men of knowing when a women was ovulating (Alexander and Noonan, 1979). The dual male mating strategy of seeking both short-term sexual partners and long-term marriage partners created an adaptive problem for women of having to discern whether particular men saw them as temporary sex partners or as potential spouses (Buss and Schmitt, in press). Sex differences in mate preferences (Buss, 1989), courting strategies (Buss, 1988; Tooke and Camire, 1991), jealousy (Buss et al., 1992), and sexual fantasies (Ellis and Symons, 1990) correspond remarkably well to these sex-linked adaptive problems. Evolutionary psychology offers the promise of providing a coherent theory of strategic sexual differences as well as strategic sexual similarities.

Strategic Individual Differences Caused By Individuals Confronting Different "Environmentally Induced" Adaptive Problems

The construction workers who are laboring on the building next door have thick callouses on their hands. My academic colleagues down the hall do not. These individual differences in callous thickness are highly stable over time. At one level of analysis, the variance can be traced solely to variance in the reliably recurring experiences of the two groups. At another level of analysis, the existence of the species-typical callous-producing mechanism is a central and necessary element in the causal explanation of observed individual differences. *Just as men and women differ in the adaptive problems they confront, different individuals within each sex face different adaptive problems over time.* Some manifest individual differences are the strategic products of species-typical mechanisms responding to recurrently different adaptive problems across individuals.

Recurrent adaptive problems can also be *socially imposed.* Consider the adaptive problems one confronts by being married to a highly Agreeable mate versus one who is highly Disagreeable. Disagreeable persons may impose on their spouses the recurrent adaptive problem of *verbal and physical abuse* (Buss, 1991), and subject their spouses to the manipulation tactics of *coercion* (e.g., He demanded that she do it) and the *silent treatment* (e.g., He did not respond to her until she did it) (Buss, 1992). Or consider Conscientiousness. In a four-year longitudinal study, we found evidence that persons low on Conscientiousness impose on their spouses recurrent acts of infidelity (Buss, 1991). Thus, the personality characteristics of one's mate can create socially imposed adaptive problems that recur over time. Stable individual differences in manifest behavior such as jealousy may flow from the individually different recurrent problems to which people are subjected.

In these examples, the individual differences are in some sense "environmental." If my academic colleagues were to trade places with the construction workers, then the manifest individual differences would reverse. If the person married to the high Conscientious mate were to trade mates with the person married to the low Conscientious mate, then the manifest individual differences in experiencing jealousy presumably would also reverse.

Nonetheless, we cannot rule out the genotype-environmental correlation processes proposed by Plomin et al. (1977) and Scarr and McCartney (1983). Some individuals, because of heritable skills, interests, or proclivities may preferentially select academic work or construction work as occupations, or high Conscientious or low Conscientious persons as mates. These selections, in turn, may create repeated exposure to friction-free or infidelity-free versus friction-prevalent or infidelity-prevalent environments, which then differentially activate the relevant species-typical mechanisms.

There are four central points in my argument thus far: (1) stable manifest individual differences can be caused by differences in the *recurrent adaptive problems* to which different individuals are exposed; (2) the personality characteristics of others inhabiting one's social environment play a critical role in determining the adaptive problems to which one is exposed; (3) the complex *species-typical mechanisms* are necessary and central ingredients in the causal explanation of individual differences because without them, the observed individual differences could not occur; and (4) the manifest individual differences are *strategic outcomes* of recurrently different input into species-typical mechanisms.

There are undoubtedly many recurrent environmental individual differences of precisely this sort. First-born children probably face recurrently different adaptive problems compared with later born children. These problems apparently trigger in first-born greater identification with the status quo, the parents, and the established scientific theories, and in later-born greater rebellion and identification with revolutionary scientific theories (Sulloway, pers. comm). Later born children apparently have less to gain by identifying with a niche that is already occupied by an older sibling.

Individuals who grow up in environments where resources are unpredictable, such as inner city ghettos, may adopt a more impulsive personality style where it would be adaptively foolish to delay gratification (Buss, 1990). In contrast, those growing up in middle class suburbs where resources and future prospects are more predictable may adopt a personality strategy involving greater delay of gratification. The resulting individual differences represent strategic solutions to the different adaptive problems encountered.

A third example involves having a long-term mate who recurrently imposes an adaptive problem on their spouse. Some mates, for example, may recurrently show signs of sexual infidelity. These signs may trigger or activate the "sexual jealousy" mechanism, which in turn may lead to recurrent tactics of mate retention. The central point of all these examples is that recurrently different social or environmental input into species-typical mechanisms can produce stable strategically patterned individual differences.

Strategic Individual Differences Caused By Individuals Confronting Different "Heritably Induced" Adaptive Problems

Recurrently different input into species-typical psychological mechanisms, of course, may come from heritable individual differences, whatever their ultimate origin (i.e., whether they originated from selection for alternative genetically based strategies, frequency-dependent selection, genetic noise, pathogen-driven selection for genetic uniqueness, or assortative mating). Individuals with an ectomorphic body type, for example, confront different adaptive problems than those who are mesomorphic. Ectomorphs may risk being at the receiving end of greater aggression than their more muscular peers, an adaptive problem that typically must be solved by means other than physical aggression. Genetic differences, in other words, pose different adaptive problems for different individuals.

Heritable dimensions of individuals -such as differences in body type, keenness of vision, oratory skills, physical attractiveness, and spatial ability- provide important input into species-typical mechanisms. These individually different inputs tell the organism about the adaptive problem it is facing. The resulting product consists of *strategic individual differences* that are stable over time. The observed strategic differences are *correlated* with genetic variance, but cannot be understood apart from the central role played by our species-typical psychological mechanisms that were "designed" to receive input -both environmentally and heritably based- about the adaptive problems confronted.

Strategic Individual Differences in Solutions to Adaptive Problems

In addition to facing different adaptive problems, some individuals experience *greater success* at pursuing certain strategies rather than others: "Selection operates through the achievement of adaptive goal states, and any feature of the world -either of the environment, *or of one's own individual characteristics*- that influences the achievement of the relevant goal state may be assessed by an adaptively designed system" (Tooby and Cosmides, 1990, p. 59; emphasis added). Individuals who are mesomorphic, for example, typically will experience far greater success at enacting an aggressive strategy than individuals who are ectomorphic (Tooby and Cosmides, 1990, call this phenomenon "reactive heritability").

Individual differences in physical attractiveness provide another example. These is evidence that physically attractive men are better able to successfully pursue a "short-term" mating strategy involving many sexual partners (Gangestad and Simpson, 1990). Physically attractive women are better able to pursue a long-term strategy of seeking and actually obtaining higher-status higher-income marriage partners (Taylor and Glenn, 1976). Relative physical attractiveness functions as "input" into species-typical or sex-typical psychological mechanisms, which then canalize the strategic solutions of different individuals in different directions.

The personality characteristics represented by the "big five" may represent (in part) individual differences in the resources individuals can draw upon to solve adaptive problems. The individual high on Surgency may be able to deploy socially dominant solutions. The person high on Agreeableness may be successful at eliciting cooperation from others in solving adaptive problems. The highly Conscientious person may solve adaptive problems through discipline, industry, and sheer hard work. The Emotionally Stable person may rely upon steadiness of nerves, inner resiliency, and the capacity to rally from setback to solve adaptive problems.

The person high on Intellectance may be adept at deploying creative cognitive solutions to adaptive problems.

In sum, this framework proposes a key role of personality in creating and solving adaptive problems: (1) personality characteristics can play a causal role in determining the adaptive problems to which one is exposed; (2) the personality characteristics of people inhabiting one's social environment can play a causal role in imposing particular problems; (3) personality characteristics influence the strategic solutions that people deploy to solve adaptive problems they confront.

I believe that the theoretical and empirical work subsumed by this general framework will have to be explored domain by domain, adaptive problem by adaptive problem (Tooby and Cosmides, 1992). A theory about how individuals differing in physical attractiveness encounter different adaptive problems and have different success with the enactment of some strategies may tell us little about the strategic consequences of individual variation in body type, oratory skills, or spatial ability. Ultimately, of course, it will be useful to integrate domain-specific theories into a more general theory of strategic individual differences.

In this brief paper, all I can do is provide one illustration of how this form of analysis might proceed. The example I use is the adaptive problem of mate retention, the complex psychological mechanisms such as jealousy involved in mate retention, and the individually different tactics people use to deal with this adaptive problem.

The Evolutionary Psychology of Jealousy - Strategic Sex Differences

Jealousy is neither a peripheral nor trivial emotion, for it is experienced in all known cultures and is the leading cause of spousal battering and homicide worldwide (Daly and Wilson, 1988). Why do humans experience jealousy? Do the sexes differ in the events that elicit jealousy? What contexts activate jealousy? And how can individual differences in jealousy *within sex* be accounted for?

Jealousy is a cognitive-emotional-motivational complex that is activated by threat to a valued relationship. It is considered "sexual jealousy" if the relevant relationship is a sexual one, but there are types of jealousy that do not involve sexual threat. Jealousy is often activated by cues to the apparent loss of key resources provided by a relationship - cues such as eye contact between one's partner and a rival, decreased sexual interest on the part of one's partner, and an increase in partner's in flirting with same-sex competitors. Jealousy channels attention, calls up relevant memories, and activates strategic cognitions. It may motivate actions designed to reduce or eliminate the threat, retain the valued relationship, and hence retain the valued resources it provides.

Because both men and women over evolutionary history have been damaged by relationship loss, both sexes have faced adaptive problems to which jealousy may have evolved as one solution. Several evolutionary psychologists have predicted that the sexes will differ in the weighting given to events that activate jealousy (Daly et al., 1982; Symons, 1979). Because fertilization and gestation occur internally within women and not men, over evolutionary history men have faced an adaptive problem not shared by women - paternity uncertainty. The reproductive threat to a man comes from the possibility of *sexual* infidelity by his partner.

In species such as ours, a woman's certainty in genetic parenthood would not have been compromised if her partner had sex with other women. Women, however, may have risked the loss of their partner's time, attention, commitment, protection, investment, and resources. This would come as a double blow, because her loss would also be an intrasexual competitor's gain if the resources were diverted from her and her children toward another woman and unrelated children. For these reasons, evolutionary psychologists have predicted that the inputs that activate jealousy for men will be biased toward cues that relate to the sex act *per se*, whereas for women they will be more biased by cues to the loss of commitment and investment from a man.

Consider this question: *What would upset or distress you more: (A) Imagining your mate having sexual intercourse with someone else, or (B) Imagining your mate forming a deep emotional attachment to someone else?* In a series of studies, we found that the overwhelming majority (85%) of women to whom this dilemma was posed found emotional infidelity to be more distressing; the majority of men (60%) reported that sexual infidelity would be more distressing (Buss et al., 1992). These sex differences were also observed in physiological arousal in response to imagining the two different scenarios. In measures of heart rate, electrodermal activity, and electromyographically recorded frowning, men showed greater physiological arousal to imagined sexual infidelity than to emotional infidelity. Women, in contrast, tended to become more physiological aroused by imagined emotional infidelity than to sexual infidelity (Buss et al., 1992). These results support the hypothesis that men's and women's psychological and physiological mechanisms are tailored to differences in adaptive problems.

Strategic Individual Differences: An Illustration Using the Adaptive Problem of Spouse Infidelity

If sex differences in jealousy are strategically patterned, are individual differences within sex also strategically patterned? I restrict my attention to two forms of strategic individual differences: (1) Do some individuals predictably experience the adaptive problem of mate infidelity more than others by virtue of (a) their own personality characteristics, and (b) the personality characteristics of their spouses? (2) Do different individuals deploy predictably different mate retention tactics, in part determined by their own personalities?

The role of personality in creating adaptive problems. To examine the role of personality in the creation of adaptive problems, we conducted a longitudinal study of 100 married couples. During their newlywed year, we assessed five major factors of personality through parallel instruments from three data sources - self-report, spouse-report, and independent interviewer reports. Four years later, subjects completed a battery of instruments that included one called "Sources of Irritation and Upset" that contained 150 previously nominated things that a member of the opposite sex could do that might irritate, anger, annoy, or upset someone. Previous factor analyses of this instrument yielded 15 major sources of problems, including a cluster labeled "Infidelity." The Infidelity factor contained the following related complaints: *He/she saw someone else intimately; He/she had sex with another person; He/she was unfaithful to me; He/she went out with another person.* Table 1 shows the correlations between personality characteristics assessed during the newlywed year and reports from the spouse of infidelity four years later.

Low Conscientious men and women, as predicted, tend to inflict this adaptive problem on their spouses more than men and women higher in Conscientiousness. An unexpected finding was that women high on Openness-Intellect tended to inflict infidelity on their spouses. Also shown in Table 1 are the links between personality of the spouse and the creation of other problems, such as abuse, insults, neglect, and inconsiderateness. These results suggest that the personality characteristics of significant others inhabiting one's social milieu play a key role in creating adaptive problems, in this case their mates.

Table 1. Personality and Problems Reported By Spouse

Problem Reported by Wife - Time 2	*Personality Characteristics of Husbands - Time 1*				
	Surgency	**Agree.**	**Conscient.**	**Emo. Stab.**	**Open.**
Abuse	.13	**-.36*****	-.15	**-.26****	-.13
Insults	**.20***	**-.37*****	-.06	-.13	.09
Neglect	.06	**-.25****	**-.30*****	.04	-.01
Infidelity	.04	-.02	**-.17***	-.13	.02
Moodiness	.09	-.13	-.10	**-.45*****	-.04
Total Problems	.13	**-.34*****	**-.24****	**-.28****	-.04
Problem Reported by Husband - Time 2	*Personality Characteristics of Wife - Time 1*				
	Surgency	**Agree.**	**Conscient.**	**Emo. Stab.**	**Open.**
Abuse	.09	**-.38****	**-.24***	**-.23***	-.04
Insults	.17	**-.29***	-.19	-.13	.04
Neglect	**.23***	-.09	**-.40*****	.06	.06
Infidelity	.12	.10	**-.31****	.01	**.34****
Moodiness	**.35****	**-.29***	**-.45*****	-.17	.16
Total Problems	.18	**-.35****	**-.40*****	-.19	.06

***p <.001; **p <.01; *p <.05

Are some people exposed to the problem of spousal infidelity because of their own personality? To answer this question, we correlated personality characteristics of persons with the degree to which they complained about spousal infidelity. *Submissive* men and women -those low on Surgency- tended to complain that their spouses were unfaithful more that those higher on Surgency. These findings are correlational, so obviously no firm conclusions can be drawn about causality. But they do suggest that submissive people may be more at risk of encountering the problem of spousal infidelity; and marrying a mate low on Conscientiousness may put one at risk of incurring this adaptive problem.

The role of personality in solving adaptive problems. Previous research has identified 19 distinct tactics that people use to retain their mates - tactics ranging from *vigilance* (e.g., he kept a close eye on her at the party) to *violence* (e.g., he hit a rival who was making moves on her)(Buss, 1988). We assessed the use of these tactics in the same sample of couples at two time periods (newlywed year and forth year of marriage) using two data sources (self-report and spouse-report). Table 2 highlights some of the links between personality characteristics and the tactics that people use to retain their mates.

Table 2. Five Personality Factors and Mate Retention Tactics

Factors	*High*	*Low*
Surgency	Resource Display	Debasement
Agreeableness	Display Love	Derogate Mate
Conscientiousness	Possessive Ornamentastion	Threaten Infidelity
Emotional Stability	Physical Signals of Possession	Derogate Competitor
Intellect-Openess	Affection and Sex	Threaten Violence

Men high on Surgency tend to retain their wives by frequent acts of *Resource Display - He spent a lot of money on her; He bought her an expensive gift; He took her out to a nice restaurant.* Men low on Surgency tended to use *Debasement* as a mate retention tactic - *He told her that he would change in order to please her; He became a "slave" to her; He gave in to her every wish.* Men high on Agreeableness tend to *Display Love and Care - He told her that he loved her; He went out of his way to be kind, nice, and caring; He was helpful when she really needed it.* In contrast, men low on Agreeableness tended to *Derogate their Mate - He told other guys terrible things about her so that they wouldn't like her; He told other guys that she was not a nice person; He told other guys that she was stupid.*

Men low on Conscientiousness tend to *Threaten Infidelity - He flirted with another woman in front of her; He went out with other women to make her jealousy.* Men low on Emotional Stability tend to *Derogate Competitors - He cut down the appearance of other males; He told her the other guy was stupid.* Men low on Intellect-Openness tend to *Threaten Violence - He yelled at other guys who looked at her; He stared coldly at the other guy who was looking at her; He threatened to hit the guy who was making moves on her.*

These findings are correlational, so we cannot draw conclusions about causality. They do suggest, however, that personality characteristics such as the Big Five are linked with the alternative tactics that men use to solve the problem of mate retention. Personality traits, as traditionally assessed, are linked in coherent ways with the tactics people use to accomplish goals and solve adaptive problems. An essential part of personality, in other words, consists of the recurrent strategies people use to solve adaptive problems.

Implications

Taken together over these studies, the evidence shows promise for an analysis of strategic individual differences that conjoins: (1) species-typical psychological mechanisms as solutions to adaptive problems (e.g., mechanisms of psychological jealousy adapted to the problems of infidelity and relationship loss); (2) individually different input into these mechanisms (e.g., differences in likelihood that spouse shows cues to infidelity), (3) individual differences in strategies used to solve species-typical adaptive problems; (4) a key role played by personality in both creating and solving adaptive problems. We have the beginnings of a model of strategic individual differences in one domain and one class of adaptive problems.

Units of Analysis - Psychological Mechanisms and Attendant Behavioral Strategies

The evolutionary psychological framework proposed here suggests two related units of analysis for modern personality psychology, evolved psychological mechanisms and behavioral strategies. These mechanisms are the processes inside the heads of people that (1) exist in the form they do because they (or other mechanisms that reliably produce them) solved specific problems of individual survival or reproduction; (2) take only certain classes of internal input; (3) specify to the organism the particular adaptive problem it is facing; (4) transform that information into output through procedures (e.g., decision ruled), which (5) produce physiological activity, manifest action, or information to other mechanisms, and (6) solve a particular adaptive problem (Buss, 1991, p. 464).

This definition undoubtedly over-simplifies an extraordinarily complex set of processes. The mate retention mechanisms discussed above, for example, would probably include a large number of contingent decision rules that gauged the probability of spouse defection, weighted the resources available, identified characteristics of the rival, assigned likely effectiveness values to the alternative mate retention tactics, and so on. This definition also highlights the fact that psychological mechanisms cannot be properly understood without identifying the manifest behavioral solutions that are the *raison d'etre* of the internal machinery.

These new units of analysis contribute to a growing trend in the field toward goal-oriented units (Pervin, 1989), such as the work on life tasks and personal strategies (Cantor, 1990), personal projects (Little, 1989), and personal strivings (Emmons, 1989). The current units differ from those of other personality approaches in placing these goal-oriented units in functional or adaptive context by viewing them as part of the evolved problem-solving machinery of our species.

Integrating "Biological" and "Social" Determinants of Personality

The term "biology" means "the study of life and living processes." Thus, all studies of human psychology are by definition "biological." According to the current evolutionary psychological framework, contrasting "biological" with "social" determinants is a confusing and misleading way to understand the causal processes of personality. Our framework rejects this dichotomy, along with other false dichotomies such as "nature/nurture" or "innate/learned" (see Tooby and Cosmides, 1992). We require more refined concepts to understand the precise nature of human causal processes.

First, psychological mechanisms, as defined above, are (a) evolved, (b) species-typical, and (c) instantiated in neurological or physiological substrates, much as my word-processing program is instantiated on machine language on IBM hardware. Complete understanding of evolved mechanisms, of course, will require identifying psychophysiological substrates (e.g., Stelmack, 1990; Zuckerman, 1990). No matter how much progress is made in understanding these "wetware" foundations, however, it is my view that the "psychological level of description" (in information-processing or other terms) will always be a useful and even necessary level of analysis and description. Thus, if the term "biological" is retained at all in our professional vocabulary, it is incumbent upon users of the term to specify what sort of "biology" is being referred to (e.g., heredity, adaptation, physiology, psychological mechanism). My preference would be to do away with the term "biology" altogether, except when used to refer to that branch of science that deals with life processes.

A second key point is that many of the most important adaptive problems that humans confronted over evolutionary history were inherently "social." Humans are an intensely group-living species. Existing evidence suggests that humans evolved in small bands ranging in likely number from 50 to 100 (Tooby and DeVore, 1987). Group-living undoubtedly conferred survival and reproductive benefits above and beyond solitary living. Groups afford protection from predators, and perhaps more important, protection from other groups of marauding males. Groups afford the possibility of cooperative hunting of large game, which could not be hunted alone. Group living, therefore, was involved in solving several critical survival problems caused by food shortages, predators, and conspecific aggressors.

Groups also contain unusual concentrations of resources that must have played a key role in reproductive success - mates, dyadic reciprocal allies, cooperative coalitions, and kin (Buss, 1991). It is a telling fact about human evolutionary psychology that one of the most devastating forms of punishment is ostracism, being cast out from the group (Gruter and Masters, 1986; Zippelius, 1986). Other people have become an integral part of our survival and reproductive strategies.

With group living, however, came an intensification of competition. Because natural selection operates by *differential* reproductive success, *relative* access to resources became pivotal. Desirable mates, for example, were always in short supply, and a selective advantage accrued to those who were able to secure the best (Buss, 1989). On the mating market, one person's gain is another's loss.

Other humans, therefore, can cripple our survival and reproductive success. Other humans are our primary source of strategic interference, as well as our primary source of strategic confluence. Other humans are our primary "hostile force of nature" (Alexander, 1987). Other humans are one of the primary "environments" that selected for our psychological mechanisms and behavioral strategies. Other humans define our adaptive landscape (Buss, 1989b).

Evolutionary personality psychology, therefore, must be inherently "social" in two senses (see also Kenrick, in press). First, many of our mechanisms and strategies have evolved to deal with social adaptive problems involved in key social relationships such as mates, reciprocal alliances, and members of the hierarchy. Second, social input is necessary for the activation of many of our evolved psychological mechanisms.

Evolutionary personality psychology also is inherently "biological" in the sense that it deals with living processes that owe their existence to evolution by natural and sexual selection. Thus, evolutionary personality psychology offers the possibility of eliminating the false dichotomy between "social" and "biological" determinants of personality, and providing an integrative framework for understanding human functioning. Perhaps through this integration, we can start bridging the traditions that historically have isolated the study of individual differences from the study of human nature since the days of Galton and Darwin.

References

Alexander, R.D. (1987). *The biology of moral systems.* New York: Aldine.

Alexander, R.D., and Noonan, K. (1979). Concealment of ovulation, parental care, and human social evolution. In N.A. Chagnon and W. Irons (Eds.), *Evolutionary Biology and Human Social Behavior.* North Scituate, MA: Duxbury Press. Pp. 402-435.

Brown, D.E. (1991). *Human Universals.* Philadelphia, PA: Temple University Press.

Buss, D.M. (1988). From vigilance to violence: Mate retention tactics. *Ethology and Sociobiology, 9,* 291-317.

Buss, D.M. (1989a). Sex differences in mate preferences: Evolutionary hypotheses tested in 37 cultures. *Behavioral and Brain Sciences, 12,* 1-49.

Buss, D.M. (1989b). *A theory of strategic trait usage: Personality and the adaptive landscape.* Paper delivered at the Invited Workshop on Personality Language, University of Groningen, The Netherlands.

Buss, D.M. (1990). Evolutionary social psychology: Prospects and pitfalls. *Motivation and Emotion, 14,* 265-286.

Buss, D.M. (1991a). Conflict in married couples: Personality predictors of anger and upset. *Journal of Personality, 59,* 663-688.

Buss, D.M. (1991b). Evolutionary personality psychology. *Annual Review of Psychology, 42,* 459-491.

Buss, D.M. (1992). Manipulation in close relationships: Five personality factors in interactional context. *Journal of Personality, 60,* 477-499.

Buss, D.M., Larsen, R., Westen, D. and Semmelroth, J. (1992). Sex differences in jealousy: Evolution, physiology, and psychology. *Psychological Science, 3,* 251-255.

Buss, D.M. and Schmitt, D. (in press). Sexual Strategies Theory: The evolutionary psychology of human mating. *Psychological Review.*

Cantor, N. (1990). From thought to behavior: "Having" and "doing" in the study of personality and cognition. *American Psychologist, 45,* 735-750.

Cronbach, L.J. (1957). The two disciplines of scientific psychology. *American Psychologist, 12,* 671-684.

Cosmides, L. (1989). The logic of social exchange: Has natural selection shaped how humans reason? Studies with the Wason selection task. *Cognition,* 31, 187-276.

Daly, M., Wilson, M. and Weghorst, S.J. (1982). Male sexual jealousy. *Ethology and Sociobiology, 3,* 11-27.

Daly, M. and Wilson, M. (1988). *Homicide.* New York: Aldine.

Darwin, C. (1859). *The origin of species.* London: Murray.

Ellis, B.J. and Symons, D. (1990). Sex differences in sexual fantasy: An evolutionary psychological approach. *Journal of Sex Research, 27,* 527-556.

Emmons, R.A. (1989). Exploring the relations between motives and traits: The case of narcissism. In D.M. Buss and N. Cantor (Eds.), *Personality Psychology: Recent Trends and Emerging Directions.* New York: Springer Verlag. Pp. 32-44.

Galton, F. (1865). Hereditary talents and character. *Macmillan's Magazine, 12,* 157-166, 318-327.

Gangestad, S.W. and Simpson, J.A. (1990). Toward and evolutionary history of female sociosexual variation. *Journal of Personality, 58,* 69-96.

Gruter, M. and Masters, R.D. (1986). Ostracism as a social and biological phenomenon: An introduction. *Ethology and Sociobiology, 7,* 1-10.

Kenrick, D.T. and Keefe, R.C. (1992). Age preferences in mates reflect sex differences in human reproductive strategies. *Behavioral and Brain Sciences*, 15, 75-133.

Kenrick, D.T. (in press). Evolutionary social psychology. In L. Berkowitz (Ed.), *Advances in Experimental Social Psychology*. New York: Academic Press.

Little, B.R. (1989). Personal projects analysis: Trivial pursuits, magnificent obsessions, and the search for coherence. In D.M. Buss and N. Cantor (Eds.), *Personality Psychology: Recent Trends and Emerging Directions*. New York: Springer-Verlag, pp. 15-31.

Pervin, L. (Ed.). (1989) *Goal concepts in personality and social psychology*. Hillsdale, NJ: Erlbaum.

Plomin, R., DeFries, J.C. and Loehlin, J. (1977). Genotype-environment interaction and correlation in the analysis of human behavior. *Psychological Bulletin, 84*, 309-322.

Scarr, S. and McCartney, K. (1983). How people make their own environment: A theory of genotype --> environment effects. *Child Development, 54*, 424-435.

Silverman, I. and Eals, M. 1992. Sex differences in spatial abilities: Evolutionary theory and data. In J. Barkow, L. Cosmides and J. Tooby (Eds.), *The Adapted Mind*. New York: Oxford University Press.

Stelmack, R.M. (1990). Biological bases of extraversion: Psychophysiological evidence. *Journal of Personality, 58*, 293-312.

Symons, D. (1979). *The Evolution of Human Sexuality*. New York: Oxford University Press.

Taylor, P.A. and Glenn, N.D. (1976). The utility of education and attractiveness for females' status attainment through marriage. *American Sociological Review, 41*: 484-498.

Tooby, J., and Cosmides, L. (1990). On the universality of human nature and the uniqueness of the individual: The role of genetics and adaptation. *Journal of Personality, 58*, 17-68.

Tooby, J., and Cosmides, L. (1992). Psychological foundations of culture. In J. Barkow, L. Cosmides, and J. Tooby (Eds.), *The Adapted Mind: Evolutionary Psychology and the Generation of Culture*. New York: Oxford University Press.

Tooke, W., and Camire, L. (1991). Patterns of deception in intersexual and intrasexual mating strategies. *Ethology and Sociobiology, 12*, 345-364.

Williams, G.C. (1966). *Adaptation and natural selection*. Princeton: Princeton University Press.

Zippelius, R. (1986). Exclusion and shunning as legal and social sanctions. *Ethology and Sociobiology, 7*, 11-18.

Zuckerman, M. (1990). The psychophysiology of sensation seeking. *Journal of Personality, 58*, 313-345.

EVOLUTIONARY PSYCHOLOGY: WHERE TO?

Commentary on the chapter by David M. Buss

Alois Angleitner
Department of Psychology
University of Bielefeld
P.O. Box 8640
4800 Bielefeld 1
Germany

Let me first express my admiration for the thought provoking contribution David Buss has offered. In general, I think that looking at personality from an evolutionary perspective is a very promising way for generating new hypotheses which are testable using cross-cultural studies. The insight we gain from an evolutionary view in psychological mechanisms like sexual jealousy, mate selection and retention can not be provided by other theories of psychological functioning. The most promising way using the evolutionary perspective, however seems to me the new look it provides on sex differences seen as strategic individual differences. The examples of sexual and emotional jealousy and tactics of mate retention are convincing. I may add that the differences in sexual and emotional jealousy between the sexes have recently been replicated also with German data (Krehmeier and Oubaid, 1992). Now let me come to some more detailed questions.

Evolutionary Psychology and Personality Traits

In the current theorizing as well as in empirical work, there is a strong emphasis in personality psychology on personality traits. However, in Buss's presentation the term trait is not mentioned. It is not clear for me how dispositions as understood within an evolutionary context are different from a trait conceptualization. The reference he gives to the "Big Five", which may be considered as the major dimensions of the "adaptive landscape" as expressed in perceiving, relating and adapting to other people -important life tasks for enhancing reproductive fitness- seems to confirm that Agreeableness, Conscientiousness and eventually Emotional Stability are important for the formation of stable social bonds in marriages and groups.

Reproductive Fitness and Personality Traits

More intriguing may be the question what the relationship of reproductive fitness to personality dimensions or traits is. As Loehlin (1992) states, there is remarkably little information about this question. He cites a study by Eaves et al. (1990) based on data from the Australian twin sample. The females were asked to report their number of children and if they still have menstruations. The number of children reported was then correlated with Extraversion (E) and

J. Hettema and I. J. Deary (eds.), Foundations of Personality, 191–196.

Neuroticism (N) measured with the EPQ. It turned out that Extraversion and Neuroticism alone were not correlated, however the combinations of E and N were correlated. Women, who reported having the most offspring tended to come from two groups: high Extraversion plus low Neuroticism, and Introversion plus high Neuroticism. The smallest number of children were reported by the following two groups: Introversion plus low Neuroticism and Extraversion plus high Neuroticism. I am not able to interpret this finding. It needs replication by asking both sexes in a cross-cultural research effort. However, including this simple question would provide us with some data (a) concerning the reproductive values of personality dimensions and traits, (b) and concerning the selective trends in operation.

Psychological Mechanisms and Temperament

Psychological problem-solving mechanisms for adaptive problems may be special-purpose and domain-dedicated. However, if we compare the definitions of personality and temperament, we may suggest that the most basic important individual differences are probably temperamental. They regulate the stimulation from inside the organism and from outside. Temperamental characteristics like activity and reactivity, are largely formal and stylistic and not as content-saturated as personality characteristics. Temperamental traits may therefore be considered as possible candidates for such evolved psychological mechanisms.

Psychological Mechanisms and Adaptional Problems

Buss is correct in stating that recurrent problems can also be socially caused. Psychological solutions to adaptational problems are usually listed as different ways of coping. Personality characteristics of one‘s partner (disagreeable) may also create socially imposed problems, that recur over time for an individual. These psychological solutions may also be unsuccessful in some cases and may lead to psychopathological behaviors. This points to a more general aspect: How can evolutionary psychology deal with psychopathology ?

Psychological Mechanisms as new units for Personality Psychology?

Buss suggests the following new units for personality psychology: (a) evolved psychological mechanisms and (b) behavioral strategies. The most prominent units in the last 50 years have been traits, states, drives and motives (needs). Theorists have also introduced as basic units roles, personal constructs, personal projects (Little, 1983), and personal strivings or goals (Emmons, 1986). Buss added behavioral acts to this list and, from an evolutionary frame of reference (with regard to enhancing the reproductive success of individuals), he now labels such actlists tactics and strategies. I have no problems with the category entitled "behavioral strategies", however I feel uneasy with the evolved psychological mechanisms as new possible units. After reading the list of human universals, which I understood as a list of possible candidates for such evolved *psychological mechanisms* I would like to ask the following questions:

1. How many of such mechanisms should be postulated: Hundreds, thousands, tens of thousands?
2. Are they specifically human or are they extended to other species also?
3. What are the criteria for identifying such mechanisms ? Is cross-cultural invariance the only necessary and sufficient criterion?
4. Are these mechanisms unrelated to each other? Is it possible to understand some of them as correlated or even integrated to form some systems? In particular the proposed approach of searching for special-purpose mechanisms may lead to the adding of one mechanism to another.
5. Furthermore, is it likely that each evolved psychological mechanism solves only *one* unique adaptive problem?

Considering the concepts used in personality psychology it seems to me that especially the needs as postulated by Murray (1938) are coming close to these proposed psychological mechanisms. Murray considers physiological needs, psychological needs, latent needs and so on. But especially important are his thoughts about the interactions of these needs as well as his ideas about the criteria for detecting needs in people's daily life. Such interaction principles are prepotency, fusion, subsidiation and conflict. Evolutionary psychology may learn by considering if and how these principles are also applicable to evolved psychological mechanisms.

References

Emmons, R.A. (1986). 'Personal strivings: an approach to personality and subjective well-being', *Journal of Personality and Social Psychology, 51*, 1058-1068.

Krehmeier, J. and Oubaid, V. (1992). Sexueller Konflikt - Deutsche Übertragung und empirische Anwendung des SCQ von David M. Buss (Sexual Conflict - German adaption and empirical test of the SCQ by David M. Buss). Unpublished master thesis, University of Bielefeld.

Little, B.R. (1983). 'Personal projects: a rationale and method for investigation', *Environment and Behavior, 15*, 273-309.

Loehlin, J.C. (1992). *Genes and environment in personality development*. Sage Publications, Texas.

Murray, H.A. (1938). *Explorations in Personality*. Oxford University Press, New York.

REPLY TO A. ANGLEITNER

David M. Buss

Professor Angleitner raises several important questions that require attention. I address these in order.

Are Current Correlations with Reproductive Success Relevant to our Evolved Psychology?

Professor Angleitner cites a study showing that, although Extraversion and Neuroticism are not directly related to reproductive success, a particular and puzzling interaction between the two is related. From the perspective of evolutionary psychology, current correlations with fitness or reproductive success are informative only about *current* selection pressures, and are not terribly informative (if at all) about prior selection pressures. The key issue is: What has the history of evolutionary selection produced that we now carry around with us today? Since selection operates slowly over thousands of generations, and the creation of complex mechanisms or adaptations typically requires thousands of generations for their formation, indices of selection pressure in the current generation are like drops of water in a rainfall - they pale in comparison to the products produced over more extensive evolutionary time. The central issue for evolutionary psychology concerns the nature of the mechanisms we carry around with us today - mechanisms created over vast expanses of evolutionary time to deal with adaptive problems in ancestral environments.

How Can Evolutionary Psychology Deal with Psychopathology?

Professor Angleitner asked why people would marry disagreeable partners, given the costs that they impose. More generally, how does evolutionary psychology deal with psychopathology? This is a fascinating and complex question that can be answered only in cursory fashion in this brief addendum.

On the issue of why anyone would marry a disagreeable person, one must keep in mind several points. First, disagreeable people may conceal the extent of their cost-imposing tendencies, which emerge only later in the relationship. Courtship is a time when people present themselves in the most positive light, highlighting their resources and benefit-bestowing abilities and proclivities and minimizing the costs they carry and the costs they might eventually inflict.

Second, not all individuals have sufficiently high "mate value" to command mates of high value. Just as only a select few can obtain the most physically beautiful, the most intelligent, the most resource-laden mates, so only some people can attract the most agreeable mates. The key point is that mating decisions are not made in a vacuum, but rather result from a confluence of dozens of factors, of which one's preferences are only one set.

More generally, psychopathology can be handled in a number of different ways by evolutionary psychology. Some psychopathologies are simply evolved mechanisms gone awry, just as hearts sometimes malfunction, the thyroid gland is sometimes over- or under- active, and so on. All mechanisms are susceptible to genetic defects and environmental insults.

More complexly, what constitute a "psychopathology" itself needs re-examination from an evolutionary perspective. For example, from the perspective of European or American middle-class values, it sometimes appears "maladaptive" for people to be impulsive, aggressive, or to fail to delay gratification. However, in environments where resources are scare and highly variable, such as in inner city ghettos, such strategies may be the most "adaptive" ones to deploy. What appears to be pathological from one perspective may turn out to be functional from another. The key point is that the very definition of psychopathology requires re-examination. A central part of this re-examination will entail distinguishing between pathology at the level of manifest behavior (e.g. a particular mating decision) and pathology at the level of the function of the underlying psychological mechanisms.

What is the Nature of Psychological Mechanisms? How Many Are There? How Are They Related? And How do we Identify Them?

Professor Angleitner raises an extremely important set of issues about the nature of evolved psychological mechanisms. Evolutionary psychologists believe, in contrast to the dominant thinking in social science over the past century, that the number of evolved mechanisms are likely to number in the hundreds or thousands, depending on how they are defined. Consider the human eye. At one level of analysis, it is one mechanism. But from an evolutionary viewpoint, different modules within the eye serve different functions. There are specific edge detectors, motion detectors, and so on. Each of these modules can be considered evolved mechanisms because each solves a distinct adaptive problem. Analogously, our psychological mechanisms are likely to be large in number, because the number of distinct psychological problems we have confronted over evolutionary time is likely to be dozens of times larger than the number of visual problems we have confronted.

Some mechanisms we share with other species, others are unique to humans. The complex collection of language mechanisms, for example, is unique to humans, even if extremely rudimentary forms can be discerned in chimps (Pinker and Bloom, 1991).

The criteria for identifying mechanisms include *precision, reliability, efficiency, specialization*, and *economy* of functioning tailored to solving a particular adaptive problem (see main text). Identifying mechanisms requires developing models or hypotheses about their nature, generating testable empirical predictions from these models, and then testing the predictions. For example, there is currently strong evidence for evolved mechanisms such as specific mate preferences (Buss, 1989a; Kenrick and Keefe, 1992) cheater-detection mechanisms in reciprocal relations (Cosmides, 1989), sex-linked jealousy mechanisms (Daly et al., 1982; Buss et al., 1992), sex-linked sexual variety-seeking mechanisms (Symons, 1979; Ellis and Symons, 1990), and many others.

These mechanisms are undoubtedly organized and linked in various complex ways, rather than being isolated. Furthermore, to achieve the complex situation-contingent functioning we observe, humans almost certainly have super-ordinate governing mechanisms that regulate in hierarchical or sequential fashion which mechanisms are activated and in what sequence. Future research is needed to identify these regulatory mechanisms.

Can Evolutionary Psychology Be Applied to Individual Differences?

Professor Hofstee (this volume) argues that evolutionary psychological analyses to date have been far more compelling in accounting for human nature and sex differences than in accounting for individual differences. He is correct. In the domain of sex differences, for example, we have powerful evolutionary theories to guide us, such as Triver's (1972) theory of parental investment and sexual selection. Individual differences within species are more challenging - not just for human researchers, but for scientists who study fish, birds, and non-human primates.

Although reasonable scientists may reasonably differ in their projections about future promise, I believe that the study of individual differences will become increasingly tractable within an evolutionary psychological framework. The major approaches to understanding such differences will include: (1) differences due to occupancy of different environments, just as stable differences in callous-thickness across individuals are due to occupancy of environments differing in friction-prevalance; (2) differences due to experiencing different environments during development that channel individuals toward one strategy or another (e.g. differences in resource variability linked with differences in proclivities to delay or not to delay gratification); (3) differences due to genetic differences where some individuals can more successfully deploy one strategy rather than another because of heritable proclivities, abilities, or assets (e.g. mesomorphs can more successfully pursue an aggressive strategy than ectomorphs).

Cross-cutting these three modes of analysing individual differences will be the sorts of analyses illustrated in this paper: (1) links between personality and the sorts of adaptive problems one confronts (e.g. submissive people confront different adaptive problems than surgent people); (2) links between personality and the adaptive problems inflicted upon others (e.g. low conscientious people inflict more extramarital affairs on their spouses); (3) links between personality and the tactics used to solve adaptive problems (e.g. agreeable people use different mate retention tactics than emotionally instable people).

Through analyses of these sorts, evolutionary psychology offers the promise of moving the field of personality from a purely descriptive science of individual differences to a broader and more encompassing explanatory science of human psychological functioning.

References

Buss, D.M. (1989). Sex differences in mate preferences: Evolutionary hypotheses tested in 37 cultures. *Behavioral and Brain Sciences, 12*, 1-49.

Buss, D.M., Larsen, R., Westen, D. and Semmelroth, J. (1992). Sex differences in jealousy: Evolution, physiology, and psychology. *Psychological Science, 3*, 251-255.

Cosmides, L. (1989). The logic of social exchange: his natural selection shaped reason? *Cognition, 31*, 187-276.

Daly, M., Wilson, M. and Weghorst, S.J. (1982). Male sexual jealousy. *Ethology and Sociobiology, 3*, 11-27.

Ellis, B.J. and Symons, D. (1990). Sex differences in sexual fantasy: An evolutionary psychological approach. *Journal of Sex Research, 27*, 527-556.

Kenrick, D.T. and Keefe, R.C. (in press). Age preferences in mates reflect sex differences in reproductive strategies. *Behavioral and Brain Sciences*.

Pinker, S. and Bloom, P. (1990). Natural language and natural selection. *Behavioral and Brain Sciences, 13*, 707-784.

Symons, D. (1979). *The Evolution of Human Sexuality*. New York: Oxford University Press.

Trivers, R. (1972). Parental investment and sexual selection. In B. Campbell (ed.), *Sexual selection and the descent of man 1871-1971*. Chicago: Aldine.

LINKING BIOLOGICAL AND SOCIAL APPROACHES TO PERSONALITY: EVOLUTION, GENDER, AND COGNITION

Douglas T. Kenrick
Department of Psychology
Arizona State University
Tempe, AZ 85287
U.S.A.

Abstract

Gender differences in mating strategies reflect evolved genetic predispositions manifested in ongoing cognition, preferences, and behavior. A summary of general principles of evolutionary psychology is followed by specific applications to gender differences. Several lines of related research are summarized. One line of research suggests that male preferences for relatively youthful females, and the converse female preference for older males, do not fit prevailing normative theories. Cross-cultural and developmental data are more consistent with an evolutionary model. Another line of research finds sex differences relatively most pronounced when the sexes differ most in potential parental investment (i.e., when considering casual sexual relationships). Under those circumstances, males' mate criteria are lower than females, and relatively uncorrelated with their own self-evaluations. The final program of research considers gender differences in ongoing social cognition. Discussion considers potential fusion of cognitive science and evolutionary psychology, and the essential importance of personality and social psychology to such an integration.

Introduction

How do men and women differ in their thoughts and behaviors about potential mates? At first blush, such a topic might seem more relevant to social psychology than to personality. However, when one considers gender differences in mating-relevant cognitions from an evolutionary perspective, it becomes clear that the topic is right at the heart of the questions being considered in this volume - the interaction of biological and social cognitive approaches to personality. In this paper I will argue that there are gender differences in behavior and cognition that appear to be based in genetic differences between males and females. It is not unreasonable to say that there are universal gender-linked "traits", in any sense of the word traits you might care to use. Those traits appear to be linked, in large part, to differential evolutionary constraints on males and females - on different sexual selection pressures faced by our male and female ancestors. Although psychologists have tended to consider evolutionary theory the domain of anthropologists or zoologists, it is becoming increasingly clear that psychologists interested in ongoing cognitive processes can profit from an understan-

J. Hettema and I. J. Deary (eds.), Foundations of Personality, 197–218.

ding of the modern developments in evolutionary theory. Without an understanding of human nature, in fact, psychologists sometimes miss some of the most important features of ongoing cognition.

During the same recent period in which there has been a resurgence of evolutionary psychology, the field has also become increasingly involved in the so-called "cognitive revolution". The papers in this volume were to address the potential integration of "biological" and "social" approaches to personality psychology. Evolutionary psychology is of course clearly linked to one of the two perspectives - the biological; whereas cognitive science is linked more closely to the other - the social cognitive. Evolutionary psychology is an interdisciplinary field - incorporating new findings and theoretical insights from behavior genetics, ethology, anthropology, and psychology (Lumsden and Wilson, 1981). Cognitive science is related to the other aspect of this conference - the social cognitive approaches, and it is also a grand integrative enterprise, encompassing developments in cognitive psychology, computer science, linguistics, anthropology, and neuroscience (Gardner, 1985; Lakoff, 1987). Thus far, evolutionary psychology and cognitive science have been largely independent. However, both interdisciplines are centrally concerned with the structure and function of the human brain, and it seems inevitable that there will be a higher level fusion. One byproduct of this fusion should be an incineration of the artificial barriers between researchers working on social cognition, interpersonal processes, and personality traits. Beyond that, a fusion between evolutionary psychology and cognitive science should also bring together researchers in social/personality and researchers in experimental, developmental, and clinical psychology.

General Principles of Evolutionary Psychology

Evolutionary theory consists, at base, of three simple assumptions, outlined by Darwin in 1859. The first assumption is that animals are engaged in a *struggle for existence* - members of the same species must compete for limited resources. The second assumption is that there is *heritable variation within a species* - animals within a species differ in many ways, and can pass some of those differences along to their offspring. The third principle of evolution - *natural selection* - follows directly from the other two. If animals must compete with one another to survive, and if animals vary in ways that can be inherited, then animals whose variations assist in the struggle for survival will have more offspring. Those offspring will, in turn, have relatively more offspring than less well-adapted strains, and so on. Over generations, the strains that are most closely suited to their particular environments will replace those that are less well-adapted.

Darwin's theory applies not only to the physical characteristics of animals, but also to behavior. Behaviors evolve in the same way as do physical features. Seals are closely related to dogs, but if a seal inherited a brain programmed to run a dog's body and tried to run down large hoofed mammals on dry land, it would not survive well; neither would a dog which attempted to swim out to sea and dive for fish fifty feet below the surface. Along with their bodies, seals, dogs, bats, giraffes, and lions inherit brains programmed to do certain things with those bodies. Thus, evolution applies to survival-related behaviors in much the same way that it applies to physical characteristics. Those animals with behavioral variations most suited to their environments (and to their bodily equipment) will survive and outreproduce those animals with less well-adapted behavioral variations.

Evolutionary theorists assume that the principles of evolution apply to all animal species, and that humans are not exempted. They assume that humans will therefore share a number of species-typical behavioral characteristics, and have consequently searched for evidence of human behavioral universals (Darwin, 1872; Ekman and Friesen, 1971; Ekman et al., 1987). One of the most exciting developments in the area of evolutionary social psychology has been the discovery of a number of such universals of social behavior that are difficult to explain using traditional sociocultural or learning approaches. For instance, Eibl-Eibesfeldt (1975) filmed natural instances of flirtation in numerous cultures, and noted that micro-analyses of the movement patterns led to the discovery of universal flirtation gestures. The sex difference in violent behavior also appears to be universal. For instance, males commit over 85 percent of all homicides in every society that has ever been investigated (Daly and Wilson, 1988a,b). I will also discuss a number of seeming universalities in mate preferences below.

INDIVIDUAL DIFFERENCES WITHIN SPECIES

Because all the members of a particular species must meet certain environmental demands, natural selection often leads to a reduction of variance on certain characteristics - most humans have two eyes and walk upright, for instance. However, it is a misconception to assume that natural selection removes all the important differences between the members of a species. If that were the case, neither artificial selection nor natural selection would have raw materials with which to work. Behavior genetic studies of twins and adoptees have supported the assumption that humans, like other animals, have many heritable variations that affect behavior (Plomin et al., 1990). As has been demonstrated by some of the contributors to this volume, variations in personality characteristics ranging from friendliness and intelligence through depression and schizophrenia appear to involve heritable components (e.g., Loehlin et al., 1988; Tellegen et al., 1988).

Only some of the individual differences within a species are due to genetic variations within species. Others are due to interactions between features of the environment and species-wide genetic proclivities keyed to those variations in the environment. There are several reasons why evolutionary pressures will not lead all animals sharing a particular environment to be identical. One reason is related to what evolutionary theorists call *density-dependent strategies* - certain behavior patterns are only adaptive when they are *not* universal. It appears that animals are programmed to vary their life strategies based on a combination of their own phylogenetically programmed predispositions and their ontogenetically programmed experiences with their social and physical environments. Such processes often lead to variations *within* a given species. For instance, there are two types of adult male blue-gilled sunfish (Gross, 1984). One type is a large territorial male whose colorful body is highly attractive to females. A second type of male is smaller and drab in appearance, resembling the less resplendent female. These smaller males are known as "sneak-copulators". Rather than investing nutritional energy in developing a large flashy physique, the sneak-copulators develop enormous sperm-producing organs. When a large territorial male is mating with a female, the smaller male will thwart his larger opponent by darting in and releasing his sperm. Obviously, the success of the smaller males' strategy depends partly on the existence of the larger males in the vicinity to attract females, and is decreased by too many other smaller sneak-copulators in the neighborhood. As another example, some animals can actually change from one physical form into another depending upon "environmental openings". In the cleaner wrasse, for instance, females are most numerous and congregate in harems around a large territoried male. When a large male dies, the largest female in his harem goes through a series of rapid physiological changes during which she grows larger and transforms into a male (Warner, 1984). Thus, the success of a particular combination of body type and behavior is linked to variations in the environment, and some species have evolved to change body types as the environment changes. Such processes might be relevant to the intra-familial behavioral differences between identical twins discussed by Plomin and Daniels (1987). There are a number of interesting questions about personality that arise from considering the extent to which early niche-picking experiences might lead to physiological changes between differences members of the same family, members of the same peer group in a school or neighborhood, and so on (Kenrick, 1987).

GENDER DIFFERENCES WITHIN SPECIES

The most prevalent within-species variations are based on gender - males and females of a given species commonly differ in size and in behavior. Many of these differences are unique to particular species: the nature of the differences between a peacock and a peahen are not the same as those between a bull and a cow. However, some wide generalizations in sex differences can be found across a wide range of vertebrate species. Darwin noted in his *Origin of Species* that males tend to be relatively larger and more showy. If one member of a fish species has more decorative fins (as in the Siamese fighting fish), if one member of a bird species has more colorful plumage (as in the peacock), or if one member of a mammalian species has larger antlers (as in the elk), it tends to be the male. There are fairly general differences in behavioral traits as well. For instance, male vertebrates tend to be more aggressive and more inclined toward dominance competitions. These sex differences lead to some common differences in mating arrangements across species. In particular, polygyny (one male, many females) is more common than polyandry (one female, many males) among vertebrates. This mammalian generalization holds for humans as well. Of 849 societies examined in Murdock's *Ethnographic Atlas*, 708 were polygynous, whereas only 4 were polyandrous (Daly and Wilson, 1983). In addition, each of the 4 polyandrous societies were *also* polygynous, whereas the reverse was not true. Two general principles are often used to explain these differences: differential parental investment and sexual selection.

DIFFERENTIAL PARENTAL INVESTMENT AND SEXUAL SELECTION

Differential parental investment refers to the fact that males and females are inherently different in the amount of resources they invest in offspring. For all animal species, eggs are generally more costly to produce than sperm. In species that utilize internal fertilization, as do most mammals, this difference is enhanced considerably. To produce a single offspring, a mammalian female must carry a fetus that requires a large amniotic sac and that takes first priority on her nutritional intake for several months. It is generally believed that the higher ratio of body fat to muscle in the human female originally stemmed from the need to ensure survival of the costly fetus in nutritionally uncertain times (Frisch, 1988). Following birth, the female mammal nurses the newborn, again sacrificing her own nutritional intake to provide nutrition for her progeny. In some species, such as humans, the offspring must be fed and cared for even after they are weaned. Therefore, the minimum female parental investment is quite large.

Males, on the other hand, could father a child with a very low investment - the amount of energy required for one act of intercourse. The record number of legitimate children recorded for one man is 899 (Daly and Wilson, 1983), and the actual number could conceivably be higher than that. On the other hand, it is biologically difficult for a woman to have more than 24 children, regardless of the number of husbands she has. For example, among the Xavante, a hunter gatherer group, the average number of offspring for males and females is 3.6 (logically, the mean has to be the same). However, the variance for women is 3.9, whereas for men it is 12.1. In other words, some Xavante men have quite a few offspring, some have very few. Only 1 of 195 Xavante women are childless at age 20; whereas 6 percent of men are still childless by age 40. One man in the group had 23 children; whereas the highest number of children for a woman was 8.

Some of the physical differences between males and females are due simply to natural selection based on these differences in parental investment. Females need a different body to produce eggs, and in the case of mammals, other specialized physical features are required to nurture the fetus and the newborn baby. However, those different parenting requirements are not enough to explain the vast differences between the sexes. One might expect that a slightly larger body would be of more use to a female mammal who must contribute her physical resources directly to the offspring (Ralls, 1976). However, males tend to be larger in most vertebrate species. Males are also more likely to have decorative features like antlers and peacock's feathers; and to use some of those features (such as antlers) to compete with one another. Darwin used the concept of *sexual selection* to explain such differences. Sexual selection consists of two separable processes. *Intrasexual selection* refers to the selection pressure that the members of one sex exert on one another via competition. In a species in which males compete for access to females by butting their heads, those individuals with the boniest heads, the strongest shoulder muscles, and the largest antlers will be more likely to win dominance competitions and survive. *Epigamic selection* is the other part of sexual selection. If one sex selects sexual partners on the basis of a certain feature, such as the possession of large antlers or bright displays of feathers, those features will be more characteristic of one sex than the other.

Darwin suspected that epigamic sexual selection applies more to female choice of males, and female choice could explain why male vertebrates tend to be larger, showier, and more dominance-oriented. The reason that females are more likely to be choosy relates back to the concept of *differential parental investment.* Because females have an initially higher investment in their offspring, they are better served to be selective about their partners. An ill-chosen mating partner is, on average, less likely to be costly for a male. A number of human sex differences seem to fit with this general mammalian pattern in which females are relatively more selective about mating partners and males are relatively less selective (Daly and Wilson, 1983; Hinde, 1983; Kenrick, 1987; Kenrick et al., 1990).

MALE PARENTAL INVESTMENT IN HUMANS

The model of differential parental investment and sexual selection cannot be applied to humans without some important qualifications. Human males invest heavily in their offspring. In a monogamous relationship, the man may provide resources for one wife and her offspring for all of his adult life. Because human males invest heavily in their offspring, they should, like females, be selective about choosing a mating partner. In fact, sex differences tend to be diminished in species in which the males invest in their offspring. Male and female humans are relatively similar in size and decoration in contrast to highly polygynous species. For instance, male baboons are twice the size of females.

Although human males are therefore expected to be selective in choosing mates, they would not be expected to choose females along the same dimensions that females use for choosing males. Men and women invest different resources, and so one would expect the two sexes to value different characteristics in a mate. Males invest indirect resources (such as food, money, protection, and security). Thus, women would be expected to value men who show the ability to provide those resources. On the other hand, women directly invest their bodily resources in the offspring. Males would therefore be expected to value women who demonstrate signs of fertility and health.

From an evolutionary perspective, human males and females are also expected to differ in the age of mates they find most attractive. Aging limits a woman's reproductive potential, and around age 50, ends it through menopause. Males do not undergo menopause, and older males are still capable of fathering offspring (Nieschlag and Michel, 1986). Male health does decrease with age, but the features which females find attractive, such as economic resources and social status, continue to accrue well past a male's physical prime. These differential changes, decreasing female fertility and increasing male resources, would lead males and females to place different value on signs of youth and physical health. In fact, data from a number of cultures has suggested that males tend to value relatively younger partners whereas females value relatively older partners (Buss, 1989; Symons, 1979). Males across cultures also report placing relatively more value on physical attractiveness. Evolutionary theorists have argued that judgments of female physical attractiveness are closely related to signs of youth and physical health.

Thus, human males and females are expected to be similar in that both may invest heavily in offspring and both are often selective about mating partners. However, males and females are expected to be selective along different dimensions. Females are expected to place more emphasis on characteristics related to resources and social status; males are expected to place more emphasis on characteristics related to youth and physical attractiveness.

Age Preferences in Mates Suggest Cross-cultural Universal

During the time that social psychologists were denying the existence of traits, they were also finding overwhelming support for the principle that people tend to mate with others whose traits are similar to theirs (Antill, 1983; Byrne, 1971). The so-called "matching principle" is one of the few laws of social psychology, and it is rarely violated. One consistent violation, however, was just alluded to. Females are attracted to older males, while males are attracted to younger females (Harrison and Saedd 1977, Cameron et al., 1977). Social psychologists have tended to attribute this age preference complementarity to culturally arbitrary gender discrepancies in social value. Females in our society presumably exchange youth and physical attractiveness for economic security.

Why do females not value youth as much as males do? The prevailing models have generally explained this and other features of mate selection in terms of cultural norms. Presser (1975) offers a tentative explanation in terms of the "norm" that "...a husband should be, or at least appear to be, mentally and physically superior to his wife. Not only should he be taller than she (for the appearance of superiority) but also older (which gives him the advantage of more time to become better educated and more experienced" (p. 202). In a similar vein, Cameron et al. (1977) explain their findings that females prefer older, taller, high status males as due to "traditional sex-role specifications ... frequently valued as sex appropriate *in American society*" which specify that women should "look up to" their male partners. These quotes reveal a tacit assumption that there are other societies in which gender differences in mate preference are fundamentally different from those found in our society.

Evolutionary theorists have noted the same gender difference in age preference, but explained it in terms of the universal features of human nature I discussed earlier. However, very different predictions follow from the evolutionary and normative explanations. Those who have adopted the normative perspective have assumed that the rule is a fairly simple one - analogous to the rule about differential height: men should be a few years older (as they should be a few inches taller; e.g., Cameron et al., 1977; Presser, 1975). None of the normative accounts I came across indicated that the size of the difference should vary with the age of the individuals involved. Since there is evidence that younger people (particularly teenage males) are particularly sensitive to sex-role norms (Deutsch et al., 1986), it might be expected that the difference would be slightly more pronounced among younger males.

If one adopts an evolutionary perspective, and considers the issue of lifespan changes in the reproductive value of males and females, however, one is led to a different set of predictions. Because a female's fertility reaches a peak in her mid-twenties and then declines more rapidly than does a male's, a male should *change* the preferred age difference between him and his partner as he ages. For a teenage male, the choice of a similar partner results in choosing someone with maximum remaining reproductive years, but low fertility. A woman slightly older than him, in her early 20s, on the other hand, still has many reproductive years, and higher fertility. So a teenage male should make little or no discrimination against women a few years older than him. For a man in his forties, however, a woman's remaining reproductive years should be a more important consideration which should act against his using similarity as his only criterion. Since females mature earlier, it makes sense for teenage females to be attracted to older males. Older males also lose physical resources (like health and sexual arousability), but at the same time, they may gain in the indirect resources and social status that females generally look for in a mate. As we indicated earlier, an older man is fully capable of fathering offspring. Those factors may simply cancel one another out. Leonard (1989) argued that, from an evolutionary perspective, a woman would optimize her reproductive potential by mating with a man ten years older. Such a male will have more resources and status than someone her own age. However, he is not so old that he will be likely to die while their children are still young. Thus, the evolutionary perspective makes a similar prediction to the normative perspective regarding female age preferences, but differs from the normative perspective in predicting that males will change their preferences as they grow older. Very young males will have no preference for younger women, and no bias against older women. As males grow older, they will progressively prefer females who are dissimilar in age, and increasingly younger than themselves.

The other major difference between the two perspectives is related to our earlier discussion of cross-cultural universals. From an evolutionary perspective, one would be led to expect that the gender difference in age preference will be relatively cross-culturally robust. Male and female differences in lifetime fertility no doubt vary across cultures, and are influenced by nutrition, age of parenting onset, and number of offspring. However, females in all societies bear the young and undergo menopause. Thus, the preference of older males for relatively younger females should be cross-culturally robust.

To examine the evolution-based predictions about age preferences, Kenrick and Keefe (1992) examined a number of data sources, including marriage records and mate advertisements. The data spanned the decades from 1913 to 1989, and also represented several different cultures, including Western Europe, India, and a small fishing village in the Philippines. Figure 1 summarizes the American data. The marriage data is based on 1,206 American marriages that took place during the late 1980s in Phoenix, Arizona, and Seattle, Washington. The preference data is based on 584 lonely hearts advertisements published in Phoenix, Arizona, and in Washington, D.C. (based on Kenrick and Keefe, 1992).

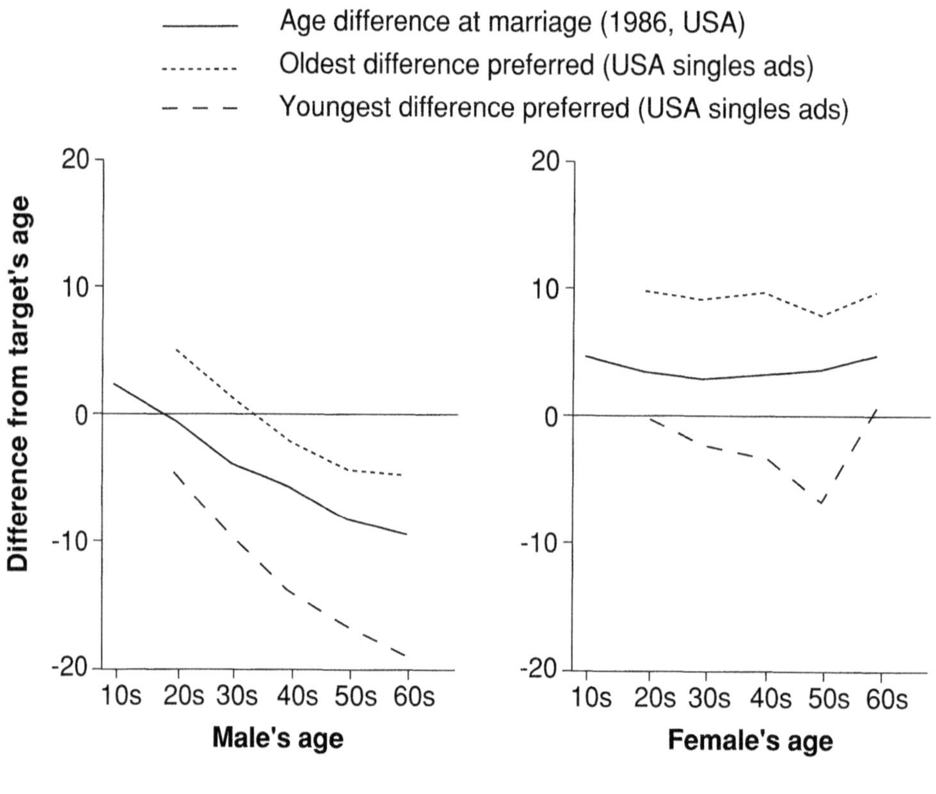

Figure 1: Age differences between mates in American singles advertisements, and American marriages (from Kenrick and Keefe, 1992).

As you can see in the figure, female specifications remained fairly constant throughout the age range. Males, on the other hand, changed their preferences in a systematic fashion as their own age increased. Males in their twenties were equally attracted to women above and below their own age, specifying partners ranging, on average, from 5 years younger to 5 years older. As males got older, however, their preferences increasingly diverged from those of females in the equivalent age group. Among males in their 50s and 60s, the maximum acceptable age was several years below their own age, and the minimum specified was almost a generation discrepant from their own age. As the marriage data indicate, people's actual partners mapped nicely onto the preferences expressed in singles advertisements.

These results match the predictions of the evolutionary model, but do not fit well with normative models. If there is a "norm" for men to prefer younger women it should show up in younger men, who tend to be more concerned with behaving in a stereotypically sex-typed manner. Yet the interesting feature of younger men's preferences is that they extended equally above and below their own ages. Thus, this data provide no evidence that a preference for younger women is a consistent feature of the normatively defined role for American males.

Cross-cultural data provide a stronger test of the normative versus the evolutionary model. Guus van Heck of the University of Tilburg provided data from singles advertisements in Holland; and Ute Hoffman and Kirstin Schaefer of the University of Bielefeldt provided some similar German data. The patterns were, in both instances, the same as those obtained in American singles advertisements. Of course, there are important similarities between American and European cultures (see Kenrick and Keefe, 1992). A more interesting comparison came from an examination of Indian marital advertisements. In comparison with American singles advertisements, Indian marital advertisements indicate very different cultural criteria. Most ads include strict limitations by caste and religion, and they commonly ask for horoscope information. An example is listed below:

> Wanted: A non-Bharadwaj smart good-looking preferably employed Kerala lyer girl below 25 for a Kerala lyer boy 29. Chemical engineer. Contact with horoscope.
> (Times of India, Bombay, Sunday, January 29, 1989)

We obtained a sample of such Indian advertisments and again used those advertisements that included the advertiser's age, and stipulated a minimum and/or a maximum age for preferred partners. There were no advertisements for women above age forty, and an Indian informant explained that women who do not marry by their thirties are not considered marriageable. The results (depicted in Figure 2) follow same pattern as the American and European data. One could still argue that modern residents of India and Europe share some cultural influences with modern Americans. However, Figure 2 also depicts data obtained from the village of Poro - a small fishing village on a remote Philippine island, far removed from any urban area, and largely isolated from any European contact to this day. The marriage data in figure 2 describe all the marriages recorded on Poro during the years 1913-1939.

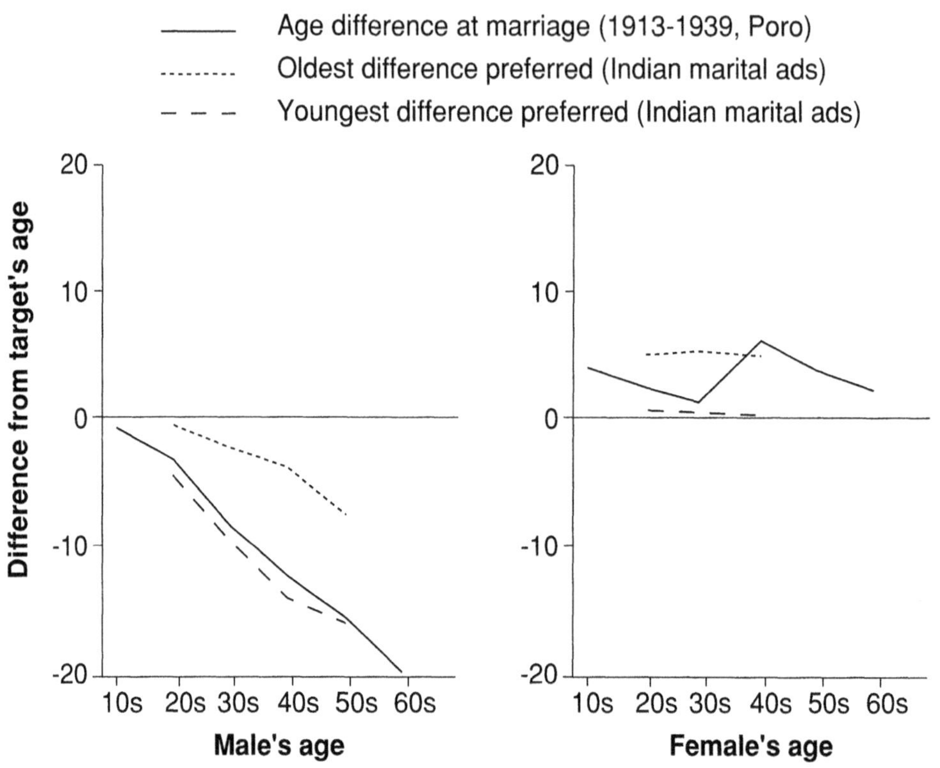

Figure 2: Age preferences indicated in marital advertisements from India, and age differences in marriages on the island of Poro between 1913 and 1939 (from Kenrick and Keefe, 1992).

In commentary published along with the original findings, two anthropologists examined data from several traditional African cultures, and found the same sex-differentiated pattern to hold there (Broude, 1992; Harpending, 1992). Other anthropological commentators argued that the pattern was *more* sex-differentiated in non-American cultures (Symons, 1992; Thornhill and Thornhill, 1992; Townsend, 1992; van den Berghe, 1992). In addition, marriage data from a diverse sample of cultures summarized by the United Nations yields sex differences in marriage precisely in line with those we found (see Figure 3 for one example, and Kenrick and Keefe, 1992 for further discussion).

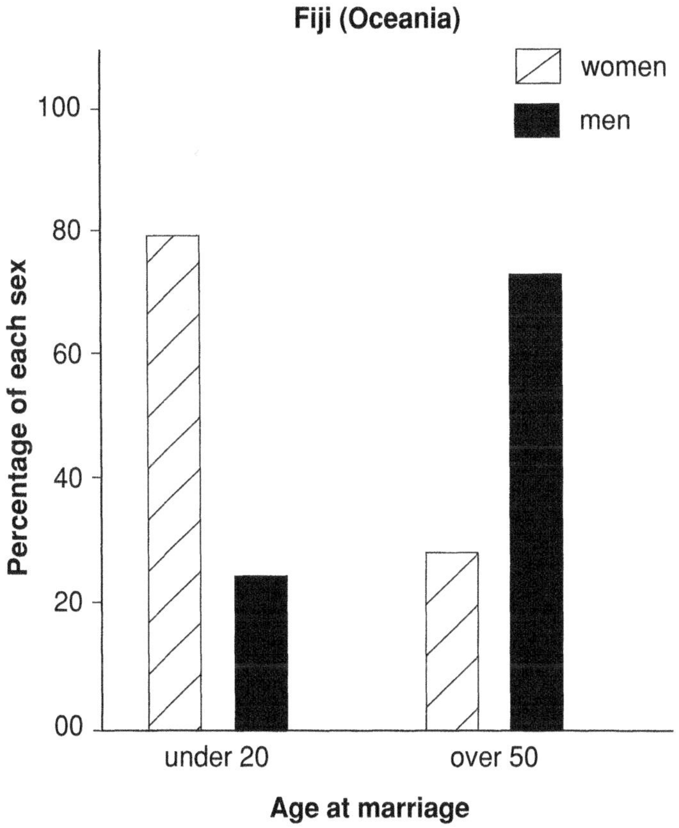

Figure 3: Ratio of males to females in marriages at different ages (based on Kenrick and Keefe, 1992).

In addition to its problems with the cross-cultural data, the normative explanation also fails to explain why young males were interested in older as well as younger females. Most of the data Kenrick and Keefe collected includes males in their twenties as the youngest group (perhaps because younger males are not attractive marital partners). It is worth noting, however, that the average age of marriage partner for the small number of teenage males in the Phoenix sample was actually older than their own age. We have recently obtained preference data from a sample of teenagers (Kenrick et al., 1992). Subjects ranging from 12 to 18 years of age were asked to indicate the youngest and oldest partners they would consider dating, and to indicate the most desirable age of a partner. Teenage females were similar to older females in specifying partners ranging from around their own age to several years older. However, when the preferences of teenage males are placed in the context of the adult singles advertisements, they indicate a particularly interesting lifetime trend. Teenage males were very different from older males in specifying ages ranging from a few months younger to *several years older*.

Most interestingly, teenage males' preferences extended much farther *above* their own age than below. Further, teenage males indicated that the age of an *ideal* partner would be *older* than their own age. This preference not only goes against the likely reward structure determined by female preferences, but it also goes against the supposed social norm (which should operate most strongly on males during the highly sex-typed teenage years). On the other hand, it is quite consistent with the evolutionary model - for a teenage male, the most fertile females are older, not younger.

Gender Differences in Trait Preferences Depend on Level of Investment

As discussed earlier, females are assumed to have an initially higher investment in potential offspring, and to therefore be more selective about the partners with which they will mate. Males, on the other hand, provide less in the way of direct resources for any given offspring, and have less to lose from mating with any given available partner. As part of the process of sexual selection, Darwin suggested that females often may mate preferentially with males who have either won the struggle for dominance over others of their sex, or who have "special weapons" suggesting their "general vigour" (Darwin, 1859). Over generations, a preference for males who display obvious signs of superiority over their competitors will lead to a steep dominance hierarchy, and at same time, select for greater size and weaponry among males. This would explain the sort of dramatic sex differences one sees among baboons, where males are perhaps twice the size of females, and also more dominant. Male baboons are that way, at least in part, because of past mating choices made by female baboons.

Consistent with this model, there is evidence that human females, like other mammalian females, are relatively more attentive to a male's position in the dominance hierarchy (Hill, 1984; Mealey, 1985; Sadalla et al., 1987; Townsend and Levy, 1990). On the other hand, Buss and Barnes (1986) also applied the differential parental investment model to human mate preferences, but found less prominent gender differences. They used a survey design and asked students to rank the characteristics they would prefer in a mate. Consistent with Buss and Barnes' evolutionary perspective - females ranked traits related to resources as more important, and males ranked physical attractiveness as more important. But the most striking feature of Buss and Barnes' data was the similarity of male and female preferences - 7 out of 10 of the highest rated preferences were identical for the two sexes. Buss and Barnes explained the lack of strong sex differences as due to the fact that differential parental investment and sexual selection are diminished in species like *homo sapiens* - which are monogamous and in which most mating age individuals pair off. If one explains sex differences in mate selection strategies in terms of evolutionary theory, and also argues that a lack of sex differences is consistent with the evolutionary perspective, a critic might justifiably complain that evolutionary hypotheses are untestable tautologies. To avoid this criticism, it is necessary to delimit our predictions. In this case, it is necessary to answer the question: "When do humans act like polygynous species, with females being more selective (as in the Sadalla et al., data), and when do we act like monogamous species, with both sexes showing high selectivity (as in the Buss and Barnes data)?

To answer this question, we turned to psychological models of mate selection. Social psychologists have long argued for a consideration of longitudinal factors in relationships, noting that what is considered desirable at one phase of a relationship may be less important at another phase. In fact, Buss and Barnes had asked their subjects about desirable characteristics in a marriage partner, whereas Sadalla et al. (1987) had asked people to rate the sexual attractiveness of a stranger. Although evolutionary theorists had not previously considered the longitudinal perspective, it is consistent with an evolutionary perspective that sex differences in selectivity will interact with the amount of time and resources partners have invested in one another.

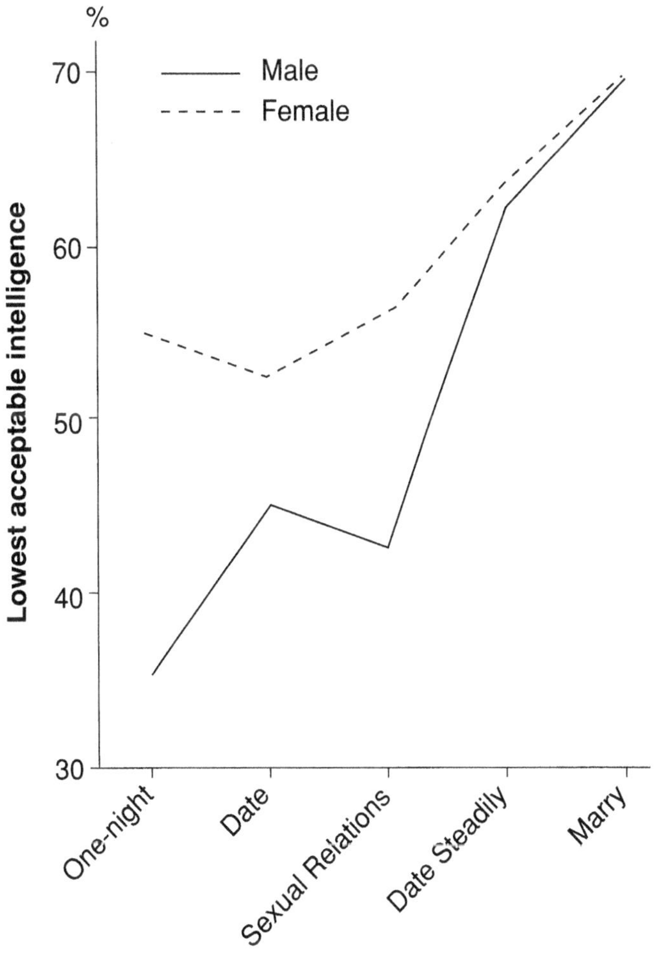

Figure 4: Minimum intelligence criteria for males and females at different levels of involvement, including "one night stand" (based on Kenrick, Groth, Trost and Sadalla, 1992).

Monogamous species typically have a lengthy courtship that precedes mating - this presumably allows both members to evaluate one another's value as a potential parent. Polygynous species, on the other hand, have brief courtships.

If the above distinction is worthwhile, one would expect that sex differences would be most prominent in casual mating liaisons, for which males stand to invest less time and resources in any resulting offspring than females do. In choosing a marriage partner, on the other hand, a male may invest resources for a lifetime - thus males and females should both be selective.

To examine the distinction suggested above, we asked college men and women about their minimum standards in a partner at four levels of involvement. Figure 4 depicts the results for the minimum intelligence that subjects demanded in a date, a sexual partner, a steady dating partner, a marriage partner, and in a partner for a "one night stand", a sexual partner whom the person would never see again. Note that females and males showed the largest differences in criteria for sexual partners.

SELF-APPRAISAL AND MATE CRITERIA

Although social psychologists have sometimes denied the existence of personality traits, they have dedicated a great deal of research to various aspects of the "self". In the relationship area, social psychological models have emphasized the importance of one's *self-evaluation* as a mediator of one's trading value in the heterosexual marketplace. Economically based models adopt a view of potential partners as searching for the best bargain in exchange for their own social assets. According to this approach, one evaluates oneself with regard to social amenities such as beauty, intelligence, charm, wealth, and social status searches for a partner with a similar net value. Evolutionary models of mate selection have tended to ignore an individual's self-evaluation, although such models actually share some of the basic assumptions of economic theories (Cooper, 1987; Frank, 1991; Kenrick and Keefe, 1992; Kenrick and Trost, 1987). It is quite consistent with evolutionary models that self-evaluation should be important. Given the assumption that individuals compete for positions in dominance hierarchies to gain access to more desirable mates, individuals should be very aware of their position in those hierarchies when considering what they can expect in a mate. In fact, personality theorists such as Hogan (1982) and Goldberg (1981) have argued that the major dimensions of personality judgment are cognitive schemas that evolved in large part to assist us in appraising our position in our social group. Interestingly, these same dimensions appear across widely divergent cultures, with "dominance versus submissiveness" always emerging as one of the two major dimensions (White, 1980). Based upon this reasoning, it could be predicted that a person's self-evaluation would be highly predictive of the minimum criteria he or she would find acceptable in a partner.

Evolutionary models suggest gender differences in the use of such self-evaluations at different levels of involvement. In general, women's self-evaluations should be highly predictive of the criteria desired in a mate for any type of sexual relationship. Males, on the other hand, should show differential selectivity depending upon the level of investment in the relationship. In choosing a partner for a more committed relationship, men's self-appraisals should match their criteria for a partner fairly closely. Given men's comparatively lower potential investment in casual sexual liaisons, however, a man's self-appraisals should be relatively less predictive of his minimum criteria for such relationships.

In the most recent series of studies in the program of research on relationship development, we accordingly asked subjects to rate themselves. When those self-ratings were correlated with criteria for a partner, they indicated an interesting pattern in line with the notion of differential parental investment. Figure 5 depicts correlations between subjects's self-rated social status and their minimum criteria for partner's status (from Kenrick et al., in press).

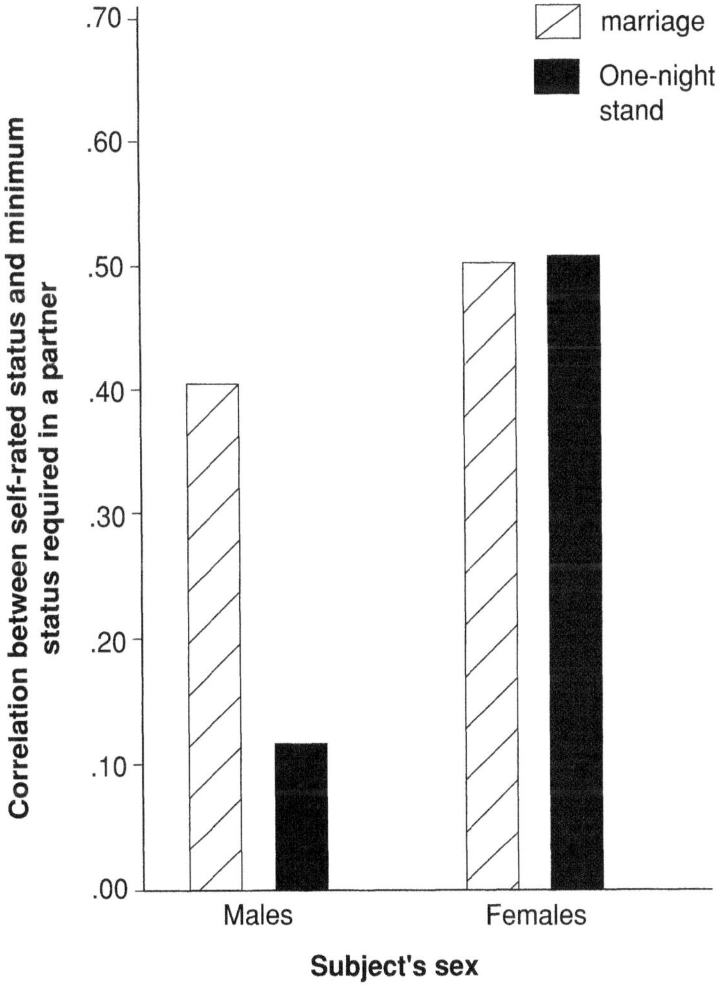

Figure 5: Correlation between self-rated status and minimum criteria for partner's status (based on Kenrick, Groth, Trost and Sadalla, 1992).

Both males' and females' criteria for status in a marriage partner correlate significantly with self-ratings - the higher a subject's self-perceived status, the higher his or her criteria for a spouse. The same holds true for female's ratings of partners for a one-night stand - the higher a woman's self-perceived status, the higher her criteria for a one-night stand. However, males' self-rated status is only weakly correlated with criteria for a one-night stand. These studies indicate inherent interaction between biologically based gender-linked traits, socially-based self-conceptions, and interpersonal behaviors. The nature of those interactions are precisely in line with an evolutionary perspective.

Evolved Gender Differences in Cognition

It has often been assumed that evolutionary theory concerns itself with "ultimate" (historically distant) explanations of behavior, whereas psychologists are more interested in "proximate" explanations that stress the interaction between the present environment and what is going on in a subject's head right now. However, proximate and ultimate levels of explanation cannot be separated. By failing to consider the evolutionary context of human behavior, psychologists sometimes miss the significance of proximate process and structure. For instance, regardless of the task at hand, subjects in laboratory experiments on social perception spontaneously attend to the categories of gender, age, race, and attractiveness. Experts on social cognition frequently note this without any mention of the potential evolutionary significance of such categorizations (e.g., Brewer and Lui, 1989; Hastie and Park, 1986; Markus and Zajonc, 1985). From an evolutionary perspective, psychologists should go beyond the description of proximate processes to ask *why* people attend to certain things more than others, *why* people encode some things more things more easily than others, and *why* people remember some social events more easily than they remember others.

Part of the reason that social cognition researchers have paid relatively little attention to the content of cognition is that it has been assumed that the same *general cognitive processes* are used to think about different types of content (Markus and Zajonc, 1985). In adopting this assumption, social cognition researchers follow cognitive psychologists in the traditional experimental areas (e.g., Glass and Holyoak, 1986). Traditional information-processing approaches to the study of cognition have focussed predominantly on "general processes" rather than specific content. Researchers who adopt an evolutionary perspective, on the other hand, have tended to argue against the existence of general all-purpose cognitive processes. Instead, they have argued that cognitive processes differ depending upon the specific content of the information being processed (e.g., Hansen and Hansen, 1988; Sherry and Schacter, 1987; Tooby and Cosmides, 1989). Sherry and Schachter (1987) reviewed extensive evidence suggesting that animals have qualitatively different memory systems to deal with different types of information. For example, a bird learns to sing the song of its species during an early sensitive period, often before it is capable of performing the song adequately, and later practice may only serve to improve the fidelity of performance. By contrast, memory for food stores is frequently erased within a few days and is "re-programmed" repeatedly throughout life. Memory for sickening foods shows a completely different pattern, in which the animal may learn an aversion on one trial and store its memory permanently with little or no input from experience. Support for content-specific cognitive processing also comes from psychophysiological research with humans, which has indicated that specific patterns of brain damage may lead to very specific and limited cognitive deficits. For instance, patients with one pattern of bilateral damage to the occipital-temporal areas suffer from a disorder called prosopagnosia (Damasio et al., 1982). Individuals with this disorder are unable to recognize specific faces, even their own, although they can recognize friends and relatives from their voices, they can find emotional expressions in photographs, and they can even pick out features such as noses, lips, and eyes. They can also distinguish the age and sex of the people they see, but not their identity. More recent research with these same patients indicates that although they are unable accurately to provide a verbal identification of which of a series of faces are familiar and which are unfamiliar, they do demonstrate marked changes in skin conductance in response to familiar faces (Tranel and Damasio, 1985). These findings are illustrative of a large body of literature that questions the assumption of a monolithic process of "memory" or "cognition" that applies uniformly to all types of content.

The alternative to the view of a monolithic cognitive processor is the view of the brain as organized into a confederation of relatively specialized modules that operate in parallel (e.g. Martindale, 1991). We are already accustomed to thinking of the brain as modular to the extent that we differentiate between auditory and visual areas of the cortex; and at a finer level, between areas such as Wernicke's area (related to understanding speech) and Broca's area (related to production of speech). Instead of considering the brain as a single organ, we can think of it as a group of interrelated organs - each somewhat specialized for particular tasks. From an evolutionary perspective, it makes sense to ask about the different kinds of tasks that a human brain would have evolved to perform. It is to be expected that, in addition to modules designed to process information about the direction of sounds, the color patterns on potentially dangerous insects, and the sweetness of fruits, the human brain also contains modules designed to solve problems in social living (Lumsden and Wilson, 1981; Tooby and Cosmides, 1989).

As psychologists, it is not necessary to study the neurophysiology underlying the operation of different brain areas. However, it is important to be aware of the functional consequences. In particular, there are marked qualitative differences in the processing of the same sorts of problems when the stimuli do and do not involve people (DeSoto et al., 1985; Tooby and Cosmides, 1989). In the past we have tended to import paradigms from cognitive psychology, asking whether the *same* sorts of processes that apply to word recognition (such as priming) might also apply to the processing of social stimuli. From the evolutionary perspective, a more fruitful approach would be to ask which types of problem social situations our ancestor's brain would have had to solve, and to look for evidence of qualititive and quantitative *differences* in the way stimuli related to those situations are processed. Given that differential reproduction is the central task of natural selection, it might be especially fruitful to consider social cognition in light of human reproduction. As I have already discussed, a consideration of the gender differences in parental investment and sexual selection has already led to a number of hypotheses about human mate choice. These same gender differences would be expected to lead to predictable variations in ongoing cognitive processes.

GENDER DIFFERENCES IN SCHEMATIC PROCESSING

An evolutionary perspective would lead us to expect that males and females differ in their chronic use of certain "schemas". The term schema is used in several ways, but it can be usefully thought of as a cognitive structure which acts as a filter guiding the selection of information for attention, encoding, and retrieval from memory. Cognition researchers note that it would be impossible to attend to every aspect of every situation in which we find ourselves. Schemas guide the selection of information in ways that are consistent with ongoing motives (Fiske and Neuberg, 1990). Most of the studies on schema activation have tried to artificially manipulate people's information processing. There has been little investigation of default schemas. What do people pay attention to and remember from social situations when they are operating on automatic, and have not been specifically primed by the experimenter?

An evolutionary theorist would assume that we automatically process for information that is relevant to survival and reproduction. Before even processing the cut of someone's clothes or their regional accent, we should notice the person's gender, his or her sexual attractiveness, and whether this person poses a potential threat or a mating opportunity. Given these considerations, it makes eminent sense that gender, attractiveness, age and race are used as spontaneous categorizations in social cognition experiments. Once again, it is possible to use the evolutionary perspective to make predictions about gender differences in such "default schemas". In line with the differential parental investment hypothesis, which sees males as actively competing for the attention of choosier and more reticent females, males should be more likely to spontaneously process information about potential mates. Lidia Dengelegi and I obtained some suggestive data regarding this type of "default processing" in a free recall study. We asked subjects to look through a yearbook, under the pretense that they were to develop an idea of the different types of students who attended different schools. After they had examined the yearbook we unexpectedly asked them if they could bring any particular face to mind, and then we asked them to locate that face. We also collected information on subjects' relational status: single, dating steady or married. Uninformed judges later rated the attractiveness of the faces that were and were not spontaneously chosen by members of each sex. When a subject recalled someone of the opposite sex, that person tended to be more physically attractive than when a subject remembered someone of the same sex. That finding itself is consistent with the idea that people "automatically" attend to mating relevant people (or at least spontaneoulsy recollect them). However, there was also a sex difference in spontaneous recall that fits with the differential parental investment hypothesis discussed above. Females were about equally likely to bring a male as a female to mind, whether they were attached or not. Males, on the other hand, were more likely to recall a female than a male. This tendency appears to be reduced when the males were attached, but even attached males were more likely to selectively recall an attractive member of the opposite sex. This finding suggests that the reproductive constraints on males and females, discussed earlier, have implications at the level of "on-line" processing. When males and females are asked spontaneously to recall a face, males' recollections indicate that they are more likely to have their schematic antennae tuned for a mating opportunity.

Another line of research indicates further differences in spontaneous schema use by males and females. This research began with a series of studies unconnected to an evolutionary model, in which we found that subjects exposed to physically attractive people later rated average looking targets as less attractive (Kenrick and Gutierres, 1980). A similar effect was found when males were asked to rate their mates after exposure to attractive centerfold photographs (Kenrick et al., 1989). However, a parallel effect was not found for women exposed to male centerfolds - these women did not show reliable decreases in attraction for their male partners. That sex difference might be consistent with the evolution-based studies which have found that women's judgments of partners are, compared with men's, based less upon physical attractiveness and more upon social status and dominance. A consideration of this literature led my colleagues and I to conduct a study in which subjects were asked to rate their commitment to their current relationships after exposure to targets who varied not only in attractiveness, but also in descriptions of their dominance (Kenrick et al., 1992). The results again indicated that exposure to attractive males had little effect on females' levels of commitment to their current mates, but did significantly affect males' commitment. On the other hand, female's judgments were influenced by the dominance of the men to whom they were exposed: more dominant men undermined commitment to a woman's current relationship.

These findings suggest gender differences in ongoing cognition in line with the evolutionary models of sexual selection and differential parental investment. They also have implications for cognitive processes at all levels. Males and females should differ in processes ranging from sensation through perception to selective memory (see Kenrick, 1993, for further discussion of these issues).

Conclusion

Given that Darwin's scheme is built on individual differences, it should come as no surprise that the perspective assumes inherent connections between social processes and personality structure (Buss, 1990; Gangestad and Simpson, 1990; Hogan, 1982; Kenrick et al., 1985; Simpson and Gangestad, 1991). Beyond these connections within the social/personality area, an evolutionary perspective places our work in the context of work in other areas of psychology. As new findings emerge on the structure and function of the human brain, experimental psychologists are increasingly placing their findings within an evolutionary context, asking how the physical and functional constraints of the brain are designed to solve adaptive problems. Evolutionary theorists believe that the human brain is designed largely to facilitate survival in social groups and, like all organic systems, to foster successful reproduction. Reproduction is, of course, an explicitly social task in all vertebrate species, and it is a particularly social task in species like ours, where two parents form long-term bonds. From this perspective, it is essential not only that personality and social psychology are closely interconnected, but also that the other subdisciplines of our field, including neurophysiology and cognitive psychology, be integrated with personality and social psychology.

References

Antill, J.K (1983). Sex-role complimentarity versus similarity in married couples. *Journal of Personality and Social Psychology, 45*, 145-155.

Brewer, M.B. and Lui, L.N. (1989). The primacy of age and sex in the structure of person categories. *Social Cognition, 3*, 262-274.

Broude, G.J. (1992). The May-September algorithm meets the 20th century actuarial table. *Behavioral and Brain Sciences, 15*, 94-95.

Buss, D.M. (1989). Sex differences in human mate preference: Evolutionary hypothesis tested in 37 cultures. *Behavioral and Brain Sciences, 12*, 1-49.

Buss, D.M. (1990). Toward a biologically informed psychology of personality. *Journal of Personality, 58*, 1-16.

Buss, D.M. and Barnes, M.F. (1986). Preferences in human mate selection. *Journal of Personality and Social Psychology, 50*, 559-570.

Byrne, D. (1971). *The attraction paradigm*. San Diego: Academic Press.

Cameron, C., Oskamp, S. and Sparks, W. (1977). Courtship American style-newspaper ads. *Family Coordinator, 26*, 27-30.

Cooper, W.S. (1987). Decision theory as a branch of evolutionary theory: A biological derivation of the savage axioms. *Psychological Review, 94*: 395-411

Daly, M. (1989). Parent-offspring conflict and violence in evolutionary perspective. In R.W. Bell and N.J. Bell (Eds.) *Sociobiology and the Social Sciences*. Texas Tech University Press.

Daly, M. and Wilson, M. (1983). *Sex, evolution, and behavior*. (2nd ed.) Belmont, CA: Wadsworth.

Daly, M. and Wilson, M. (1988a). *Homicide*. New York: Aldine deGruyter.

Daly, M. and Wilson, M. (1988b). Evolutionary social psychology and family homicide. *Science, 242*, 519-524.

Damasio, A.R., Damasio, H. and Van Hoesen, G.W. (1982). Prosopagnosia: Anatomic basis and behavioral mechanisms. *Neurology, 32*, 331-341.

Darwin, C. (1859). *The origin of species by natural selection or the preservation of favoured races in the struggle for life*. London: Murray.

Darwin, C. (1872/1965). *The expression of emotions in man and animals*. Chicago: University of Chicago Press.

Deutsch, F.M., Zalenski, C.M. and Clark, M.E. (1986). Is there a double standard of aging? *Journal of Applied Social Psychology, 16*, 771-775.

DeSoto, C.B., Hamilton, M.M. and Taylor, R.B. (1985). Words, people, and implicit personality theory. *Social Cognition, 3*, 369-382.

Eibl-Eibesfeldt, I. (1975). *Ethology: The biology of behavior* (2nd Ed.). New York: Holt, Rinehart and Winston.

Ekman, P. and Friesen, W.V. (1971). Constants across cultures in the face and emotion. *Journal of Personality and Social Psychology, 17*, 124-129.

Ekman, P., Friesen, W.V., O'Sullivan, M., Chan, A., Diacoyanni-Tarlatzis, I., Heider, K., Krause, R., LeCompte, W.A., Pitcairn, T., Ricci-Bitti, P.E., Scherer, K., Tomita, M. and Tzavaras, A. (1987). Universals and cultural differences in the judgments of facial expressions of emotion. *Journal of Personality and Social Psychology, 53*, 712-717.

Fiske, S.T. and Neuberg, S.L. (1990). A continuum of impression formation, from category-based to individuating processes: Influences of information and motivation on attention and interpretation. In M.P. Zanna (Ed.) *Advances in experimental social psychology*. Vol. 23, pp. 1-74. New York: Academic Press.

Frank, R.H. (1991). Economics. In M. Maxwell (Ed.) *The Sociobiological Imagination*. Albany, NY: SUNY Press.

Frisch, R.E. (1988, March). Fatness and fertility. *Scientific American*, pp. 88-95

Gangestad, S. and Simpson, J.A. (1990). Toward an evolutionary history of female sociosexual variation. *Journal of Personality, 58*, 1, 69-96.

Gardner, H. (1985). *The mind's new science: A history of the cognitive revolution*. New York: Basic Books.

Glass, A.L. and Holyoak, K.J. (1986). *Cognition* (2nd Ed.). New York: Random House.

Goldberg, L.R. (1981). Language and individual differences: The search for universals in personality lexicons. In L. Wheeler (Ed.) *Personality and social psychology review* (Vol. 2). Beverly Hills, CA: Sage.

Gross, M. (1984). Sunfish, salmon, and the evolution of alternative reproductive strategies and tactics in fishes. In G. Potts and R. Wooton (Eds.) *Fish reproduction: Strategies and tactics*. New York: Academic Press.

Hansen, C.H. and Hansen, R.D. (1988). Finding faces in the crowd: An anger superiority effect. *Journal of Personality and Social Psychology, 54*, 917-924.

Harpending, H. (1992). Age differences between mates in southern African pastoralists. *Behavioral and Brain Sciences, 15*, 102-103.

Harrison, A.A. and Saedd, L. (1977). Lets make a deal: analysis of revelations and stipulations in lonely hearts advertisements *Journal of Personality and Social Psychology*, 35: 257-264

Hastie, R. and Park, B. (1986). The relationship between memory and judgment depends on whether the judgment task is memory based or on-line. *Psychological Review, 93*, 258-268.

Hill, J. (1984). Prestige and reproductive success in man. *Ethology and Sociobiology, 5*, 77-95.

Hinde, R.A. (1983). Why do the sexes behave differently in close relationships? *Journal of Social and Personal Relationships, 1*, 471-501.

Hogan, R. (1982). A socioanalytic theory of personality. In M. Page (Ed.), *Nebraska symposium on motivation*. Lincoln, NE: University of Nebraska Press, pp. 55-89.

Kenrick, D.T. (1987). Gender, genes, and the social environment: A biosocial interactionist perspective. In P. Shaver and C. Hendrick (Eds.), *Review of Personality and Social Psychology* (Vol. 7). Newbury Park, CA: Sage.

Kenrick, D.T. (1993). Evolutionary social psychology: From sexual selection to social cognition. In M.P. Zanna (Ed.) *Advances in Experimental Social Psychology*. San Diego: CA: Academic Press.

Kenrick, D.T., Engstrom, C., Keefe, R.C. and Cornelius, C. (1992). Age preferences for dating partners among teenagers: Support for an evolutionary model. Unpublished manuscript: Arizona State University.

Kenrick, D.T., Groth, G., Trost, M.R. and Sadalla, E.K. (1993). Integrating evolutionary and social exchange perspectives on relationships: Effects of gender, self-appraisal, and involvement level on mate selection. *Journal of Personality and Social Psychology*, (in press).

Kenrick, D.T. and Gutierres, S.E. (1980). Contrast effects and judments of physical attractiveness: When beauty becomes a social problem. *Journal of Personality and Social Psychology, 38*, 131-140.

Kenrick, D.T., Gutierres, S.E. and Goldberg, L. (1989). Influence of popular erotica on judgments of strangers and mates. *Journal of Experimental Social Psychology, 25*, 159-167

Kenrick, D.T. and Keefe, R.C. (1992). Age Preferences in mates reflect sex differences in human reproductive strategies. *Behavioral and Brain Sciences, 15*, 75-133.

Kenrick, D.T., Montello, D. and MacFarlane, S. (1985). Personality: Social learning, social cognition, or sociobiology? In R. Hogan and W. Jones (Eds.), *Perspectives in Personality* (Vol. I). Greenwich, CT: JAI Press.

Kenrick, D.T., Neuberg, S.L, Zierk, K. and Krones, J. (1992). Evolution and social cognition: Contrast effects as a Function of Sex, Dominance, and Physical Attractiveness:. Manuscript submitted for publication.

Kenrick, D.T., Sadalla, E.K., Groth, G. and Trost, M.R. (1990). Evolution, traits, and the stages of human courtship: Qualifying the parental investment model. *Journal of Personality, 58*, 97-117.

Kenrick, D.T. and Trost, M.R. (1987). A biosocial model of heterosexual relationships. In: K. Kelly (Ed.) *Males, females ans sexuality: Research and theory*. Albany, NY: SUNY Press.

Lakoff, G. (1987). *Women, fire, and dangerous things: What Categories reveal about the mind.* Chigago: Univ. of Chicago Press.

Leonard, J.L. (1989). Homo sapiens: A good fit to theory, but posing some enigmas. *Behavioral and Brain Sciences*, 12: 26-27.

Loehlin, J.C., Willerman, L. and Horn, J.M. (1988). Human behavior genetics. *Annual Review of Psychology, 39*, 101-133.

Lumsden, C.J. and Wilson, E.O. (1981). Genes, mind, and culture: The coevolutionary process. Cambridge, Mass.: Harvard.

Markus, H. and Zajonc, R.B. (1985). The cognitive perspective in social psychology. In G. Lindzey and E. Aronson (Eds.) *Handbook of Social Psychology,* (Vol. 1, pp. 137-230). New York: Random House.

Martindale, C. (1991). *Cognitive Psychology: A neural-network approach.* Pacific Grove, CA: Brooks/Cole.

Mealey, L. (1985). The relationship between social status and biological success: A case study of the Mormon religious hierarchy. *Ethology and Sociobiology, 6,* 249-257.

Nieschlag, E. and Michel, E. (1986). Reproductive functions in grandfathers.. In L. Mastroianni, Jr. and C.A. Paulsen (Eds.) *Aging, reproduction, and the climacteric,* Plenum Press.. Pp. 59-71.

Plomin, R. and Daniels, D. (1987). Why are children in the same family so different from one another? *Behavioral and Brain Sciences, 10,* 1-16.

Plomin, R., DeFries, J.C. and McClearn, G.E. (1990). *Behavioral genetics: A primer* (2nd edition). W.H. Freeman.

Presser, H.B. (1975). Age differences between spouses: Trends, patterns, and social implications. *American Behavioral Scientist,. 19,* 190-205.

Ralls, K. (1976) Mammals in which females are larger than males. *Quarterly Review of Biology, 51,* 245-276.

Sadalla, E.K., Kenrick, D.T. and Vershure, B. (1987). Dominance and heterosexual attraction. *Journal of Personality and Social Psychology, 52,* 730-738.

Sherry, D.F. and Shachter, D.L. (1987). The evolution of multiple memory systems. *Psychological Review, 94,* 439-454.

Simpson, J.A. and Gangestad, S. (1991). Individual differences in sociosexuality: Evidence for convergent and discriminant validity. *Journal of Personality and Social Psychology, 60,* 870-883.

Symons, D. (1979). *The evolution of human sexuality.* Oxford University Press.

Symons, D. (1992). What do men want? *Behavioral and Brain Sciences, 15,* 113-114.

Tellegen, A., Lykken, D.T., Bouchard, T.J., Jr., Wilcox, K.J., Segal, N.L. and Rich, S. (1988). Personality similarity in twins reared apart and together. *Journal of Personality and Social Psychology, 54,* 1031-1039.

Thornhill, N.W. and Thornhill, P.A.A.. (1992). The preferred age of a potential mate reflects evolved male sexual physiology. *Behavioral and Brain Sciences, 15,* 114-115.

Tooby, J. and Cosmides, L. (1989). Evolutionary psychology and the generation of culture, Part I: Theoretical considerations. *Ethology and sociobiology, 10,* 29-49.

Townsend, J.M. (1992). Measuring the magnitude of sex differences. *Behavioral and Brain Sciences, 15,* 115-116.

Townsend, J.M. and Levy, G.D. (1990). Effects of potential partners' physical attractiveness and socioeconomic status on sexuality and partner selection. *Archives of Sexual Behavior, 19,* 149-164.

Tranel, D. and Damasio, A.R. (1985). Knowledge without awareness: An autonomic index of facial recognition by prosopagnosics. *Science, 228,* 1453-1454.

Van den Berghe, P.L. (1992). Wanting and getting ain't the same. *Behavioral and Brain Sciences, 15,* 116-117.

Warner, R.R. (1984). Mating behavior and hermaphrodotism in coral reef fishes. *American Scientist, 72,* 128-134.

White, G.M. (1980). Conceptual universals in interpersonal language. *American Anthropologist, 82,* 759-781.

CAN BIOLOGY HELP PERSONALITY?
Commentary on the chapter by D.T. Kenrick

Willem K.B. Hofstee
University of Groningen
Department of Psychology
Grote Kruisstraat 2/1
9712 TS Groningen, The Netherlands

In the present context, disciplines like biology and subdisciplines like social psychology are auxiliary sciences. The psychology of personality has outgrown the stage in which traits were a dependent variable to be explained as an artifact of social perception, or to be reduced to biological differences. Upon considering personality as a topic in its own right, instead of an epiphenomenon, the question is what other disciplines can contribute to the study of personality. With respect to evolutionism, I shall defend the position that its contribution can be modest at best. In the process, I shall clarify the conception of personality psychology that underlies my argument.

Grand Theories of Human Nature

To anticipate an advance objection, there *are* conceptions of personality to which the evolutionary point of view is eminently relevant. One is personality in the sense of human nature - the wiring of the human species. Another is differential psychology, specifically, gender psychology. In both these areas, the evolutionary point of view offers grand theories of personality. I am not using 'grand' in any derogatory or ironical sense, but as a technical term to denote that the theory applies to common traits, not to individual differences (see also Buss, 1990).

Biological narratives of human nature have great intellectual significance. They provide sobering thoughts. In Western culture, voluntaristic interpretations of human existence have become dominant; opinion-makers celebrate the makeability of the person and deny or at least overlook any natural constraints. The popular psychology of gender, for example, emphasizes cultural determinants of sex differences. In the face of this emphasis, it is a sobering thought that males commit over 85 percent of all homicides irrespective of culture (Daly and Wilson, 1988), and that age preferences in mates are not explained by cultural variation but obey biological laws (Kenrick, this volume). The intellectual significance of such sobering thoughts is in the reduction of *hubris*: They teach us tolerance for own and others' limitations, and warn us against overoptimism with respect to the changeability of human nature.

Moreover, the come-back of biological psychology has had an important spin-off for personality psychology as an intellectual enterprise. For several decades, social psychologists have been trying to explain personality traits away by emphasizing situational determinants, attribution error, or both. Behavior-genetic evidence has liberated personality psychology from

J. Hettema and I. J. Deary (eds.), Foundations of Personality, 219–226.

this foreign occupation, and evolutionary psychology is playing an auxiliary role in the background.

However, all this is not to say that grand theory should be the aim of personality psychology. There is a very practical objection against such a view. The specific task for personologists is to explain behavior in terms of individual differences. No other branch of science does that, so if personality psychology did not exist, it would have to be invented. Not only do grand theories not serve that purpose; through their emphasis on generalities, they can only deflect the attention of personologists from their specific task.

Another way to phrase the argument is that a conception of personology as the provider *par excellence* of theories of human nature is presumptuous. Developmental psychology, social psychology, not to speak of general psychology, are at least equally entitled to exercise that claim. By virtue of its specific task, which is emphasizing individual differences in contrast to common traits, personology might even be the last to credibly claim any monopoly on the study of human nature.

The Psychometric Conception of Personality

The above argument can serve as a stepping stone to articulate a psychometric conception of the psychology of personality (see also, e.g., Hofstee, 1984; Wiggins, 1973). There is nothing essentially new here, so the exposition can be brief. On the other hand, there are times and contexts in which it is appropriate to apply oneself at such a restatement, and this may be one of those.

In the psychometric or trait conception of personology, the focal task of the subdiscipline is to predict individual differences in socially important behavior - the fact that Mary does well at a particular job or school whereas John does not; the fact that one reacts positively to a particular treatment whereas the other does not, and so on. For the prediction of these socially important individual differences in behavior one logically needs a predictor. Scores on a predictor variable are not socially important in and of themselves - there is nothing exciting at all about some number that is based on solving inconsequential puzzles or answering fairly superficial questions - but derive their relevance from their predictive power. Logically again, predictor variables and criterion behaviors must have something in common if the prediction is to be effective. This common content may be called a trait. We may theorize about traits in an attempt to make prediction more effective and efficient. However, the proof of the personality pudding is in predictive validities.

Personality textbooks usually list the trait approach along with the psychodynamic and social learning approaches to personality. However, this juxtaposition obscures an important difference between the psychometric view and other approaches. Psychodynamic and social learning theories try to explain how traits come about. In this perspective, traits are the dependent variable of developmental antecedents. The application that ensues from these paradigms is treatment, or trait change. The underlying reasoning is experimental, or manipulative. In the psychometric paradigm, test scores are predictor variables. The application perspective is selection (if that term is taken in a wide sense). The reasoning is correlational or observational. In sum, the explanatory paradigms like the psychodynamic and social-learning are retrospective, looking at the antecedents of traits; the psychometric paradigm is prospective, proceeding from measured trait variables onward (Hofstee, 1992). It is no use presenting the paradigms as alternatives or competitors. They span different worlds.

Evolutionary Theory and Individual Differences

In the context of the psychometric perspective, the question with respect to evolutionary theory is whether it can indicate where to look for promising individual differences. My tentative answer is that this is asking too much.

Naturally, individual differences have a place in evolutionary theory. When circumstances change, variety within the species is a condition for its survival. For example, if the ozone layer is further affected and if dramatic changes in climate materialize, it may be beneficial to the human kind not to be composed of Caucasians only, as they seem to be ill-equipped for such a climate. However, as these environmental changes are accidental, all that the species can do is keep an open mind, so to speak. The evolutionary principle does not dictate which traits should show large variance and which should not. Even on hindsight, it is not altogether clear why little variation is observed, for example, with respect to the number of fingers. Conceivably, our facility at binary calculation would have benefited if some of us would be born with two fingers on each hand. So all the evolutionary point of view has to say is that there be variety.

Going a step further, the viewpoint would seem to imply that existing variability is largely inconsequential for adaptive purposes, given the circumstances. For example, if extraversion were important for impressing potential mates, we would expect a highly negatively skewed distribution after so many generations. The large observed spread of a trait like extraversion therefore testifies to its unimportance. There is an interesting contradiction here with the lexical hypothesis which implies that spread of a trait goes together with its social importance. The primary explanation of individual differences in evolutionary terms is genetic noise. Therefore from that point of view, the big five factors of personality (Digman, 1990; Goldberg, 1992; John, 1990) would represent the summit of behavioral luxury and inconsequentialness.

In the evolutionary dialectic, some auxiliary hypotheses are found for explaining existing variety. One is frequency-dependent selection. For example, by the time the sweet young things would get almost completely overwhelmed by extraverted suitors, they would find rest in the arms of the few remaining introverted outcasts and give them the time of their evolutionary life. The other is the parasite script in which the extraverted suitors would warm up more potential mates than they can serve, the remainder being taken care of by introverts who have kept their powder dry. With respect to personality psychology, it is difficult to evaluate these notions. There is no doubt about their potential to serve in ad-hoc explanations, but it is quite a different thing to have such mechanisms predict the major dimensions of personality.

In one respect the evolutionary perspective may appear to be directly relevant to personality theory. It has been noted that the big five factors of personality are differentially associated with the biological sex variable. Due to differences in the rotational positions of the axes of the five-space, it is not altogether clear what factor or factors carry most of the sex variance. For example, in the American-English Big Five solution (Goldberg, 1992) the Extraversion or Surgency factor comes out pretty macho, whereas the Agreeableness factor is a bit on the soft side. In the Dutch solution, biological sex is most associated with Factor IV, Emotional Stability. A good idea would be to have one factor coincide with biological sex, that is, to rotate towards that variable. Only social taboos seem to have refrained taxonomists from using this anchoring point.

Ideally, four more such anchors would be needed to fix the positions of the five major dimensions of personality. In the absence thereof, I consider all interpretations of the Big Five to be premature: These interpretations are bound to be based upon a rotational position that can only be arbitrary to some extent. One may hope that some day biological variables, probably genetic in kind, will help to secure the axis positions. In the meantime, these positions should be kept flexible. The AB5C model (Hofstee et al., 1992) is a step towards this requirement.

The Abridged Big-Five Circumplex (AB5C) Model

Indeterminacy of rotational positions of axes of a trait space is incorporated in circumplex models (Wiggins, 1980) of personality. In these two-dimensional models, there is no presumption of a simple structure that would fix the positions of the two axes. Instead, the angular positions of the trait variables are thought to spread more or less evenly. Usually, the plane is divided into octants, clustering traits that are close together. However, that is an arbitrary abridgement. The emphasis in circumplex models is upon the freedom of rotation; the axes of the plane, and the ensuing positions of the octant boundaries, form no part of the unabridged circumplex model depicting only the angular positions of the trait variables.

A full five-dimensional circumplex is insufficiently parsimonious as a taxonomic model. The AB5C representation contains two abridgements. First, trait variables are represented by their projections on the two-dimensional slice of the trait sphere to which they are closest. There are ten such slices (circumplexes), one for each pair out of the five factors. This abridgement is justified by the empirical finding that very few variables have substantial loadings on more than two factors (Hofstee and De Raad, 1991; Hofstee et al., 1992). Second, the planes are divided into twelve segments of 30 degrees (duodecants), and traits within one such segment are considered to be equivalent. As a result, traits are represented by the factor poles on which they have their highest, and their second highest loading. An AB5C solution is economically presented by a 10x10 table, with the ten poles of the five factors as rows and as columns. The cells of the table contain trait variables. The column represents their primary loading, the row, their secondary loading. Of the 100 cells in the table, ten represent combinations of the positive and negative poles of the same factor and are thus empty by definition; other cells may be empty empirically.

Taxonomic representations such as the AB5C model are to be judged on their emergent properties. The following properties may be considered:

Trait clustering. The AB5C approach is a case of soft modelling, as opposed to simple-structure models which are relatively harsh. In a simple-structure model, variables are assigned to the factor on which they load highest. Simple structure may thus be considered to be an abridgement of a circumplex, using quadrants rather than octants or duodecants: All traits are clustered together that are less than 45 degrees away from a particular factor pole. With five factors, there are just ten such clusters or segments. Within a cluster, two trait variables may be nearly orthogonal to each other. Generally, a simple-structure approach produces very heterogeneous clusters. (Usually, factor analysis is opposed to cluster analysis; upon interpreting factors, however, marker variables are drawn together, so that implicitly a cluster analysis is taking place).

The greater subtlety of the AB5C approach is best illustrated by considering marker variables for a factor. A simple-structure solution would take the highest loading variables, without systematic attention to their loadings on other factors. In an AB5C solution, the primary loading of a factor-pure variable is guaranteed to be at least 3.73 times as high as its loading on any other factor, 3.73 being the cotangent of an angle of 15 degrees: A variable with a greater angle gets assigned to a segment representing a mixture of factors. An investigator interested in finding marker variables for a simple-structure solution would therefore be well-advised to carry out the AB5C-procedure.

Assignment to circumplexes. A similar argument applies with respect to two-dimensional circumplex solutions. The Interpersonal Circumplex (Wiggins, 1979) is now generally considered to consist of the Big Five Factors I and II (McCrae and Costa, 1989; Trapnell and Wiggins, 1990). However, if only one slice of the five-dimensional space is considered, trait variables may be included that have a higher projection on one of the nine other slices (Hofstee et al., 1992). Thus even if theoretical reasons would lead one to single out a particular circumplex for consideration, the advice would be to run a complete AB5C analysis first.

Both simple-structure and two-dimensional circumplex solutions are special cases of the AB5C model. Not only is the latter model superior in that technical sense, it also corrects certain shortcomings of the former models by taking their wider context into account.

Facets. Through the use of duodecants rather than octants in dividing the circumplexes, facets of factors are formed through the admixture of other factors. For example, the segment containing trait variables with their highest loading on the positive pole of Factor I and non-negligible secondary loadings on the positive pole of Factor II (the I+II+ segment) forms a facet of Factor I; the II+I+ segment forms a facet of Factor II. The six most representative trait adjectives (American-English AB5C solution, see Hofstee et al., 1992, p. 156/7) for these two facets, together with the factor-pure (I+I+ and II+II+) segments, are listed below:

I+I+	I+II+	II+I+	II+II+
talkative	sociable	merry	sympathetic
extraverted	social	cheerful	kind
aggressive	enthusiastic	happy	warm
verbal	communicative	friendly	understanding
assertive	spirited	effervescent	sincere
unrestrained	energetic	jovial	compassionate

Potentially, each of the ten factor poles has eight facets resulting from the admixture of the positive and the negative pole of each of the four other factors, in addition to the factor-pure facet. Four of the remaining facets of Factor I+ are listed below:

I+III+	I+III-	I+IV+	I+IV-
active	boisterous	confident	flirtatious
competitive	mischievous	bold	explosive
persistent	exhibition.	assured	wordy
proud	immodest	uninhibited	extravagant
	gregarious	courageous	
	demonstrative	brave	

These facets do not come about by using a hierarchical model, which would violate the basic tenet that five dimensions are sufficient to cover the personality sphere (the violation consisting of postulating a separate factor for each facet): Each AB5C facet is contained within the five-dimensional space. As many of the facets that are constructed by authors using hierarchical models will easily be recognized as mixtures of the Big Five, the AB5C approach is more parsimonious.

Chiasmic configurations. Finally, an intriguing taxonomic configuration usually referred to as the Peabody plot, appears to be subsumed and generalized by the AB5C model. Peabody (1967) constructed fourfold tables such as the following:

generous	thrifty
extravagant	stingy

The left-right distinction is interpreted as substantive (in this case, ease of spending versus close-fistedness), whereas the vertical distinction is interpreted as social desirability versus undesirability. Opposite terms are found along the diagonals of the Table (generous versus stingy and thrifty versus extravagant).

If one of the Big Five factors is taken to represent social desirability (or is made to do so through a target rotation), these chiasmic configurations could be constructed automatically. For example, pitting Factor I against Factor II (which tends to congrue most with social desirability) results in:

	I+	I-
II+	sociable social enthusiastic	timid unaggressive submissive
II-	dominant domineering forceful	unsociable uncommunicative seclusive

Again, the left-right distinction is substantive, the common denominator being surgency versus lack of it; the vertical distinction may be interpreted in terms of social desirability (however, see below); opposite terms are found along the diagonals (e.g., sociable vs. unsociable; dominant vs. submissive). For each of the three remaining Factors, a similar set of chiasms could be constructed in principle.

However, interpreting one of the five factors as social desirability is probably not a good idea, if only because purely evaluative terms are usually removed from the item set. Target rotation toward maximal congruence with the vector of item desirabilities will not work in the manner of rotating the factors toward biological sex: Social desirability is not a person variable, so it is not possible to have one of the factors coincide with it and remove all correlation with the remaining factors.

Abandoning the idea of a special function for one of the factors leads to a further generalisation of the chiasmic model: Any combination of two factors can be used to set up a chiasmic configuration. For example, take I and III:

	I+	I-
III+	active competitive persistent	reserved restrained serious
III-	boisterous mischievous exhibitionistic	unenergetic uncompetitive sluggish

There are 20 ordered pairs of factors, giving so many generalised chiasms. Not all of these are well-filled. However, I have demonstrated a fundamental correspondence between the chiasmic, circumplexical, and AB5C taxonomic models. More precisely, chiasms are contained in circumplexes, which in their turn are contained in the AB5C configuration (as are simple-structure solutions).

Concluding Remarks

In addition to the properties emphasized above, the AB5C model to some extent responds to the need for keeping the description of personality flexible, given the absence of cogent anchors for fixing the positions of the axes. For example, it conceives of sociability (I+II+), often confused with extraversion, as a mixture of extraversion (I+I+) and warmth (II+II+); it writes impulsivity (III-IV-) as a blend of disorderliness (III-III-) and moodiness (IV-IV-). The implication is that the axes could equally well be drawn through these mixtures, promoting the blends to factor-pure status and degrading other segments to bastard status. In fact, some of the labels that are traditionally used to indicate the Big Five Factors appear as blends in the AB5C solution, notably, Agreeableness (II+I-), Conscientiousness (III+II+), and Intellect (V+IV+), testifying inadvertently to the large margins of the factor positions. Furthermore, the facet structure takes away the need for postulating a new dimension for each favorite concept that is not precisely captured by a simple-structure solution: the chances are that it can be represented as a mixture. The solution is thus both flexible and robust.

A price is paid for the abridgements applied to the full five-dimensional circumplex structure. If axes are rotated over angles different from 30 degrees, traits within a cluster do not stay together but are distributed over adjacent segments. Also, the sizes of the pure-factor segments differ somewhat from the sizes of the other segments. This problem is comparable to constructing a round leather ball, for which both hexagonal and pentagonal pieces are needed. However, these problems do not detract greatly from the ease with which rotations can be conceived.

The AB5C structure is thus receptive to future findings concerning the anchoring of the factor space. I have expressed some doubts as to whether evolutionary psychology will be of much help in this respect. Behavior-genetic research may be more promising.

References

Buss, D.M. (1990). Toward a biologically informed psychology of personality. *Journal of Personality, 58,* 1-16.

Daly, M. and Wilson, M. (1988). *Homicide.* New York: Aldine.

Digman, J.M. (1990). Personality structure: Emergence of the five-factor model. In M.R. Rosenzweig and L.W. Porter (Eds.), *Annual Review of Psychology: Vol. 41* (pp. 417-440). Palo Alto, CA: Annual Reviews.

Goldberg, L.R. (1992). The development of markers of the Big-Five factor structure. *Psychological Assessment, 4,* 26-42.

Hofstee, W.K.B. (1984). What's in a trait: Reflections about the inevitability of traits, their measurement, and taxonomy. In J.C.J. Bonarius, G.L. van Heck and N.G. Smid (Eds.), *Personality psycholgy in Europe: Theoretical and empirical developments* (pp. 75-81). Lisse, The Netherlands: Swets.

Hofstee, W.K.B. (1992). *Persoonlijkheidseigenschappen: de caleidoscoop van individuele verschillen* [Personality traits: The caleidoscope of individual differences]. Francqui lecture, Free University of Brussels, 22 May.

Hofstee, W.K.B. and De Raad, B. (1991). Persoonlijkheidsstructuur: de AB5C taxonomie van Nederlandse eigenschapstermen [Personality structure: The AB5C taxonomy of Dutch trait terms]. *Nederlands Tijdschrift voor de Psychologie, 46,* 262-274.

Hofstee, W.K.B., De Raad, B. and Goldberg, L.R. (1992). Integration of the big five and circumplex approaches to trait structure. *Journal of Personality and Social Psychology, 63,* 146-163.

John, O.P. (1990). The "Big Five" factor taxonomy: Dimensions of personality in the natural language and in questionnaires. In L.A. Pervin (Ed.), *Handbook of personality: Theory and research* (pp. 66-100). New York: Guilford Press.

McCrae, R.R. and Costa, P.T. (1989). The structure of interpersonal traits: Wiggins' circumplex and the five-factor model. *Journal of Personality and Social Psychology, 56,* 586-595.

Peabody, D. (1967). Trait inferences: Evaluative and descriptive aspects. *Journal of Personality and Social Psychology Monographs, 7* (Whole No. 644).

Trapnell, P.D. and Wiggins, J.S. (1990). Extension of the Interpersonal Adjective Scales to include the Big Five dimensions of personality. *Journal of Personality and Social Psychology, 59,* 781-790.

Wiggins, J.S. (1973). *Personality and prediction: Principles of personality assessment.* Reading, Mass.: Addison-Wesley.

Wiggins, J.S. (1979). A psychological taxonomy of trait-descriptive terms: The interpersonal domain. *Journal of Personality and Social Psychology, 37,* 395-412.

Wiggins, J.S. (1980). Circumplex models of interpersonal behavior. In L. Wheeler (Ed.), *Review of personality and social psychology* (Vol. 1, pp. 265-294). Beverly Hills, CA: Sage.

MARGINAL DEVIATIONS, AGGREGATE EFFECTS, DISRUPTION OF CONTINUITY, AND DEVIATION AMPLIFYING MECHANISMS[1]

Gian Vittorio Caprara
Dipartimento di Psicologia,
Università di Roma "La Sapienza"
Via Degli Apuli 8
00185 Roma, Italia

Abstract

Children at risk show a profile of traits and response dispositions that deviate from the norm (of other children) in only marginal ways. While there may be no difference at all on some dimensions, on others the differences are slight, may not even be statistically significant, but are reliably present. Nevertheless, this composite profile does predict subsequent anti-social outcomes for children who are so defined.

Over time, originally small marginal deviations may be amplified and become significantly different from the pattern typical of peer cohorts. It is hypothesised that the causal path by which marginal deviations are transformed into macro-deviations includes:

1) *aggregation of disturbing effects on teachers and peers generated by the slightly deviant reactions;*
2) *lack of awareness in both teachers and peers of what causes the source of disturbance which may break the continuity of the interpersonal relations;*
3) *deviation amplifying mechanisms such as "contrast" associated with withdrawal and rejection.*

Introduction

In these notes, I intend to develop further some previous reflections on early indicators of social relations (Caprara et al., 1988; Caprara and Pastorelli, 1989).

I intend, in particular, to consider the prominence that the notion of "marginal deviation" has come to assume in the development of a model with which to analyse and act upon the precursors and the moderators of psychological and social maladjustment. In fact, for some years now, my co-workers and I have been carrying out a range of observations and experiments involving primary school children with the aim of discovering and understanding the factors that carry the risk of producing and consolidating maladjusted interactions.

We began by focusing on aggressive behavior. Our main concern, at first, was to develop and validate tools for measuring the propensity to behave aggressively, emotional instability, and prosocial behavior orientation. Previous research activity on the process of regulating

J. Hettema and I. J. Deary (eds.), Foundations of Personality, 227–244.

aggression guided our choice both of the object of research and of a largely correlational approach.

As regards research in the laboratory and with adults, a developmental perspective seemed to offer the key to a better and deeper understanding of the webs of affects, cognitions, behaviours and relationships that underlie the building and the perpetuation of aggressive behaviour.

Previous research had shown that it was useful to distinguish, as far as possible, between displays of aggression in which a broad and often sophisticated involvement of cognitive regulatory process was apparent, as in the case of premeditated offense and vindictiviness, and displays of aggression in which, on the other hand, cognitive functioning was impaired in concomitance with the alteration of excitatory functions (Caprara, 1987). Findings of studies on emotional susceptibility and irritability had suggested investigating whether such dimensions correlated as highly with aggression disposition among children as they had done among adults.

The literature, for its part, led us to pay particular attention to the links between emotional stability and aggressive disposition, both because of the frequent positive correlations found between lack of emotional control and aggression, and because of the high predictive value attributed to both dimensions as regards subsequent deviant outcomes (af Klinteberg and Magnusson, 1989; af Klinteberg et al., 1990). The literature also suggested examining prosocial dispositions as counterpoise to aggressive disposition (Eron and Huessman, 1984; Parke and Slaby, 1983).

A series of factorial analyses enabled us to identify our three constructs and to develop three scales with which to measure them, in the form of self-assessment questionnaires. Several replications confirmed the validity and reliability of the three self-assessment tools, even for seven-years-old children at the end of their second year of primary school. Other studies confirmed our expectations of high positive correlations between emotional instability and aggressive disposition, high negative correlations between aggressive disposition and prosocial disposition, and a strong level of agreement among self-assessment, teachers' assessment, parents' assessment and peer assessment (Caprara and Laeng, 1988; Caprara and Pastorelli, 1992). It therefore seemed plausible to identify children whom the various assessments agreed had a highly aggressive disposition, were highly unstable emotionally and whose prosocial disposition was slight, as "children at risk".

This was the foundation for a new phase of research in which so-called "children at risk" were systematically compared with so-called "normally-adjusted" children in a variety of experiments designed to investigate their possession and degree of development of particular cognitive and social skills in broad sense associated with the notion of "social intelligence" (Cantor and Kilstrom, 1986) which seemed essential for achieving and maintaining adequate interactions.

To this extent we examined the following: the ability to recognize emotions from facial expression, following Ekman and Friesen's paradigm (1975); the ability to infer specific affects from specific events, following Weiner's paradigm (1986) concerning the role of attribution processes in regulating emotions and motivation; and preferred forms of social influence and those felt to be most effective, following French and Raven's paradigm (1959).

Since most of these studies have already been published, or at least reported in other published studies (Caprara et al., 1988; Caprara and Pastorelli, 1989), I do not feel that a detailed description here would be pertinent to what follows. Suffice it to note that:

a. we used Weiner's paradigm to examine the ability to recognize the importance that an individual's degree of control over an event has in relation to the emotional reactions that such event is likely to arouse in other people;
b. we used Ekman and Friesen's paradigm to examine the ability to distinguish the six basic emotions by recognizing the universal characteristics of their facial expression;
c. we used French and Raven's paradigm to examine the ability to conform to the generally shared judgment that modes of social influence or power that require some type of "surveillance", such as reward and punishment, are less effective than those that are independent of any "surveillance", such as information or example.

Marginal Deviations

As can be intuited from the above, the original interest in studying indicators and precursors of aggressive behavior has developed into a broader interest in studying those processes and exchanges which, underlying the construction of social relations, also lay the foundations for having recourse to aggressive behavior, and hence supply the conditions for consolidating and perpetuating such behavior. Indeed, I am convinced that aggressive behavior is nothing more than an epiphenomenon of the dynamic, mutually determinant interaction between the individual and the enviromnent. Behavior itself, that which we register with our senses and which becomes the object of awareness and maybe even of measurement, is nothing more than an epiphenomenon of the network of exchanges that supports and articulates it. The behavior we notice is always something "intermediate", whose beginning and final outcome generally escape our full awareness and control.

To speak of precursors implies reducing the gap between beginning and appearance. As regards aggression, if I may be allowed to link it to illness, it implies acting upon incubation processes and not ignoring processes of convalescence if we want to avoid relapses. Focusing on the phenomena implies taking into account the plot of the exchanges: who does what, and what they get in return. It implies identifying the components and the results of continual interactions, combinations and aggregations. And it implies penetrating that flow of affects, cognitions and reactions which organize behavior and mark the building of experience through the feelings that they arouse in others and the opportunities which these feelings reveal and maintain. It is therefore inevitable that we go beyond manifest behavior to the processes by which it is built.

At the same time, we have to reconsider the appropriateness of the models and units of measurement that have traditionally guided our research. This implies, as suggested by Magnusson (1990), a change of paradigm from a variable approach to a person approach and a shift of focus from responses to processes. In this regard an original sensitivity towards that which is subtle, nuanced, tenuous, indistinct and marginal becomes crucial. Marginal deviations from generally shared modes of functioning and systems of expectations may also prove crucial in a developmental process which is largely mediated by other people's expectations.

Within an interactional perspective, whereas aggressive behavior figures as a symptom of a defect in the project of adjustment, which progresses along the sequence of transactions that take shape between the individual and the environment, the rejection by others that accompanies such behavior assumes a critical role as cause and effect in its production and reproduction.

The literature points out that aggression is one of our most stable personality characteristics (Olweus, 1979): an aggressive child is more likely to become an aggressive adult than a sociable or conscientious child is to become an equally sociable or concientious adult.

Rejection, for its part, is shown to be other people's most frequent reaction to aggression. A variety of data and lines of reasoning give equal support to hypotheses which see rejection, such as that by a mother (Olweus, 1980), as primarily a cause, and those which, in contrast, see rejection, for instance by peers (Parker and Asher, 1987), as a highly common or probable effect. In the face of this dilemma, I believe that it is worth going beyond the objective manifestation of both aggressive behavior and of rejection to investigate the system of affects, representations, meanings, rules and shared expectations which constitute the subjective and relational web underlying the two behaviors, and which may explain the links between the two phenomena.

At this point, the importance of the subtle, the nuanced and the marginal comes to the fore, because the components of the plot are not of the same order of magnitude as the noticeable behaviors; they are not equally visible, and they are not so easily accessible to consciousness.

It is precisely to this web that our interest has turned: firstly, to the child's ability to become aware of the invisible rules which others share, and which it is presumed everyone should share in rejoicing, in worrying, in desiring, in inferring, in predicting; and secondly, to the extent to which the social environment is prepared to support the child's development of "social intelligence" (Cantor and Kilstrom, 1986). We thus began to examine the web of emotional experience of "children at risk", especially their knowledge and possession of "social grammar", in other words of that set of shared rules that define how one should behave and how it is reasonable to expect others to behave. The first pointers to emerge from our data, and which we intend to examine more closely, suggest that "children at risk" are "ungrammatical" children: they know the language, but the use they make of it is not wholly appropriate.

Such pointers, however, fail to emerge from a traditional-type reading of the data. Individual comparisons among groups only yielded significant differences in a limited number of experiments; and these, taken in isolation, would at most allow us to write further papers on the robustness of the effects predicted by Weiner's theory, and on the significant role that individual differences may play in extending the range of related hypotheses (Caprara, 1987). The picture changes, however, when comparisons are made cumulatively, and when we dwell upon the "added value" due to the accumulations of deviations from the norm which, among children at risk, are generally in the less favored direction.

Aggregate Effects

Children at risk prove to be less sure in indicating the emotional reaction generally associated with the attribution of specific causes; they are more uncertain in recognizing emotions from facial expression; they attach greater importance to coercive strategies. Our studies have shown that generally normal children have an extremely clear understanding of rules regarding causal controllability (responsibility, accountability) (see for more details Weiner et al., 1982; Weiner and Handel, 1985).

Having a somewhat uncertain grasp of the links between effort and failure that may prompt reactions of blame (for failure through lack of effort) or else of indulgence (toward failure despite effort) is not especially worrying in itself, even though expecting blame for failure despite effort may preclude the possibility of taking advantage of other people's willingness to help.

For the same-age children (7-9 yrs. old), our studies have also shown that normal children are in general capable of recognizing the six basic emotions (happiness, sadness, anger, fear, disgust, surprise) when presented the series of slides portraying the facial expressions (see for more details Ekman and Friesen, 1975).

Being uncertain about recognizing the facial expression of an emotion like sadness is not worrying if it only happens a few times, even though, on those occasions, it may affect our ability to help other people.

Children, like adults, are able to appreciate the different value of forms of social influence or power based upon informational, expert and legitimate means versus those of coercion and reward (see Schmidt and Raven, 1985).

Overestimating the importance of "surveillance", as in the case of reward and punishment, may prove a barrier to trying out forms of persuasion, such as convincing through exchange of information, example and authoritativeness, which are definitively more economic, longer lasting and more "socially approved" modes.

Furthermore, it is not necessary to deviate significantly from the norm to violate, however minimally, the system of shared expectations, and thereby eventually to annoy those with whom one interacts by introducing unforeseen elements of indeterminacy and hence of discontinuity into social relations.

For the most part, the nuisance generated by the accumulation of so many little imperfections, mistakes and uncertainties on the part of those who do not have a full command of the grammar of social relations is a subtle one. It is not merely a nuisance, however, for it commits the interlocutor to continual corrections and translations.

In recent years, the literature on personality has brought out the importance of aggregated measurements in counting for the stability and continuity of behavior (Epstein and O' Brien, 1985). Our own research has convinced us that equal importance should be accorded not only to the aggregation of situations, time intervals and individual reactions, but also to the aggregation of the effects that these produce in other people.

Micro-deviations along a variety of dimensions of social interaction, in none of which the individual's behavior is significantly deviant, may, when aggregated, arouse feelings and reactions in others that may eventually launch a carreer of deviancy. They introduce into relationships elements of unpredictability which violate others' expectations of continuity. It is therefore plausible that the accumulation of elements of discontinuity provokes anxiety, which in turn generates reactions of avoidance and distancing if these may serve to reduce the unease and annoyance being felt. Withdrawal and avoidance are indeed effective, socially appropriate measures with which to express one's unease and intolerance especially when it seems impossible to explain the cause of the feeling and hence to take alternative action against it.

Ordinary people are not aware of the theory that attribution processes guide emotional reactions, but they intuitively know that failure because of lack of effort by those with ability is blameworthy and provokes anger in others, whereas failure because of lack of ability and in the presence of effort deserves indulgence from others. They are amazed when this does not occur, and their amazement turns into anxiety and intolerance when it repeatedly fails to occur. Ordinary people have no difficulty distinguishing happiness from sadness in other people's facial expression, and in either case they know how to behave and how it is reasonable to expect others to behave. What, on the other hand, they do not know, or are inclined to overlook, is that it is less easy to conform to such expectations if one is uncertain about recognizing happiness and sadness, due to an interpretation deficit and not to indifference or hostility. The deficit that exposes an individual to risk is therefore not only that which underlies a deviance apparent to everyone; rather, it is also the micro-deficit that escapes most people's attention.

The micro-deviation that carries little weight in itself becomes significant the moment it is added to other micro-deficits with which it becomes confused and whose anonymity it shares.

Perversely, although the accumulation of various micro-deviations increases annoyance, it also further confuses the already difficult task of recognizing the individual elements causing it. Unfortunately, when other people's withdrawal and avoidance take the form of rejection, their effects are by no means innocuous.

Continuity and Disruption of Continuity

Within the debate on the stability/continuity of behavior, concerning the validity of the notion of traits and the possibility of depicting the personality as a relatively stable, organized configuration of traits or disposition, Caspi et al., (1989) rightly draw attention to the retroactive and reflexive influences that behavior ultimately exerts upon itself as a result of the consequences that it produces. With this in mind, they suggest making a distinction between cumulative continuity and interactive continuity. The former refers to the kind of continuity which derives from an individual interactive style that leads the individuals into situations reinforcing that very style through the progressive accumulation of its consequences. The latter refers to the kind of continuity which establishes itself when an individual style evokes reactions in others that end up perpetuating that same style. Their example of cumulative continuity is the phenomenon examined by Patterson (1988) of the child whose displays of aggression provoke aggressive reactions by the parents, which in turn lead to more aggressive displays by the child in a spiral in which the child becomes both victim and architect of a coercive system. Indeed, the aggressive reactions with which the child responds to parental punishment ultimately lead the latter to renounce any attempt at imposing discipline, thereby confirming to the child the effectiveness of his/her behavior and dissuading him/her for trying out other interactive strategies than that of aggression. Their example of interactive continuity is the phenomenon examined by Dodge (1986) of the aggressive child who expects others to be equally aggressive and accordingly acts toward them in a way that eventually provokes their aggression, thereby confirming the child's expectations and, implicitly, the validity of his/her strategy. The distinction between cumulative continuity and interactive continuity is a subtle one and by no means irrelevant. It fails, however, to explain how one type of continuity can turn into the other, as happens when a disposition is subsumed by another with different properties and connotations.

Our own data suggest that teachers and children share a tendency to confuse emotional instability with aggression. Indeed what we are really dealing with are phenomena which are different despite connected features. In fact, there are substantial differences in terms of the causes and processes that determine and regulate their various respective manifestations, in terms of the social connotations that such manifestations may take on, and, consequently, in terms of the treatments to which they may be subjected. It is only as regards their effects that instability and aggression may be treated as the same, since both are disturbing and as such they both may elicit from others the same aversive reactions. For their part, the frequent disruption of the continuity of the shared rules of social relations that the child "at risk" occasions through being "socially ungrammatical" is also disturbing.

The continuity of relations that derives from stability of mutual expectations is just as important as the continuity and stability of behavior with respect to the need to generate and to maintain a shared interactive context. The marginal deviations discussed above may thus represent repeated attacks on the continuity of relations based on a system of shared rules. In this respect, it is possible to invoke the indulgence on the part of the interlocutors, so that the need for coherence will ensure that such micro-deviations are "normalized". I nevertheless fear that beyond a certain threshold, the value added by the accumulation of micro-deviations may trigger mechanisms which push the deviation further from the norm. When this happens, it becomes even more probable that a child who disturbs because she/he is ungrammatical or emotionally unstable will become an aggressive child largely independent of whether or not she/he is actually more aggressive than the norm.

The label "aggressive" is in fact much more socially and cognitively "accessible" than the label "emotionally unstable" or other, even more sophisticated labels. Moreover, it is inevitable that the increase in the perceived deviation should raise the probability of isolation and rejection, and that these, in turn, should trigger counterreactions of withdrawal and aggression which further "confirm" the appropriateness of the label. A sort of collusion between the subject's behavioral inadequacy and the interlocutor's interpretative inadequacy therefore makes it possible for a disruption of continuity (in the relationship) to transform or subsume a dimension of individual continuity (emotional instability) into another dimension, in which it opens a career (aggression). Other people's reactions therefore assume a crucial position as mediators between an individual and his/her behavior. In the particular case of aggressive behavior at school, the processes underlying the reception and the reduction of deviations from normative standards, the processes of attributing causality and of labeling, and the reactions of acceptance or rejection that derive from such processes and accompany them, all appear to be important.

Assimilation, Contrast and the Cost of Adjustment

The problem of how marginal deviations can be amplified or attenuated in relation to the environment and in connection with the reactions they give rise to is not new. In a broad sense this may be referred to more general phenomena of negative and positive feedback in the context of cybernetics speculations on mutual causal systems (Maruyama, 1963). Whereas negative feedback between their elements lead deviation- counteracting systems to restore and maintain previous equilibrium, positive feedback between their elements lead deviation-amplifying systems to amplify unsignificant or accidental behavioral expressions, to build up deviation and to finally diverge significantly from the initial condition.

As noted by Maruyana (1963) "in contrast to the progress in the study of equilibrating systems, the deviation-amplifying systems have not been given as much investment of time and energy by the mathematical scientist on the one hand, and understanding and practical application on the part of geneticists, ecologists, politicians and psychotherapists on the other hand" (p. 164). Nor have I found any reference to the processes which cognitively may mediate the negative and positive feedback in terms of "contrast" and "assimilation". Indeed these seem to me pertinent psychological processes for providing an explanation of possible outcomes in terms of attenuation or, rather, of amplification. These are processes originally investigated in the context of perception of physical phenomena and which subsequently have been particularly extended in the context of perception of social phenomena (Helson, 1964; Pepitone and Di Nubile, 1976). In cases of series of phenomena the process of maintaining perceptual constancy by means of attenuation of differences has been called "assimilation"; and the symmetrical process of amplification of differences, "contrast". Apparently when various cases are serially submitted to judgment, the judgment which precedes provides an anchor to the judgment which comes later and the discrepancy among the former and the latter determine the threshold with which shifts are assimilated and attenuated and by which shifts are contrasted and amplified.

Dealing explicitly with social stimuli in their social judgment theory of attitude change, Sherif and Hovland (1961) have proposed that the judgment of a communicator's position on a pro-con extremity scale with respect to a given attitude issue is strongly influenced by the individual's own attitude position which serves as a standard of comparison or anchor. When there is an objectively large discrepancy between the communicator's and the individual's attitude position, i.e., when the former is outside the range of positions acceptable to the individual - the judgment of the communicator's position is displaced away from this "latitude of acceptance". When there is a small discrepancy between the communicator and individual, the former position is assimilated, i. e., judged to be within or near the latitude of acceptance.

Apparently,with respect to an attitude position to which judgment becomes anchored, the latitude of acceptance represents the border within which shifts are assimilated and attenuated and beyond which shifts are contrasted and amplified. In the case of a child's conduct, whereas the norm-anchor is likely to be what results from the average conduct of other children, the attitude position, which serves as standard of comparison, corresponds to what is held to be the prototypical child's conduct, on behalf of peers, teachers and parents in the various settings. Therefore, it is likely that similar mechanisms of contrast and assimilation also affect the perception of children's conduct on behalf of peers and teachers. However, since the child finds itself acting in various contexts it is plausible that the same shifts are differently assimilated or contrasted in the family, at school, and with playmates.

Also it is evident that this diversity of treatment has notable implications with regard to a judgment of coherence, stability or continuity of conduct, and with regard to the perpetuation or amplification of the same shifts. Whereas the child's conduct is constantly confronted with the concrete expressions of other's behavior and with the mental representations and expectations of various perceivers, the level of discrepancy which is tolerated in the various contexts differently may lead to an amplification of deviations via contrast or to an attenuation of deviations via assimilation.

It is plausible that affect and indulgence in the family provides ideal social support conditions not so much for an encapsulation of the shift, as for an attenuation of the processes of stigmatization in favour of processes of gradual recovery and correction. Moreover, mechanisms of denial cannot be excluded which end up leaving the child even more unprepared when his difficulties are noted and stigmatized outside of the family. The conditions in which relationships develop with playmates and in the classroom with the teacher and with schoolmates are likely to leave less room for indulgence. Resources in the form of the attention and patience of the educator are more limited; there is less space for gradual recovery action; the collective requests for adjustment are more pressing; risks of isolation and refusal and, thus, of an amplification of the shifts are greater.

The situation is obviously critical because in the case of marginal deviations by definition it is difficult for the family and the educator to anchor their different perceptions of the child's behavior to specific events, to find elements upon which to develop mutual understanding and accordingly to implement a common action aimed at a better support of the child.

The situation becomes dramatic when the discrepancy between family indulgence and extra-family intolerance places the child in the critical situation of not knowing what to do and to whom to pay attention to avoid mistakes, because whatever he/she does he/she ignores someones' expectations and demands. It may happen in a similar fashion to what Watzlavick et al. (1967) have referred to in terms of "paradoxical communication": the child becomes the target of incompatible pressures he/she cannot avoid on behalf of relation systems upon which he/she depends. Since what disturbs the continuity of the relationship is not understood, the efforts of the child to meet the pressures of the world of companions and adults and the efforts of the educator to free the child's potential are in vain; and it is inevitable that the accumulation of discouragement ends up well beyond marginal deviations. In these cases the cost of adjustment becomes extremely high.

Tacit Knowledge and Tacit Ignorance

To go beyond conjectures into the sphere of action, my colleagues and I designed a study using material gathered on four "at risk" children and four "well-adjusted" children. Children at risk were selected on the basis of their extreme scores on aggression as rated by themselves, by peers and by teachers, and on the bases of high scores on rejection on behalf of peers. Similarly, well-adjusted children were selected on the basis of their low scores on aggression, as rated by themselves, by teachers and by peers, and high scores on popularity as rated by their peers.

A record was prepared for each child including: a) questionnaires to assess emotional instability, prosocial behavior and physical and verbal aggression; b) the various answers provided by the children on the various tasks associated with: 1) inference of emotions, inference of anger in presence of lack of effort, inference of pity in presence of lack of ability; expected anger for controllable reasons and revealing uncontrollable causes in cases of a broken social contract, 2) recognition of facial expression of emotions (happiness, sadness, anger, fear, surprise and disgust) ; 3) choice and efficacy of different forms of social influence (informational, referential, legitimate, expert, coercive, reward). In order to avoid phenomena of uncontrolled sequence, the various stimuli were presented in random order.

Figures 1, 2, and 3 show the average profiles of "well adjusted" (popular) children and of "at risk" (rejected) children resulting from various tasks.

Fig. 4 provides a summary picture of inferences given in terms of accuracy indices according to the rationale of each task.

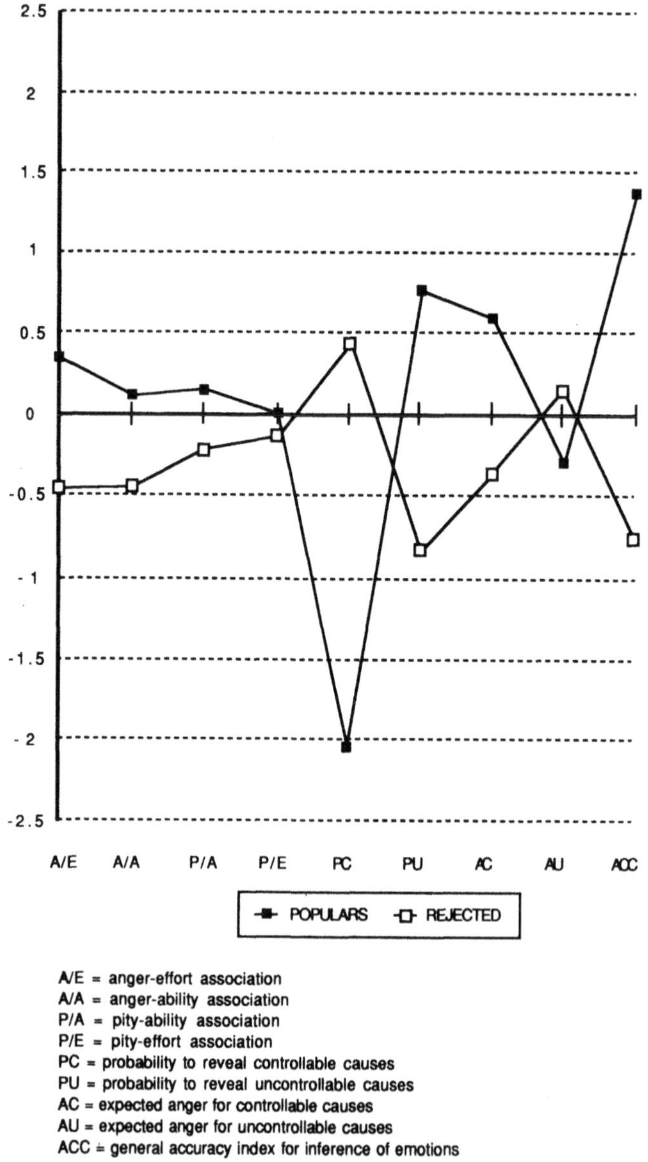

Figure 1. Inference of emotions (z scores)

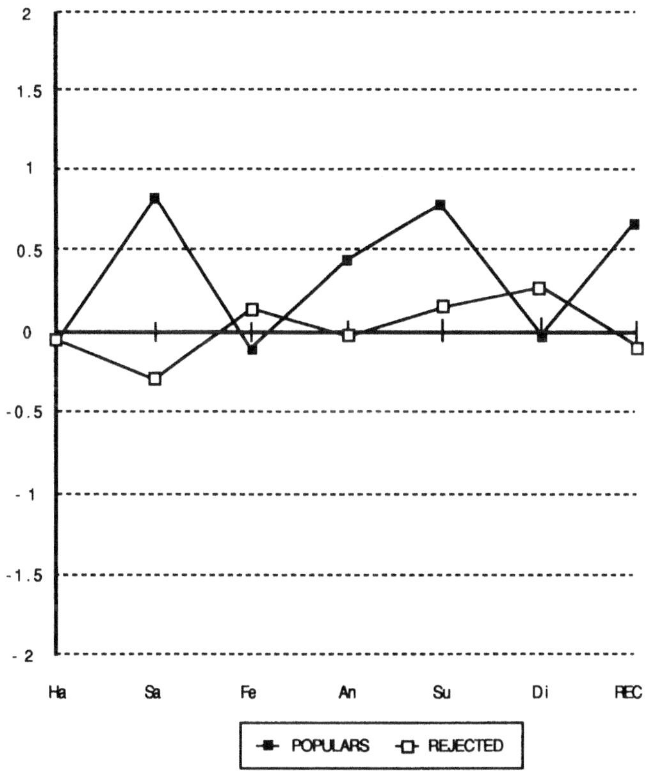

Ha = happiness
Sa = sadness
Fe = fear
An = Anger
Su = Surprise
Di = Disgust
REC = general accuracy index of recognition

Figure 2. Recognition of Facial Expressions of Emotions (z scores)

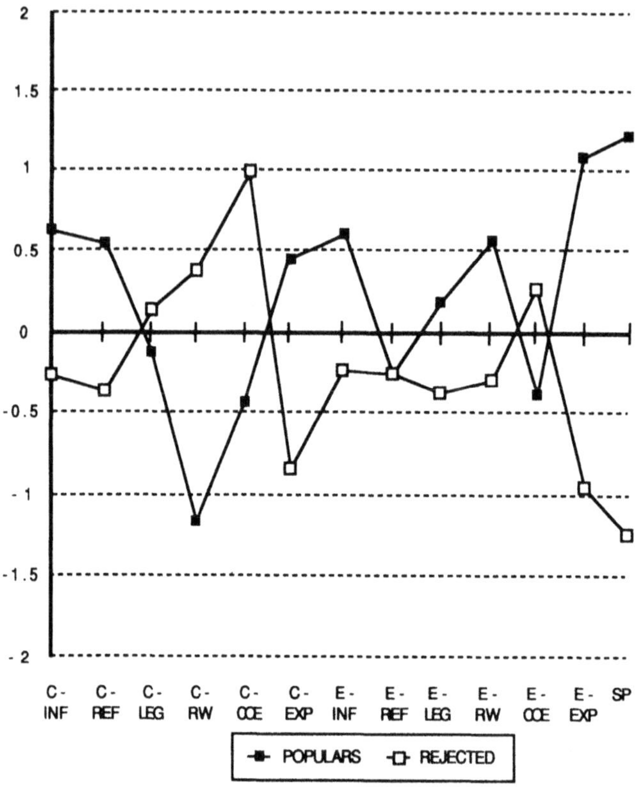

C-INF = choice for informational power
C-REF = choice for referential power
C-LEG = choice for legitimate power
C-RW = choice for reward power
C-COW = choice for coercive power
C-EXP = choice for expert power
E-INF = efficacy for informational power
E-REF = efficacy for referential power
E-LEG = efficacy for legitimate power
E-RW = efficacy for reward power
E-COE = efficacy for coercive power
E-EXP = efficacy for expert power
SP = general accuracy index for social power

Figure 3. Bases of social power (z scores)

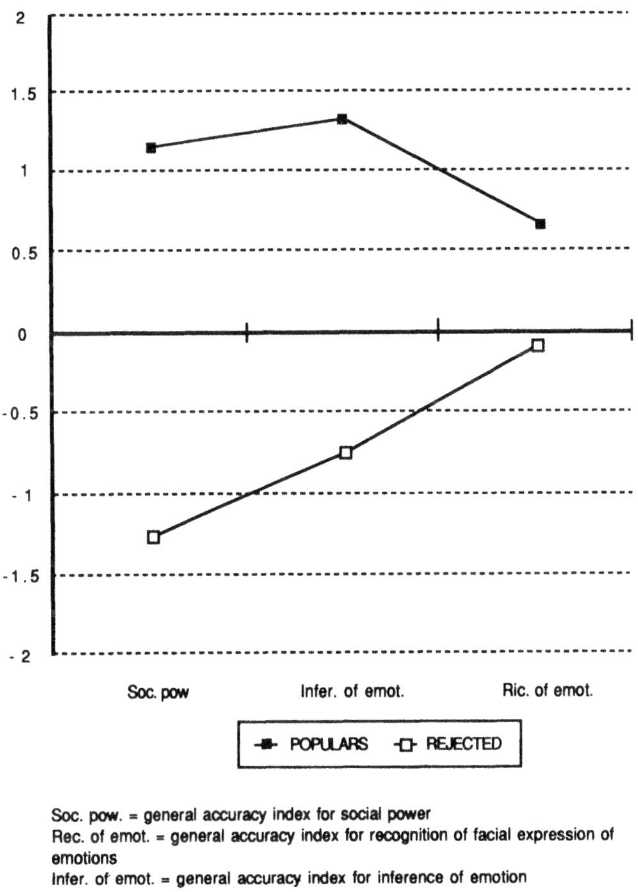

Figure 4. Accuracy index for bases of social power, inference of emotions, recognition of facial expression of emotions (z scores)

Since a detailed account of procedures and analyses are available in an other study (Caprara et al., 1992), I will not give an extended description of them here. I would only note that the 8 records were examined by various psychology students on three different occasions.

Based on the records, subjects were instructed to form an idea of the personality of each child; name, age, and class of each child appeared on the front of the record. Subjects were instructed to assign to each record two numbers, from 1 to 8; their judgment was based on two criteria: "having adjustment problems" and "creating adjustment problems".

On two occasions judgments were made by second year psychology students who individually examined the reports in a university classroom. In the first administration, 13 students examined the completed records; in the second administration, questionnaire responses and tasks covering inference of emotion, social influence and facial expression of emotions were examined separately, by 22 and 22 students respectively.

The examination of results in all situations showed that the "at risk" children were generally differentiated from the "well-adjusted" children for both criteria and in the expected direction. The differentiation was both significant in the case of complete records ($t = 5.66$; $p = .000$) and of separate records (questionnaires, $t = 5.04$; $p = .000$ and tasks, $t = 3.12$; $p = .005$).

On the third occasion, judges were 10 students in the School of Specialization in Clinical Psychology, thus presumably more expert than the former in terms of diagnostic-clinical ability. The 8 records were given to them with the same instructions regarding individual examination; they were also allowed to take the records home. Following completion of the individual examination and assignment to each child of a position regarding the two criteria, subjects were subdivided into two groups (one group composed of 4 judges, one group composed of 6 judges) and asked to come to a conclusion as a group about the rank to give to each child's record according to the two criteria. Also in this case both in the individual ($t = 5.65$, $p < .001$, $n = 10$) and in the group examination ($t = 3.02$, $p < .02$, $n = 4$; $t = 2.97$, $p < .03$, $n = 6$) the "at risk children" were clearly differentiated from the "well-adjusted" children.

For both psychology students and students specializing in clinical psychology it seemed plausible to acknowledge a form of "tacit knowledge" which led them to draw correct impressions from the records concerning the psychological picture resulting from the child's simple responses and choices in the absence of any explicit diagnostic evaluation. However, when the criteria used for arriving at the correct solution were considered, in most cases (with some exceptions in the case of the questionnaires), recognition was almost completely lacking of the underlying rationale for the various tasks as well as recognition of the implications which the various responses or solutions given by the children might have for the judgment of more or less problematic adjustment. A rationale was not lacking, but this seemed to be the result of the application of a variety of "implicit theories" of personality to a series of impressions rather than the result of the examination of objective elements that could be extracted from the various tasks for a detailed evaluation of the child's psychological-social functioning.

Thus, it seemed reasonable to acknowledge a form of "tacit ignorance" in the incongruous application to a series of impressions of a variety of judgments and labels which are completely divorced from any awareness of what determines those impressions.

If this is similar to what happens in the case of children where marginal shifting through aggregation generates non-specific impressions of discomfort and disturbance, the prognosis can only be alarming. In fact, there is the risk that the various impressions take the form of completely functional judgments for whomever makes them; but they are extremely dangerous for whoever they are destined. There is the risk of burying with a judgment feelings which warrant attention; of precluding by a false explanation the possibility of removing the elements which impede the complete psychic functioning of the subject and undermine the continuity of social relations; finally, there is the risk of participating in the initiation of a marginal career, of isolation, and deviance, by a variety of more or less rash interventions which may derive from premature judgments and false explanations.

From Risk Factors to Mechanisms of Risk

In the most recent perspective (Caprara and Laeng, 1988; Rutter, 1988; Robins and Rutter, 1990), research on mechanisms of risk represents the necessary development of traditional research on risk factors. Even if the paradigm of the risk factor, understood as the favouring condition, has represented an important step in the examination of the precursors of discomfort and maladjustment, today this paradigm appears to be surpassed by a conception which aims, on one side, at a more precise contextualization of individual risk factors and, on the other, at a more thorough examination of the processes supporting and regulating *risk* factors and protection factors. Above all it seems crucial to identify what determines and what results from the unfolding of the individual-environment relationship in various phases and various situations.

Risk mechanisms include cause and effect relationships, sequences and circularities, mutual causal processes which, at the level of basic mechanisms, set the premises and maintain the conditions for types of emotional and cognitive organization, and conduct and interactions capable of compromising the individual's development and psychological and social equilibrium. Even though the notion of psychological risk is primarily associated with specific conduct, to lose sight of the whole limits the possibility of identifying the meaning of individual behavior, singling out the factors which determine and maintain it, and the possibility of reconstructing the sequences which lead to it.

The multicausality of phenomena and circularity of causal processes lead to a revision of the traditional notions of continuity and stability of conduct, since what appears to be relevant is not so much conduct per se as the role it plays in the individual's constellation of relationships with the environment. The picture which results from the number of variables in play and from their aggregations, combinations and integrations reveals levels of complexity which impose more sophisticated interpretative models than those anchored to traditional taxonomies. Whereas complexity supports the plausibility of various theories, at the same time it denounces the limits of individual theories in providing an exhaustive explanation of the entire range of phenomena in play. This leads to a reformulation of a conception of personality whose components assume importance primarily for the organization which results from them due to pressures and social attributions.

The notion of "fitting" moves attention from the characteristics of the individual and the environment to the characteristics of the relationships and exchanges which take form and work out in time between the two in terms of compatibility and reciprocity. The person is received and influenced by an environment which, in turn, is profoundly modified and influenced by him/her. On the individual side, what happens in terms of modulation of affects, development of cognitive structures and in terms of ability to satisfy requests and expectations of the environment appears crucial. On the environmental side, what is expected, what is received and what is offered in terms of care, solicitations, and support appears crucial. Thus, all those elements which break or impede "compatibility "are elements of risk; and all those mechanisms which amplify and exasperate marginal misfits are mechanisms of risk.

Small shifts or delays with respect to an expected normative model, in the development of language, in motor development, in cognition and in mood can be elements of risk when they compromise the attachment relationship; when they break the acceptance of the child by whoever is delegated to care for him; when these are associated in others with feelings of inadequacy and refusal; and when they are accompanied by negligence, abandonment or isolation.

To adhere to the program which derives from this new conception of relationships between person, environment and conduct leads us beyond most traditional controversies and imposes a broadening of perspective to the more subtle expressions and effects of behaviors.

In this regard, the notion of marginal deviations, of aggregate effects, of disruption of continuity, of assimilation and contrast may contribute to a better and deeper understanding of mutual causal relationships in the morphogenesis of deviance.

Conclusions

The development of some types of social maladjustment in children is viewed from the perspective of a cyclical interaction over time of personality and social psychological processes.

Children, identified as being "at risk", because they exhibit high levels of aggression and other forms of social maladjustment, differ initially from their peers only marginally on selected socially relevant tasks. However, this profile of originally small, marginal deviations is a risk indicator of eventual significant deviations in social behavior. Aggregation of effects may, in fact, amplify these small deviations until they are salient enough to result in stigmatization, peer rejection, and conflict with authority systems. Lack of awareness of microdeficits on behalf of educators makes children's relations with others less easy, on the one hand; and different responses to them in different settings, on the other band, further raise the cost of child adjustment. As a consequence, prevention strategies should be implemented at any of a number of points in cycle in order to prevent the spiralling which from the aggregation of effects of marginal deviations may end into amplification of deviations, misattribution, rejection and stigmatization.

A number of issues remain to be further clarified. Of particular interest is identifying the variables involved in the transition phase from assimilation and minimalization of marginal deviance to contrast and highlighting it. Some candidates are: a) intensity and frequency of deviations; b) obvious, salient physical characteristics of the child, such as appearance in terms of his/her size, deformities, race and ethnic features; c) educator characteristics and values such as: authoritarian, rigid, controlling, "burned out"; d) characteristics of the situation, such as: excessive demands on educators, strict discipline demanded by the school system, poor moral of educators, size of the classroom, conflicts and scarce collaboration between school and family.

Author's note

1 I greatful aknowledge earlier comments and suggestions of Prof. Philip Zimbardo.
Part of these researches were supported to the author by Consiglio Nazionale delle Ricerche (grant.CNR-Strategico 89/90) and the Henry Frank Guggenheim Foundation (grant to individuals, 1990-91). Correspondence for reprints should be addressed to: Gian Vittorio Caprara, Dipartimento di Psicologia, Università di Roma "La Sapienza", Roma, Italia.

References

Caprara, G.V. (1987). The disposition-situation debate and research on aggression. *European Journal of Personality, 1*, 91-101.

Caprara, G.V. and Laeng, M. (Eds). *Indicatori e Precursori della condotta agressiva* 1988, Roma: Buizoni.

Caprara, G.V., Borgogni, L., Mazzotti, E. and Pastorelli, C. (1988). Indicatori precoci della relazione sociale. *Eta' Evolutiva*, 29, 91 - 101.

Caprara, G.V. and Pastorelli, C. (1989). Toward a reorientation of research on aggression. *European Journal of Personality, 3*, 121-138.

Caprara, G.V. and Pastorelli, C. (1992). Early indicators of emotional instability, prosocial behavior and aggression. Submitted for publication.

Caprara, G.V. and Pastorelli, C. (1992). Early emotional instability, prosocial behavior and aggression: Some methodological aspects. *European Journal of Personality*, in press.

Caprara G.V., Barbaranelli C., Pastorelli, C. and Perugini, M. (1992). Da mere impressioni ad attribuzione di disposizioni (From mere impressions to attributions of dispositions). *Rassegna Italiana di Psicologia*, in press.

Cantor N., Kihlstrom S.F. (1986). *Personality and social intelligence*. Englewood Cliffs, NJ: Prentice Hall.

Caspi, A., Bem, D.J. and Elder, G.H. (1989). Continuity and consequences of interactional styles across life course. *Journal of Personality, 57*, 375-406.

Dodge, K.A. (1986). A social information processing model of social competence in children. In M.P. Perlmutter (Ed.), *Minnesota symposium on Child Psychology* (vol. 18, pp.55-125). Hillsdale, NJ: Erlbaum.

Ekman P. and Friesen, (1975) *Unmasking the face*. Englewood Cliffs, NJ: Prentice Hall.

Eron, L.D. and Huesmann, L.R. (1984). The relation of prosocial behavior to the development of aggression and psychopathology. *Aggressive Behavior, 10*, 201-211.

Epstein, S. and O' Brien, E.J. (1985). The person-situation debate in hystorical and current perspective. *Psychological Bulletin, 98*, 513-537

Helson, H. (1964). *Adaptation level theory*. New York: Harper and Row.

French and Raven (1959). The bases of social power. In Cartwright, D. (Ed). *Studies in social power*. Ann Arbor, MI: Institute for Social Research.

af Klinteberg, B. and Magnusson, D. (1989). Hyperactive behavior in childhood and adult impulsivity: A longitudinai study of male subjects. *Personality and Individual Differences, 10*, 43-50.

af Klinteberg, B., Schalling, D. and Magnusson, D. (1990). Childhood behaviour and adult personatity in male and female subjects. *European Journal of Personality 4*, 57-71.

Magnusson, D. (1990). Personality research-Challenges for the future. *European Journal of Personality, 4*, 1-17.

Maruyama, M. (1963). The second cybernetics: deviation-amplifying mutual causal processes. *American Scientist*, 164-179.

Olweus, D. (1979). Stability of aggressive reaction patterns in males: a review. *Psychological Bulletin, 86*, 852-875.

Olweus, D., (1980). Familial and temperamental determinants of aggressive behavior in adolescent boys: A causal analysis. *Developmental Psychology 16*, 644-660.

Parke, R.D. and Slaby, R.G. (1983). The development of aggression. In: P.H. Mussen (Series Ed.) and M. Hetherington (Vol. Ed.), *Handbook of child psychology* (4th ed., Vol. 4, pp. 547-642). New York, Wiley.

Parker, J.G. and Asher, S. R. (1987). Peer relation and later personal adjustment: are low accepted children at risk?. *Psychological Bulletin, 102*, 357-389.

Patterson, G.R. (1988). Family process: Loops, levels and lies. In N. Bolger, A., Caspi, G. Downey and M. Moorhouse (Eds). *Persons in context: Developmental Processes*. Cambridge University Press, Cambridge.

Pepitone, A. and Di Nubile, M. (1976). Contrast effect in judgments of crime severity and the punishment of criminal violators. *Journal of Personality and Social Psychololy, 33*, 448-459.

Robins, L.N. and Rutter, M. (1990). *Straight and devious pathways from childhood to adulthood*, Cambridge University Press. Cambridge.

Rutter M. (1988). *Studies of Psychosocial risk: the power of longitudinal data*. Cambridge University Press, Cambridge.

Schmidt, A. and Raven, B. (1985). *Children cognition of persuasion strategies and compliance*. Paper presented at the Biennal Meeting of the Society for Research in Child Development, Toronto.

Sherif, C.W. and Hovland, C. (1961). *Social Judgement: Assimilation and contrast effects in communication on attitude change*. New Haven: Yale University Press.

Watzlavick, P., Beavin, J.H. and Jackson, D. (1967). *Pragmatics of human communication. A study of interactional patterns, pathologies and paradoxes*. New York: W.W. Norton and Co.

Weiner, B. (1986). *An attributional theory of motivation and emotion*. New York: Springer-Verlag.

Weiner, B., Graham, S., Stern, P., and Lawson, M.E. (1982). Using affective cues to infer causal thoughts. *Developmental Psychology, 18*, 278-286.

Weiner, B. and Handel, S. (1985). Anticipated emotional consequences of causal communications and reported communication strategy. *Developmental Psychology 21*, 102-107.

THE LONG ROAD FROM MARGINAL DEVIATIONS TO MAJOR EFFECTS
Commentary on the chapter by G.V. Caprara

Ivan Mervielde
Department of Psychology,
University of Ghent.
H. Dunantlaan 2,
B-9000 Ghent, Belgium.

Marginal Deviations

Caprara's contribution presents a number of important ideas for improving research outcomes in the area of early indicators of social maladjustment and social deviation. His suggestions focus on a process as opposed to a trait orientation and challenge major assumptions underlying the trait approach to personality. Caprara's conceptual framework is mainly based on social psychological theories and emphasizes the social rather than the biological approach to individuality.

Trait oriented research (Olweus, 1979) has demonstrated that childhood aggressiveness, as measured by peer nominations and ratings by parents and teachers, is a stable trait that predicts adult social maladjustment and deviation. Moreover, Caprara points out that "an aggressive child is more likely to become an aggressive adult than a sociable or conscientious child is to become an equally sociable or conscientious adult" (p. 4), and thus acknowledges that trait aggression is one of the more stable traits.

To increase our understanding and to improve detection of children at risk, Caprara argues that it is "worth going beyond the objective manifestation of aggressive behaviour and of rejection to investigate the system of affects, representations, meanings, rules and shared expectations" (p.4). More importantly, he asserts that "an original sensitivity towards that which is subtle, nuanced, tenuous, indistinct and marginal becomes crucial" (p.3).

Caprara's plea for studying marginal deviations is in tune with basic assumptions underlying developmental psychology. Development is conceived as a process of increasing differentiation from birth to adulthood. Looking at development in retrospect, it seems logical to assume that major deviations in adulthood, must have been reflected by minor or marginal deviations at some point in development. Case studies often document marginal childhood indicators that escaped the attention of teachers and parents. No matter how convincing such studies may be, it cannot be ruled out that retrospective evidence is often unreliable and blurred by hindsight (Halverson, 1988). The crucial evidence needed to establish the significance of marginal deviations has to come from prospective or longitudinal studies. Caprara suggests that assessment of certain process variables, in addition to the assessment of childhood aggressive behaviour, would increase our understanding of the underlying mechanisms and improve risk assessment and prediction of adult deviation. It is difficult to judge the value of this approach without longitudinal data. Moreover, marginal deviations are

J. Hettema and I. J. Deary (eds.), Foundations of Personality, 245–250.

by definition minor deviations and evidence about the intra-individual stability of these marginal deviations should be supplied. Furthermore some rule should be given to distinguish "marginal" from "major" and "insignificant" deviations. Only intra-individually stable marginal deviations will provide the cumulative effect that Caprara expects from the aggregation of marginal differences.

The more important question is not whether and how marginal deviations should be studied but *which* marginal deviations should be assessed. Caprara's selection of process variables is in tune with current issues in person perception research: attribution of emotions (Weiner, 1986), recognition of facial expressions (Ekman and Friesen, 1975) and perception of various forms of social influence. These variables emerge from social psychological theories that focus on effects of experimental conditions and tend to ignore inter and intra-individual differences. The validity of these variables as indicators of reliable individual differences is not yet established. Moreover, it is difficult to link these variables to taxonomies of individual differences such as the five-factor model or Eysenck's PEN-model. This of course does not impede the value of these variables but given the growing consensus about the value of such taxonomies for establishing nomological networks (Costa and McCrae, 1992), new variables are more readily accepted when they can be linked to such a conceptual network.

Aggregation of Effects

One of the more interesting assertions from attribution theory is that lay perceivers tend to match the size of cause and effect (Kelley, 1967). Minor effects are produced by minor causes and hence major effects require major causes. Caprara suggests that major effects can be caused by the cumulative effect of minor deviations. Trait approaches are inclined to explain major behavioral effects by looking for major deviations on one or a few traits. In principle, trait theories could attribute major effects to an accumulation of effects produced by minor deviations on many traits. Trait theories do not adopt this strategy because they are based on structural analysis. Factor analysis groups related "minor" traits in "major" or broad band factors. Process oriented approaches, such as the one proposed by Caprara, tend to ignore structural analysis of relationships among variables derived as process indicators. It would be useful to look at the structural relations between the person perception variables proposed by Caprara. If accuracy of attribution of emotions, decoding of facial expressions and preferred social influence style correlate to some extent, then aggregation of these variables into a broad band process variable such as accuracy of person perception is indicated and the presumed contradiction between the process and the trait approach could be reduced. Moreover, aggregation of process variables into a broad band process variable such as accuracy of person perception, would perhaps transform the observed minor deviations into a major deviation on the broad band variable. Structural analysis of these process variables together with traditional trait measures might also reveal significant relationships with Eysenck's PEN-model variables and those included in the five-factor model.

Disruption of Continuity

Caprara states that "teachers and children share a tendency to confuse emotional instability with aggression" (p.7). My own research ((Mervielde 1991, De Fruyt and Mervielde 1992) about teacher ratings of kindergarten as well as primary school children shows that teachers make independent judgements of emotional stability, agreeableness, conscientiousness and extraversion. Aggression is usually subsumed under the broad band trait unagreeableness in the five-factor model or psychoticism in the PEN model and is therefore conceived as independent from emotional instability or neuroticism. My research with peer nominations (De Fruyt and Mervielde, 1992) suggests that, at least for nominations on *negative* items, children from the third to the fifth grade do not make independent nominations for "most unfriendly, most quarrelsome, most difficult" and "most restless, most nervous and least patient". Sixth graders, however, already make the distinction. Peer nominations for *positive* traits reveal independent judgements for agreeableness and emotional stability from the fourth grade onwards.

The disruption of relations is conceived by Caprara as the result of a continuous attack on shared rules. Within the five-factor model this would correspond to the negative sides of agreeableness and conscientiousness. Agreeableness is known to be the factor that correlates most with social desirability. Many of our shared rules define socially desirable behaviours and persons that do not conform to those rules tend to be labelled with traits that cluster at the negative pole of the agreeableness factor such as uncooperative, aggressive, unagreeable, difficult, quarrelsome and unfriendly . A similar argument can be made for the broad band factor conscientiousness that refers to traits such as : orderly, persistent, hard working, careful and attentive. The psychoticism factor of Eysenck's PEN-system is known to be inversely related to both agreeableness and conscientiousness (McCrae and Costa, 1985). The disruption of the continuity of relationships by repeated attacks on shared rules therefore seems to be similar to what in traditional trait approaches is referred to as the negative side of agreeableness and conscientiousness or as psychoticism in the PEN model.

Tacit Knowledge Versus Ignorance

To substantiate his ideas, Caprara provides a summary of a series of studies on clinical judgement. The main purpose of these studies is to illustrate that people take into account marginal deviations but are unaware of the factors that influence their judgements. Questionnaire responses and answers on person perception tasks from eight cases are presented (4 popular and 4 rejected children) to students who are asked to judge to what extent the children will have adjustment problems and create adjustment problems. The average differences between the four popular and the four rejected cases, as depicted in fig. 4, are by no means marginal: more than two standard deviations for judgments of "social power", about two standard deviations for inferences of emotions and about one standard deviation for recognition of emotions. Given the rather small samples of judges (13, 22), a t-value of 5.66 ($p = .000$) for perceived differences between both groups is not a "marginal" effect. Smaller effects are observed when students are not provided with the questionnaire responses of the cases. On p. 9 Caprara states that "Children at risk were selected on the basis of their extreme scores on aggression as rated by themselves, by peers and by teachers, and on the basis of high scores on rejection on behalf of peers". The questionnaires assessed "emotional instability, pro-social behaviour and physical and verbal aggression" and hence are more related to the criteria for dividing the eight cases into "popular" and "rejected" cases. The studies summarised in this paper are not very relevant as tests of the hypothesis that aggregation of marginal deviations will improve the *prediction* of major social effects. Moreover, I disagree with Caparara's conclusion that these studies illustrate "tacit ignorance". The studies show that even small samples of judges generate different average ratings for "popular" and "rejected" case groups. If tacit ignorance cannot explain the performance of the judges how can it be explained ? Research on person perception has shown that likableness or social desirability is a major dimension of person perception. When judges have to integrate multi-dimensional information, they tend to ignore descriptive meaning and integrate evaluative meaning (Mervielde, 1986). The students' judgments can be explained by a similar process. When they have to judge the expected extent of adjustment problems, an undesirable characteristic, they encode the presented case information into "degree of undesirability". According to Anderson's (1971) information integration theory, the general impression of the case will be similar to the weighted average of the social desirabilities of the presented information. This general impression can subsequently be used as anchor for the judgment of "adjustment problems".

Conclusion

The utility of marginal deviations has to be confirmed with data from prospective, longitudinal studies. Moreover, a clear criterion should be provided to distinguish among marginal, major and insignificant deviations.

Process variables indicating marginal deviations should be subjected to structural analysis in order to test whether they are independent variables or merely different measures of a broad band "process" factor such as accuracy of person perception.

The disruption of continuity of relations, as conceived by Caprara seems rather similar to the negative poles of conscientiousness and agreeableness or to psychoticism as measured by Eysenck's EPQ-R.

The studies on clinical judgment are not directly relevant as tests of the utility of marginal deviations. Instead of providing evidence for "tacit ignorance" of the judges, the data can also be interpreted as showing the power of simple information integration strategies for making so-called clinical judgements.

References

Anderson, N. H. (1971). Integration theory and attitude change. *Psychological Review, 78,* 171-206.

Costa, P.T., Jr. and McCrae, R. R. (1992). Multiple uses for longitudinal data. *European Journal of Personality, 6,* 85-102.

De Fruyt, F. and Mervielde, I. (1992). The "Big-Five" in children's sociometric judgments (Abstract). *International Journal of Psychology, 27,* 353.

Ekman, P. and Friesen, W. (1975). *Unmasking the face.* Englewood Cliffs, NJ: Prentice Hall.

Halverson, C.F., Jr. (1988). Remembering your parents: reflections on the retrospective method. *Journal of Personality, 56,* 435-443.

Kelley, H. H. (1967). Attribution theory in social psychology. In D. Levine (Ed.), *Nebraska Symposium on Motivation, 1967* (pp. 192-238). Lincoln, NB: University of Nebraska Press.

McCrae, R. R. and Costa, P. T. Jr. (1985). Comparison of EPI and Psychoticism scales with measures of the five-factor model of personality. *Personality and Individual Differences, 6,* 587-597.

Mervielde (1986). The attribution of personality traits. In A. Angleitner, A. Furnham, and G. Van Heck (Eds.), *Personality Psychology in Europe: Current issues and controversies* (Vol. 2) (pp. 159-175). Lisse, The Netherlands: Swets & Zeitlinger.

Mervielde, I. (1991). *The Big Five in Belgian teachers' constructs and ratings of 4-12 year old children.* Paper presented at the NIAS conference on Development of Temperament and Personality from Infancy to Adulthood, Wassenaar, The Netherlands.

Olweus, D. (1979). Stability of aggressive reaction patterns in males: a review. *Psychological Bulletin, 86*, 852-873.

Weiner, B. (1986). *An attributional theory of motivation and emotion.* New York, NY: Springer-Verlag.

REPLY TO I. MERVIELDE

Gian Vittorio Caprara

In my reply to Mervielde first of all I will clarify my basic conception of personality. The assumptions from which my reasoning derives imply that personality is an open system whose main properties are to be self reflexive and self regulatory. In great part these properties emerge from the encounters of the individual open system with other analogous open systems in the realm of the social environment. In this view focusing on "process" does not imply underestimating stable dispositions or predispositions. Rather, the dynamic properties of traits as emerging characteristics of individuation are underlined. The construction of personality appears to be the result of bio-social processes which are in great part transactional and irreversible.

Whereas I agree with Mervielde on the critical importance of longitudinal data, our experience suggests we may rely upon longitudinal data to investigate the effects of aggregation of marginal deviations only in part. Experimentation is a crucial complement of longitudinal research, given the great number of factors which transform, disappear, overlap and intermingle with each other over time. Since the relevance of marginal deviations, intended as marginal perturbations, broadly depends on normative values or baselines and related latitudes of acceptance which may change over time, experimentation may extend our control over sources of variations which longitudinal research risks to miss or to confound.

The recommendation of Mervielde to examine the latent structure of the various aspects of social intelligence that have been considered is well taken. Indeed, I don't see any contradiction between trait and process oriented approaches. In this regard the present research as well as all my previous research attest to the validity of pursuing the study of invariants of personality, integrating the study of individual differences with the study of more basic processes. Whereas in previous research individual differences have often been used instrumentally to amplify and to clarify general regularities, in the present research regularities associated to what in broad sense has been referred to as social intelligence have been used instrumentally to clarify the emergence and the maintenance of individual differences.

Since Allport and Lewin we have been used to considering the study of individual differences as an indispensable avenue to the search of general regularities. It is likely that various processes associated with attribution of emotion, with recognition of facial expressions of emotions, with social influence can be reduced, at least in part, to a common latent construct such as "social intelligence". Also it is likely that both attenuation and amplification of marginal deviations can be examined, at least in part, as aspects of individual differences associated with a tendency to assimilate or to contrast.

However, I doubt that processes which emerge from transactions between individuals as the ones referred to in terms of "aggregation of marginal deviations", "disruption of continuity", "amplification of deviance" all be reducible to the notion of trait. If this were the case the notion of trait should no longer be referred to as property of persons, but as property of interactions or transactions. By all means I don't think that processes can be subsumed under traits.

Regarding aggression and emotional instability my arguments are sustained by the findings of over twenty years of research that have been amply documented, among others, in the European Journal of Personality and in Personality and Individual Differences.

By ignoring the section on "Assimilation, contrast and the cost of adjustment", concerning the phenomena of attenuation and amplification of deviance via assimilation and contrast -i.e. via negative and positive feedback respectively- Mervielde fails to capture not only the implications of my reasoning but also the essence of the study that is presented. This section is critical to bridge the section on "continuity and disruption of continuity" with the section on "Tacit knowledge and tacit ignorance" and with the section on "Risk factors and risk mechanism" which in addition seems to have been unfortunately ignored.

Since the rationale of the presented study seems to have been missed, I must recommend a closer scrutiny of the material presented and of the documentation to which reference has been made explicitly. The study offers sufficient evidence, although indirect, of how marginal deviations aggregate in the mind of the perceiver to lead to decisions which are coherent with our theoretical expectations: children who have been rejected by their peers are identified correctly as "children at risk of maladjustment" by different judges and in repeated studies; however, the rationale the judges subsequently provided to explain their correct decisions suggested that the judges were able to appraise the effects but were incapable of recognizing their causes.

PERSONALITY: AN INTERACTIONAL PERSPECTIVE

Norman S. Endler
Department of Psychology
York University
4700 Keele Street
North York, Downsview
Ontario M3J 1P3, Canada

Abstract

Various personality models are reviewed and the theoretical foundations and histories of the associated perspectives are discussed. In light of these, it is suggested that while the contributions to personality made by social and biological factors are of crucial importance, psychological factors must be incorporated in both theory and research. Mechanistic and dynamic models of interactionism are discussed, and an interactional model, which acknowledges the mutual influences existing both between and within persons and situations, is presented. It is suggested that personality researchers must recognize the inherent complexity of human experience and, relevantly, exercise caution when interpreting their field's research findings. General recommendations are offered for future research attempted under an interactional perspective, and conclusions are drawn regarding directions for scientific progress.

Introduction

Personality is probably the most complex, ambiguous, and challenging area within the field of psychology. Much of the research in this area has raised more questions than answers. Nevertheless, some progress is slowly being made. Unfortunately, the crises discussed in the 1970s and early 1980s have not been easily resolved, and criticisms about the area voiced at that time are just as relevant today in the 1990s (Carlson, 1971; Carson, 1989; Magnusson, 1990b; and Sechrest, 1976). Endler and Parker (1992a) have noted that many current personality researchers have ignored the previous unresolved crises and have simply continued doing their own research. Endler (1982) stated that "a major problem with respect to personality research during the 1950s, 1960s and 1970s is that it has been basically piecemeal, and ad hoc, rather than programmatic" (p. 215). To a great extent this was true in the 1980s and continues to be true in the 1990s.

However, at least two research programmes, one representing a longitudinal approach (cf. Magnusson, 1992b) and one representing a cross-sectional or cross-situational approach (Endler, 1983; 1988) to interactionism, are notable exceptions to this trend. Magnusson's longitudinal project on "Individual Development and Adjustment" is being conducted with his colleagues in Sweden (Magnusson, 1992b) and Endler's cross-situation project in "The Interac-

J. Hettema and I. J. Deary (eds.), Foundations of Personality, 251–268.

tion Model of Stress, Anxiety and Coping" is being conducted with his colleagues in Canada (Endler, 1988; Endler and Parker, 1990, 1992a). Both research programmes, which will be discussed shortly, are derived from theoretical perspectives on personality.

Personality Models

Historically, both personality research and theory have been directed by four major models: traits or the study of individual differences, situationism, psychodynamics and interactionism (Endler, 1983; Endler, in press; Endler and Magnusson, 1976; Endler and Parker, 1991, 1992a; Magnusson, 1990b).

THE TRAIT MODEL

For the most part, and beginning with Galton (1884), personality psychology has been a psychology of individual differences, and has been dominated by the trait model. The trait model postulates that *traits*, or stable and latent dispositions, are the primary determinants of behaviour. Cattell (1957), among other trait theorists, believed that traits are predispositional bases for response-response consistencies of behavior in various situations. Allport (1937), a major North American trait psychologist, conceptualized traits as tendencies (predispositions) to respond in a particular fashion, and these predispositions were *not* linked to specific stimuli. Rather, they were general and enduring tendencies. The various trait theorists mentioned did not agree exactly on what a trait is, nor did they agree as to the exact number, names and kinds of traits. However, most trait theorists agree that traits are internal factors and that they account for behavioural consistency across a variety of different situations. What was true in the 1930s is true in the 1990s, as evidenced by the controversy between the big five (Costa and McCrae, 1992a, 1992b, 1992c) versus the big three (Eysenck, 1992a, 1992b, 1992c). Although a discussion of the big five (openness, conscientiousness, extraversion, agreeableness, and neuroticism) or of the big three (psychoticism, extraversion, neuroticism) is beyond the scope of this chapter, we will return to a discussion of the role of traits in personality theory, later in this chapter.

The historical roots of the investigation of personality traits in North America stem from at least two sources, both of which had a practical function: the intelligence testing movement (Danziger, 1990; Parker, 1990, 1991), and the investigation, via personality inventories, of emotionally disturbed Army recruits during World War I (Hilgard, 1987; Nunnally, 1984). Both intelligence tests and personality tests evolved out of practical concerns. Binet's intelligence test was initially developed to identify mentally retarded children. The first personality inventory aimed to identify World War I Army recruits in the United States who were emotionally disturbed. Woodworth (1918) developed an "adjustment inventory" called the *Personal Data Sheet*. The U.S.A. Armed Forces wanted to weed out persons who were emotionally unfit, before they were sent overseas. Prior to this, screening had been conducted by psychiatric interviewers to assess everyone; in a sense, Woodworth had each person "interview himself". The development of intelligence testing had a significant impact on personality assessment, which in turn profoundly influenced the definition of personality research as a field. As Endler and Parker (1992a) note, starting in the 1920s, psychologists began investigating personality traits in order to improve the prognostic ability of existing intelligence tests. Attempts were made to determine those traits that were most important for success in schools, colleges, and/or in business or occupational areas.

During the 1920s, investigators conducting research on various traits were *not* especially interested in developing personality theory; they were more interested in focusing on technical questions related to the development of useful tests with practical applications (Danziger, 1990). Psychologists attempted to collect empirical data from as many diverse samples and populations as they could (cf., Symonds, 1924). During the 1920s and early 1930s, most researchers were not concerned with the conceptual and theoretical confusion revolving around the various trait concepts (cf., Allport, 1927; Krout, 1931).

However, during the late 1930s a number of personologists attempted to develop trait theories (e.g., Allport, 1937; Cattell, 1945; Stagner, 1937), and some of them (e.g., Cattell, 1946) also attempted to develop associated personality trait measures. The trait theories were about personality structure and said nothing about personality dynamics or personality development. They focused on personality description rather than explanation. To a great extent this is still true today, in the 1990s.

Personality theories developed from the trait model tend to focus on specific constructs, most frequently those associated with particular scales (cf. Endler and Parker, 1992a). For example, Eysenck's Extraversion-Introversion and Neuroticism-Stability scales (Eysenck, 1947), first developed in the late 1940s, and Costa and McCrae's (1992c) recent NEO-PI-Revised Scales measuring the Big Five are examples of trait scales designed to assess limited sets of variables, and generative of numerous empirical papers. These empirically "robust" constructs (e.g., anxiety or neuroticism) are usually significantly related to many other constructs (e.g., depression, anger).

Endler and Parker (1992a), in a crude sampling of research published in the 1989 volume of *Personality and Individual Differences*, found that for studies using either the EPI or the EPQ, the extraversion construct correlated with such other variables as "purpose of life", "internal locus of control", "affect intensity", "depression", "emotional inhibition", "alexithymia" and with "self-monitoring", with the correlations ranging from -0.37 to 0.51. It would tax the imagination to develop a theoretical model, with extraversion as the focus, that would offer an explanation or integration of such a diverse set of variables. The construct's practical and theoretical utility thus seems rather limited. Nevertheless, investigators will continue to conduct research, with such "robust" constructs, explaining and sharing modest amounts of variance with other variables, measures, and scales.

Current personality research focuses on the study of single traits or personality dimensions, and sometimes on a few traits (cf., Rorer and Widiger, 1983; Wiggins, 1980). The decision as to which traits to use is not commonly determined by theoretical considerations. Rather choices are often made, to a great extent, by the availability of personality measures assessing specific traits and by the zeitgeist as to "the flavour of the month", or what traits are currently in fashion (cf. Endler and Parker, 1992a). The choice of which cluster or pattern of personality traits to assess is a complex matter. Deciding what traits to study was a problem in the 1930s (Allport and Odbert, 1936; Vernon, 1933) and continues to be a problem in the 1990s. What traits should one include? What trait should be excluded?

During the late 1980s and early 1990s, many personality psychologists (e.g., Angleitner et al., 1990; Cochran, 1984; Peabody, 1987) have proposed developing taxonomies of traits which could serve as a focal point of personality research and theory. Some personality researchers believe that they have discovered the "holy grail" embodied in the big five personality dimensions. That is, a number of investigaors have noted that the *natural language* for describing personality can best be summarized in terms of five basic dimensions, with different terms used by different investigators (for reviews see Digman, 1990; John, 1990) . Nevertheless, the five dimensions are often referred to as "openness", "conscientiousness", "extraversion", "agreeableness" and "neuroticism", which readily form the acronym OCEAN.

Parallel to the research on the big five based on the use of *natural language*, a number of researchers (cf. McCrae, 1989; McCrae and Costa, 1986), reviewing the literature on factor analytic studies of various personality measures and scales, have suggested that five basic factors can be derived from commonly used personality measures, namely the same five factors named above - OCEAN. McCrae and Costa (1986) thus proposed that the "big five" model "offers a universal and comprehensive framework for the description of individual differences in personality" (p. 1001). Digman (1990) has noted that the big five model presents a theoretical basis for personality research. We would suggest that the big five provides an important starting point for personality structure, but is not a theory *per se*. Hofstee (1990) has noted that it is an open question "whether the natural language of personality provides an adequate point of departure for the construction of a scientific system of personality categories" (p. 78). It is also not clear at this point whether personality structure can best be described by the big five or by Eysenck's (1992a, 1992b, 1992c) big three. The generalizability and stability of the big five (or for that matter the big three) to and for diverse cultural groups is still an open question (cf. Briggs, 1989; Hofstee, 1990). An important theoretical advance has been made by Hofstee et al. (1992) in their attempt to integrate 5-dimensional simple structure and 5-dimensional circumplex models of personality.

Nevertheless, the big five does not provide a comprehensive personality theory. As Endler and Parker (1992a) have noted, the solution to problems associated with traits is neither better classification systems for existing traits nor the development of more elegant and sophisticated research and/or statistical procedures for assessing these traits. The solution would seem to lie in the direction of developing comprehensive, meaningful and relevant personality theories. At present, personality research seems to be trait driven (in terms of currently fashionable traits) rather than theory driven.

Most of the research, theory and crises have been dominated by the trait and situationism models, fuelled initially by the person vs. situation controversy (see Mischel, 1968). Although psychodynamics (primarily psychoanalysis) has had a profound influence on Western society and culture, especially in literature and the performing arts, it has had remarkably little sustained impact on personality research and theory (Bakan, 1977). A discussion of the psychodynamic model is beyond the scope of this chapter.

THE SITUATIONISM MODEL

The trait model has focused on internal person factors. Situationism proposes that external environment factors are the major determinants of personality and behaviour. Modern situationists or social learning theorists, however, are cognizant of the role of individual differences in personality.

During the 1920s, concurrent with the expansion of research on individual differences, there were a number of investigations of the effects of environmental or situational factors on both animal and human behaviour (Endler and Parker, 1992a). According to Buckley (1989), although this research was initially disjointed, isolated and fragmented, under the direction of such prominent psychologists as Thorndike, Watson, Tolman and Hull, it evolved into the systematic body of knowledge and theory regarding *learning* in the 1930s and 1940s. Parker

(1990, 1991) has noted that during the 1920s and 1930s, *personality researchers* focused on studying individual differences, while *learning* researchers, using a different methodology, collected reams of data on the systematic effects of environmental variables on behaviour. Danziger (1990) has pointed out that the experimental methods fostered by learning theorists were imitated by most psychologists. During the late 1930s and early 1940s, learning theorists such as Guthrie (1944) proclaimed that "laws of learning", derived from experimental learning studies of behaviour, created a solid foundation for the understanding of personality.

On the basis of research on learning, a number of investigators developed systematic theories about personality (e.g., Dollard and Miller, 1950; Miller and Dollard, 1941; Tolman, 1932; Thorndike, 1940). These personality theories, derived from learning experiments obviously had a broader scope than those proposed by trait theorists (e.g., Cattell, 1946) in the 1940s and early 1950s.

During the 1950s the classical social learning theorists (Dewey and Humber, 1951; Dollard and Miller, 1950; Rotter, 1954), who represented a heterogeneous set of viewpoints, developed theories, which included organismic variables, that had an impact on personality research. During the 1970s behaviour theorists (e.g., Bandura, 1971; Mischel, 1973; Rotter, 1975) proposed theories that were basically concerned with the person's behaviour rather than with traits, attributes and motives. Nevertheless, these theorists included person factors in their theories. In the 1980s, Bandura's (1986) revised cognitive social learning theory, which included the concept of self-efficacy, originally proposed in the 1970s, was basically a transactional or interactional theory.

It should be noted that by the mid-1960s the dominant position of the trait personality model started to be challenged by situationism. The amount of research, at least in North America, on the effects of situational factors on personality rivalled research in the personality trait area (Endler, 1983). In the 1970s, there were actually suggestions by some investigators (cf. Mischel, 1973) that the trait or individual differences personality research area might be in trouble. This prediction decline, however, was premature. Not only did the trait area *not* decline; it thrived during the 1970s and 1980s, and continues to do so in the 1990s. The person vs. situation debate, a conflict of two major traditions - viz. trait psychology and social learning theory - was more than a disagreement about how to interpret research results. As Endler and Parker (1992a) note, the debate was about the primacy of particular research theories and methodologies. It was also about different theoretical perspectives.

With respect to the resolution of the person-situation controversy, Kenrick and Funder (1988) have commented that "as with most controversies, the truth finally appears to lie not in the vivid black or white of either extreme, but somewhere in the less striking gray area" (p. 31). Although many personality researchers believe that this crisis has largely been resolved, and proclaim an adherence to an interaction perspective to personality, this is true in the abstract, but not in practice. Trait theorists, in their research, rarely examine the role of situations systematically; situationists rarely examine the role of individual differences variables systematically in their research. Furthermore, situationist research usually has limited generalizability. While multivariate analyses are more frequent today than in the 1950s, Cronbach's (1957) *American Psychological Association* Presidential address is still relevant today. "No experimenter would deny that situations and responses are multifaceted, but rarely are his procedures designed for a systematic multivariate analysis. The typical experimental design and the typical experimental law employs a single dependent variable" (p. 676). Although situationists have reformulated their theories and research, much of it does not capture the dynamic quality of interactionism, and should be differentiated from the type of programmatic research proposed by interactionists (cf. Endler and Parker, 1992a).

THE INTERACTIONISM MODEL

As Endler and Edwards (1986) have noted, the concept of interactionism can be traced back to the time of Aristotle (cf. Shute, 1973), Descartes and other philosophers. Furthermore, early examples of interaction models can be found in the physical sciences, i.e., Hooke's law which states that "within the elastic limit, strain is proportional to stress" (Harris and Levey, 1975, p. 845); elasticity has a different meaning for fluids and gases. This suggests that the elasticity of any substance is an interactive function of the degree of situational stress, and the nature of the material. Is there a Hooke's law for personality?

Interactionism, which has a long history (cf., Ekehammar, 1974; Endler and Edwards, 1986), began to have some impact on the field in the mid-1970s (Endler, 1983; Endler and Parker, 1992a). Researchers on interactionism (Bowers, 1973; Endler, 1988; Endler and Magnusson, 1976) attempted to reconcile the trait and situationism models. Nevertheless, as Endler and Parker (1992a) have noted, the influence of interactionism in the personality field has been primarily rhetorical. Personality researchers have paid "lip-service" to interactionism, but in practice much of the research has been of little relevance to this approach, with a few notable exceptions (e.g., Endler and Parker, 1990; Magnusson, 1992a). Prior to discussing interactionism in more detail, let us review biological, psychological and social perspectives in personality.

Biological, Psychological and Social Perspectives

The NATO Advanced Research Workshop on Biological and Social Approaches to Individuality (Personality) on which this book is based is timely, but a third factor should have been included: psychological approaches. I will return to this shortly. Although, as I have noted above, there has traditionally been a great deal of research on social approaches (e.g., situationism) to personality (e.g., Mead, 1934; Miller and Dollard, 1941), the role of biological and genetic factors has been relatively neglected (cf. Endler, 1989; Eysenck and Eysenck, 1985) until the 1980s, especially in North America. Recently, it has been suggested that in order for personality theories to be heuristically useful they must explicitly include both biological and social factors (Endler, 1989; Endler and Parker, 1992a; Magnusson, 1990a; 1992a; Van Heck, 1991). "Obviously no behavior can be independent of an organism's heredity, by the same token no behavior can be independent of an organism's environment" (Endler, 1989, p. 158). Analogously, no behavior can be independent of either biological or social factors; these factors are not additive but are interactive.

BIOLOGICAL FACTORS

In the late 1980s and early 1990s there have been a number of books that have focused on the role of biological factors in personality (e.g., Eaves et al., 1989; Zuckerman, 1991). Amongst other types of constructs, biological factors include genetic, constitutional, physiological and biochemical variables. Although genetic factors obviously play a role in personality we do *not* inherit behaviours. We do, however, inherit genes, and they regulate the proteins endemic to the structure and function of the nervous system. Genes may serve as a predisposition to traits and consequent behaviours via the control of neural growth and development, but there are many factors that modify and regulate both the genetic mechanism, the traits, and their behavioural expression. In short, "there is many a slip twixt the cup and the lip".

Nevertheless, there is both direct and indirect evidence for the role of biological factors in both personality development and regulation. Bouchard et al. (1990) in their "Minnesota Twins Reared Apart" Study, have reported "evidence for the strong heritability of most psychological traits" (p. 223). However, they recognize the value of "parenting, education and other propadeutic interventions" (p. 223). Eaves et al. (1989) have concluded that there is a strong genetic component to the psychoticism, extraversion and neuroticism personality dimension of the Eysenck PEN model. Zuckerman (1991), going beyond the contribution of genetic factors, presents evidence for the additional role of biochemical and biophysical processes in personality development and functioning. None of these investigators ignore the role of environmental or social factors in personality; they merely conclude that biological factors play a greater role. It should be noted that biological factors include physiological, biochemical, and genetic variables.

Stelmack (1990) has provided electrodermal and electrocortical activity as empirical evidence to indicate that introverts manifest greater physiological reactivity to sensory stimulation than extraverts. Zuckerman (1991) has provided evidence for relationships among endocrines, pharmacological and other physiological factors and individual differences or personality. Cox et al. (1983), in discussing the psychological and endocrine responses related to stressful experiences, indicate the complexity of the field. For example urinary adrenaline (a catecholamine) secretion generally increases as a function of stress. However, physical activity, posture, gender, personality variables and other factors may affect the adrenaline level. We would suggest that excreted adrenaline levels may be affected by many factors including diet, steroids, drugs, alcohol, and genetic factors.

There is also a difference between basal levels and transient response measures, especially in indices of arousal; as exemplified by the difference between trait and state physiological response (cf. Lader, 1983; Stelmack, 1990). Thus, excreted adrenaline levels may not always be a reliable index of autonomic reactivity and one should be cautious about making conclusions about psychological variables and personality on the basis of physiological measures.

SOCIAL FACTORS

Social factors have been strongly implicated in personality development and functioning (e.g., Bandura, 1986; Endler, 1988; Miller and Dollard, 1941). Although biological factors play a crucial role in personality, the role of stimuli, situations and environments, the social factors, cannot be ignored. As I have stated elsewhere, "the environment is the general and persistent background or context within which behavior occurs; whereas the situation is the momentary or transient background. Stimuli can be construed as being the elements within a situation" (Endler, 1981, p. 364). Note that this distinction is analogous to the trait-state differentiation, i.e., environments as traits, and situations as states. For a more detailed exposition of the differences among stimuli, situations and environments, the reader is referred to Pervin (1978).

Social factors, particularly situations, have an effect on personality. Such factors may have a profound effect in extreme cases as witnessed by the survivors of Hiroshima, the Holocaust and the children in Ethiopia and Somalia. Although individuals react to situations they also affect the situations with which they interact, there is a constant and continuous interaction between persons and situations. Furthermore, in addition to having situations or social factors imposed on us, we often select the situations or stimuli we encounter. All of us experience complex situations in our daily lives, at all levels of functioning. As Endler (1983) has noted "stimulation affects behavior both in terms of information being processed at the moment, and also by interacting with information that has previously been stored" (p. 170). Magnusson (1978) has noted that the total environment with which we interact includes a very complex system of social, cultural and physical-geographical factors. He also maintained that "the total environment influences individual development and behavior [but] the influence of environment is always mediated via the actual situations" (Magnusson, 1978, p. 1).

Usually, when one uses the concept *differential psychology* it is in reference to individual differences. Both Endler (1983) and Magnusson (1978) have suggested the need for a complementary *differential psychology* of situations, wherein one systematically scales and classifies situations, and determines their important dimensions. But this has to be founded on a sound theoretical basis. In fact, no one has found a major impact of situations on personality, for evidence derived from studies of twins reared apart (Bouchard et al., 1990) does not necessarily mean that personality is basically due to hereditary factors. It may mean we have not yet developed the tools to systematically and coherently investigate the role of situation (environmental) factors for twins reared apart. Thus, we need to develop strategies for systematically investigating situations. Although we have a fairly advanced differential psychology of individual differences, our differential psychology of situations is still in the dark ages. Nevertheless, it is readily apparent that social factors do have an impact on personality. The nature of this impact is not yet clearly known.

PSYCHOLOGICAL FACTORS

The perception of a situation can be an important determinant of personality. One can consider perceptions as psychological factors involving the interactions of persons and situations. That is, in addition to biological and social factors, psychological factors must be addressed by personality research and theory. These include cognitive and phenomenological factors. The meaning one attributes to a situation or how one perceives one's physiological reactions are important determinants of behaviour (Endler, 1988; Magnusson, 1992b). As Magnusson (1990b) has noted "the cognitive-interpretation of events in the present situation guides thoughts and actions and evoke physiological systems that, in turn, influence psychological events, thoughts, and emotions" (p. 6).

With respect to the psychological significance or meaning of situations, it is extremely important to distinguish between the *situation perception* attribute and the *situation reaction* attribute. Two persons may *perceive* the same situation as threatening, yet one person may *react* by *attacking* it, and the other individual may *react* by *withdrawing* from it. As Endler (1983) has noted, temporal factors are important: at time A a person may react to perceived stress by attacking, and later at time B the same person may react by withdrawing. Motivational and contextual factors also play a role in mediating the relationship between perception and reaction. Information processing and cognitions are also important aspects of personality.

Thus, biological, social and psychological factors, and their interactions all play important roles in personality development and functioning.

Interactionism: Between and Within

Prior to discussing longitudinal and cross-situational approaches to interactionism in detail, let me present the various meanings of interactionism as well as an interactional model, within which, I think, longitudinal and cross-sectional approaches can be incorporated.

VARIOUS MEANINGS OF INTERACTIONISM

There are two *basic* different types of interactions, namely *mechanistic* (or structural) interaction and *dynamic* (or process) interaction. Nevertheless, the concept of interaction has been used in numerous ways. For example, Carson (1991) in discussing a social-interactional perspective, is essentially referring to interpersonal relationships, i.e., interactions with other persons. In our schema, this would be a subset of person by situation interactions, where *other* persons are part of one's situation. Olweus (1977) has discussed four meanings of the interaction concept: (1) in a *general sense*, conceptualizing how situations and persons *combine*, or "unidirectional interaction"; (2) in terms of the *interdependency* of persons and situations; (3) in terms of *reciprocal* action, and (4) in terms of its use in the *analysis of variance*. Basically, Olweus (1977) is presenting three types of interaction: (a) unidirectional (mechanistic); (b) reciprocal (dynamic); and (c) a third type where it is not possible (methodologically), at present, to separate situations from persons. Thus, we are left with two meaningful and testable conceptions of interaction: mechanistic and dynamic (cf. Endler, 1983).

Mechanistic Interaction. The mechanistic model of interaction focuses on the interaction of main effects (e.g., person and situation) on behaviour. It uses the analyses of variance as its measurement model, is unidirectional, and makes clear and precise distinctions between independent variables (e.g., person and situational factors) in determining behaviour (a dependent variable). It is not concerned with the interaction between independent and dependent variables. The "interaction is between causes and not between cause and effect" (Overton and Reese, 1973, p. 78).

Dynamic Interaction. The dynamic model of interaction focuses on the reciprocal interaction between behaviour and both situational events and person factors. Furthermore, situations affect persons and vice-versa; a reciprocal relationship exists (cf. Endler, in press). Dynamic interaction is multidirectional, process oriented and refers to interactions between independent variables, as well as interactions between independent and dependent variables. "*Reciprocal Causation* means that not only do events affect behavior of organisms but the organism is also an active agent influencing environmental events" (Endler and Magnusson, 1976, p. 969).

Let us now discuss the "between and within" interaction model of personality that we mentioned above. Figure 1 presents a "between and within" person by situation interaction model. Note that there are four phases in this Figure A, B, C and D, and there are both squares and circles. The squares represent large categories of phenomena. Therefore, if we examine phase A, we note that there are two categories of phenomena, person variables and situation variables. The interaction of person variables and situation variables, which affects the perceptions of situations in phase B, we refer to as *between* interactions. The sub-categories within the person variables and situation variables in phase 1 (the circles) can also interact with the biological variables to influence perceptions. In addition, within the circle of biological variables, various systems can affect one another. These we call *within* interactions. For example, the cardiovascular system can affect the autoimmune system or the respiratory system can affect the number of red blood cells. Biological factors can affect psychological factors and vice versa. Therefore, one may identify various levels of interactions; *between* (represented by squares) and *within* (represented by circles). The more molecular one's interests, the greater the number of within-subjects subcategories that are required.

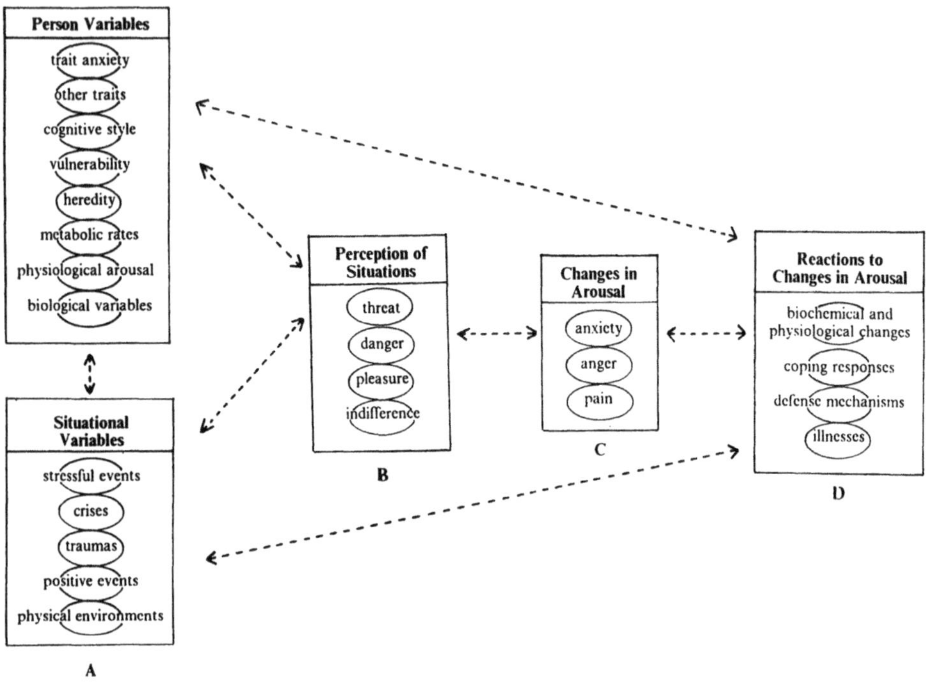

Figure 1. Between and Within Person by Situation Interaction Model.

Note that phase A refers to person variables, such as trait anxiety and heredity, as well as situation variables as stressful events, phase B to perceptions or cognitions, phase C to changes in arousal level, and phase D to reactions to these changes in arousal. This may be conceptualized as a feedback loop, so that phase D can affect the person and situation variables in phase A. Phase D may also be seen as affecting phases B and C. Note that this is a process oriented model, and we are dealing with "movies" not with still "snap shots", it is an ongoing multidimensional process. If one wanted to take a life-span developmental approach, as has Magnusson (1992b), one would add antecedents to phase A and consequences to phase D in Figure 1. Block (1971, 1977) has focused on a longitudinal trait approach.

Magnusson's (1992b) research has focused on the longitudinal approach; my own research (Endler, 1983, 1988; Endler and Parker, 1992b) has focused on a cross-sectional or cross-situational approach. I believe that these are complementary orientations.

Longitudinal Approaches to Interactionism

Magnusson (1992b) presents a general overview of the Orebro or "Individual Development and Adjustment (IDA)" project which he and his colleagues have been working on for almost 30 years - since 1965. One of this project's major studies focuses primarily on the relationship between autonomic reactivity (as measured by adrenaline excretion based on urine samples) and antisocial behavior (as indexed by registered criminal activity, i.e., the number of officially registered instances of law-breaking in Sweden). Magnusson (1986) has discussed this phase of the IDA project in greater detail.

The theoretical orientation, which presents an interactional perspective, has much merit and it should be pointed out that this labour intensive project has both important theoretical and practical implications. Furthermore, the investigators in this project or research programme are among the minority of personality researchers who have investigated biological as well as psychological and social variables, and their interactions.

One important aspect of this research program has focused on the role of the sympathetic nervous system (a biological marker) in both social and antisocial behaviour. The general assumption has been that low autonomic physiological reaction in males (as indexed by catecholamines, specifically adrenaline levels in urine samples) is related to antisocial behaviour. However, Magnusson (1986, 1992b) found that the relationship between adrenaline levels at age 13 and antisocial behaviour at age 30 was more complex. The relationship of low adrenaline levels in urine and antisocial behaviour was tested in three groups of males: (1) non-crime (N=48), teen crime but non-crime in adulthood (N=13) and teen crime plus adult crime (N=11). The low levels of adrenaline secreted in urine at age 13, was related to antisocial behavior at age 30, but only for the teen plus adult crime group. It did not hold for the teen crime only group, thus indirectly suggesting the possibility for an interactive relationship between biological and social factors.

This is quite a robust and impressive finding, considering some of the possible methodological and empirical problems involved, including the small size of the samples and possible reliability and validity difficulties regarding the two major variables measured: antisocial behaviour and autonomic reactivity. As indicated earlier in this chapter, excreted adrenaline levels (the measure of "autonomic reactivity") may be affected by many factors including diet, steroids, drugs, alcohol, and genetic factors. The "antisocial behaviour" variable, defined in terms of "registered criminality" includes a heterogeneous set of possible antisocial activities, such as: (a) offense against property; (b) offenses against persons; and (c) moral offenses (e.g., drug and alcohol offenses). Are these offenses functionally and psychologically equivalent? Are there differences in "autonomic reactivity" between individuals registered for *one* offense as compared to those registered for *multiple* offenses? Are there differences between murderers and those charged with breaking and entering? At what age was the crime committed? Furthermore, attempting to predict criminal behaviour at age 30 from urine samples collected at age 13 is like attempting to predict how fast a feather dropped from a height will fall during a hurricane!

Despite the methodological and measurement problems, discussed above, which would tend to attenuate the results, the findings are quite impressive and compelling. Nevertheless, a more direct test of interactionism, within a longitudinal perspective, would be useful. However, the theory is sound and the IDA project (Magnusson, 1986, 1992b) has made very important contributions to our understanding of personality.

The relevance of a longitudinal approach has also been demonstrated by Block (1971, 1977) who, in a longitudinal study from a trait perspective, has found consistency in personality (stability) between early adolescence and middle to late adulthood. Nevertheless, there are two additional interpretations beyond the one of consistency - (1) since people select and maintain those situations they find rewarding, stabilities may also be influenced by the similarities of situational encounters at various stages of life; and (2) if everyone changes in the same way throughout life, e.g., if everyone "mellows", then stability coefficients would be high and significant. These interpretations supplement the consistency or stability hypothesis, and in no way negate Block's (1971, 1977) impressive findings.

Cross-Situational Approaches to Interactionism

In our own research, within the context of an interaction model of stress, anxiety, and coping (Endler, 1983, 1988; Endler and Parker, 1990), we have evaluated interactionism from a cross-situational or cross-sectional perspective. Elsewhere (Endler, 1983; Endler et al., 1991; Endler et al., 1989) we have summarized the research on stress and anxiety, and we have recently extended this to include our research on coping (Endler and Parker, 1990; 1992a). Basically, as detailed in the next section, we have assessed persons in both non-stressful and stressful situations (measuring their perception of the situations), and we have found that persons (as assessed by multidimensional measures of that anxiety) interact with situational stressors to induce increases in state anxiety. Using real life situations (e.g., actors and actresses in plays, athletes performing, classroom examinations, etc.), we have found confirmation for the interaction model of anxiety (Endler, 1983; Endler et al., 1989, 1991; Endler and Parker, 1990, 1992a). We have recently extended this model to examine the consequences of increases in anxiety, such as coping (Endler and Parker, 1990, 1992a), and this will be discussed in the next section.

Like longitudinal studies, cross-sectional studies also have problems in terms of representativeness of situations, paucity of statistical procedures, for analyzing processes, few studies using biological markers, and representativeness of sampling, etc. Some of these problems are explicated in the next section.

Future Directions for Research on Interactionism

One of the most difficult problems with research on the interaction model of personality involves the inadequacies of the methodologies used to study the constructs, the variables, and their interrelationships (Endler, 1991; Endler and Parker, 1992a). This is especially true in assessing biological factors, as Magnusson (1992a), and Plomin (1990) have noted. We have to go beyond correlational techniques and simple analysis of variance (ANOVA) designs. As Magnusson (1990a) has stated elsewhere, an essential feature of the interaction model of personality is its assumption "that the characteristic functioning of an individual in the dynamic person-environment interaction process depends on and influences the continuous reciprocal process of interaction among psychological and biological subsystems" (p. 202). Endler (1983, 1989) has pointed out that it is extremely important to assess both the interaction *between* systems (e.g., persons vs. situations or biological vs. psychological) and the interactions *within* systems (e.g., the various biological subsystems). Depending on the level of analysis a within system can be conceptualized as a between system. For example, biological and psychological systems are *within* the person variable (Phase A in Figure 1). However, we can discuss interactions *between* the biological vs. psychological systems or *within* the biological system (respiratory vs. cardiovascular) or within the psychological systems (emotions vs. cognitions). There have been recent attempts by others (in addition to those by Magnusson and myself) to develop new interaction models. The biosocial interaction model developed by Kenrick (1987), and the open-systems adaptation model proposed by Hettema (1989a; 1989b) are illustrative. For both these heuristic models, personality research is conceptualized in terms of the study of behavior as a function of three potentially interrelated basic factors: cognitive processes, physiological reactions and situational factors (Hettema and Kenrick, 1989). Specific studies may give differential weights to the role of these three factors. However, valid, reliable, and theoretically sound models must include biological, social, and psychological factors.

In our own work (Endler, 1991; Endler et al., 1991; Endler and Parker, 1992b; Endler et al., 1992) we have been conducting programmatic research on an interaction model of stress anxiety and coping. Using Figure 1 as a guide, the person variables we have been studying are trait anxiety interacting with stressful situations (Phase A) to affect perceptions of danger or threat (Phase B). This leads to increases in state anxiety, or changes in arousal, (Phase C) with consequent coping reactions, including coping with illness, and physiological and biochemical reactions (Phase D). These in turn affect the person variable and stressful situations (Phase A).

In effect we have a feedback loop and continuous processes. We can assess various aspects of the model or the model as a whole (Endler, 1988; 1991). As our research program continues, the model will have to be modified and the theory revised. We see this as a perpetual rough draft wherein we have to constantly juggle many changing variables simultaneously. However, this may be regarded as a challenge rather than as a problem.

Conclusions Regarding Scientific Progress

As Endler and Parker (1991) have noted, "The study of personality is influenced by theories and models, existing methodologies, and sociopolitical factors" (p. 258). Nevertheless, the dynamic yeast for growth and progress in any science, especially the complex and ambiguous one of personality psychology, stems from competing ideas, theories, and honest disagreements. No one point of view has a monopoly on the truth and thus intellectual conflict, honest disagreements and resulting experiments and research based on theories can yield progress. Conant (1947) has maintained that "science emerges from the other progressive activities of man to the extent that new concepts arise from experiments and observations, and the new concepts in turn lead to further experiments and observations (p. 24)". Conant (1947) also noted that the test of any new idea is not only in relating existing facts, but more importantly its role in stimulating further research. "This dynamic quality of science viewed not only as a practical undertaking but as a development of conceptual schemes seems to me to be close to the heart of the best definition" (Conant, 1947, p. 24).

The field of personality is very complex and there are no simple answers. Obviously biological, social and psychological factors all play a role in personality development and functioning (cf. Endler, 1989). To paraphrase what a colleague once told me, "To every *complex* problem, there is a simple answer. And it is wrong!"

Author's Note

The writing of this paper was partially supported by Research Grant No. 410-91-1150 from the Social Sciences and Humanities Research Council of Canada (SSHRC).

The author wishes to thank Gordon Flett, James D.A. Parker, and Laura J. Summerfeldt for their comments and suggestions.

Reprint requests should be sent to Norman S. Endler, Department of Psychology, York University, North York, Ontario, Canada, M3J 1P3.

References

Allport, G.W. (1927). Concepts of trait and personality. *Psychological Bulletin, 24*, 284-293.

Allport, G.W. (1937). *Personality: A psychological interpretation*. New York: Holt, Rinehart and Winston.

Allport, G.W. and Odbert, H.S. (1936). Trait names: A psycho-lexical study. *Psychological Monographs, 97*, 211.

Angleitner, A., Ostenfdorf, F. and John, O.P. (1990). Towards a taxonomy of personality descriptors in German: A psycholexical study. *European Journal of Personality, 4*, 89-118.

Bakan, D. (1977). Political factors in the development of American psychology. *Annals of the New York Academy of Sciences, 291*, 222-232.

Bandura, A. (Ed.) (1971). Psychological modelling: Conflicting theories. New York: Aldine-Atherton.

Bandura, A. (1986). *Social foundations of thought and action: A social cognitive theory*. Englewood Cliffs, NJ: Prentice Hall.

Block, J. (1971). *Lives through time*. Berkeley, CA: Bancroft Books.

Block, J. (1977). Advancing the psychology of personality: Paradigmatic shift or improving the quality of research. In D. Magnusson and N.S. Endler (Eds.), *Personality at the crossroads: Current issues in interactional psychology*. Hillsdale, NJ: Lawrence Erlbaum Associates, pp. 37-63.

Bouchard, T.J. Jr., Lykken, D.T., McGue, M., Segal, N.L. and Tellegen, A. (1990). Sources of human psychological differences: The Minnesota study of twins reared apart. *Science, 250*, 223-228.

Bowers, K.S. (1973). Situationism in psychology: An analysis and a critique. *Psychological Review, 80*, 307-336.

Briggs, S.R. (1989). The optimal level of measurement for personality constructs. In D.M. Buss and N. Cantor (Eds.), *Personality psychology: Recent trends and emerging directions*. New York: Springer-Verlag, pp. 246-260.

Buckley, K.W. (1989). *Mechanical man: Joan Broadus Watson and the beginnings of behaviorism*. New York: Guilford Press.

Carlson, R. (1971). Where is the person in personality research? *Psychological Bulletin, 75*, 203-219.

Carson, R.C. (1989). Personality. *Annual Review of Psychology, 26*, 393-414.

Carson, R.C. (1991). The social-interactional viewpoint. In M. Hersen, A.E. Kazdin and A.S. Bellack (Eds.), *The clinical psychology handbook*. Toronto: Pergamon Press, pp. 185-199.

Cattell, R.B. (1945). Principal trait clusters for describing personality. *Psychological Bulletin, 42*, 129-161.

Cattell, R.B. (1946). *Description and measurement of personality*. New York: World Book Co.
Cattell, R.B. (1957). *Personality and motivation structure and measurement*. New York: World Book.
Cochran, L. (1984). On the categorization of traits. *Journal for the Theory of Social Behaviour, 14*, 183-209.
Conant, J.B. (1947). *On understanding science: An historical approach*. New Haven: Yale University Press.
Costa, P.T. and McCrae, R.R. (1992a). Four ways five factors are basic. *Personality and Individual Differences, 13*, 653-665.
Costa, P.T. and McCrae, R.R. (1992b). Reply to Eysenck. *Personality and Individual Differences, 13*, 861-865.
Costa, P.T. Jr. and McCrae, R.R. (1992c). Revised *NEO Personality Inventory (NEO-PI-R) and NEO Five-Factor Inventory (NEO-FFI): Professional Manual*. Odessa, Florida: Psychological Assessment Resources, Inc. (PAR).
Cox, T., Cox, S. and Thirlaway, M. (1983). The psychological and physiological response to stress. In A. Gale and J.A. Edwards (Eds.), *Physiological correlates of human behaviour: Vol. 1. Basic Issues*. London: Academc Press, pp. 255-276.
Cronbach, L.J. (1957). The two disciplines of scientific psychology. *American Psychologist, 12*, 671-684.
Danziger, K. (1990). *Constructing the subject: Historical origins of psychological research*. New York: Cambridge University Press.
Dewey, R. and Humber, N.J. (1951). *The development of human behavior*. New York: Macmillan.
Digman, J.M. (1990). Personality structure: Emergence of the five-factor model. *Annual Review of Psychology, 41*, 417-440.
Dollard, J. and Miller, N.E. (1950). *Personality and psychotherapy: An analysis in terms of learning, thinking and culture*. New York: McGraw-Hill.
Eaves, L.J., Eysenck, H.J. and Martin, N.G. (1989). *Genes, culture and personality: An empirical approach*. London: Academic Press.
Ekehammar, B. (1974). Interactionism in personality from a historical perspective. *Psychological Bulletin, 81*, 1026-1048.
Endler, N.S. (1981). Situational aspects of interactional psychology. In D. Magnusson (Ed.), *Toward a psychology of situations: An interactional perspective*. Hillsdale, NJ: Lawrence Erlbaum Associates, pp. 361-373.
Endler, N.S. (1982). Interactionism comes of age. In M.P. Zanna, E.T. Higgins and C.P. Herman (Eds.), *Consistency in social behavior: The Ontario Symposium*. Volume 2. Hillsdale, N.J.: Lawrence Erlbaum Associates, pp. 202-249.
Endler, N.S. (1983). Interactionism: A personality model, but not yet a theory. In M.M. Page (Ed.), *Nebraska Symposium on Motivation 1982: Personality - Current theory and research*. Lincoln, Ill.: University of Nebraska Press, pp. 155-200.
Endler, N.S. (1988). Hassles, health and happiness. In M.P. Janisse (Ed.), *Individual differences, stress and health psychology*. New York: Springer, pp. 24-56.
Endler, N.S. (1989). The temperamental nature of personality. *European Journal of Personality, 3*, 151-165.
Endler, N.S. (1991). The Interaction Model of Stress, Anxiety and Coping. Invited address presented at the International Congress on Stress, Anxiety and Emotional Disorders. University of Minho, Braga, Portugal, July 1-3, 1991.

Endler, N.S. (in press). Personality. In D. M. Atrens (Ed.), *Psychology: Concepts and Issues*. Markham, Ont.: Pippin Publishing Ltd.
Endler, N.S. and Edwards, J.M. (1986). Interactionism in personality in the twentieth century. *Personality and Individual Differences, 7*, 379-384.
Endler, N.S., Edwards, J.M. and Vitelli, R. (1991). *Endler Multidimensional Anxiety Scales (EMAS): Manual*. Los Angeles, CA: Western Psychological Services.
Endler, N.S., Edwards, J.M., Vitelli, R. and Parker, J.D.A. (1989). Assessment of state and trait anxiety: Endler Multidimensional Anxiety Scales. *Anxiety Research: An International Journal, 2*, 1-14.
Endler, N.S. and Magnusson, D. (1976). Toward an interactional psychology of personality. *Psychological Bulletin, 83*, 956-974.
Endler, N.S. and Parker, J.D.A. (1990). *Coping Inventory for Stressful Situations (CISS): Manual*. Toronto: Multi-Health Systems.
Endler, N.S. and Parker, J.D.A. (1991). Personality research: Theories, issues and methods. In M. Hersen, A.E. Kazdin and A.S. Bellack (Eds.), *The clinical psychology handbook* (2nd ed.) New York: Pergamon Press, pp. 258-275.
Endler, N.S. and Parker, J.D.A. (1992a). Interactionism revisited: Reflections on the continuing crisis in the personality area. *European Journal of Personality, 6*, 177-198.
Endler, N.S. and Parker, J.D.A. (1992b). Toward a reliable and valid method for the multidimensional assessment of coping with health problems. Paper presented at the annual meeting of the Canadian Psychological Association, Quebec City, Quebec, Canada, June.
Endler, N.S., Parker, J.D.A. and Summerfeldt, L.J. (1992). Coping with health problems: Developing a reliable and valid multidimensional measure. *Department of Psychology Reports*, York University, No. 204.
Eysenck, H.J. (1947). *Dimensions of personality*. London: Routledge and Kegan Paul.
Eysenck, H.J. (1992a). Dimensions of personality: 16, 5, or 3? - Criteria for a taxonomic paradigm. *Personality and Individual Differences, 13*, 773-790.
Eysenck, H.J. (1992b). Four ways five factors are not basic: *Personality and Individual Differences, 13*, 667-673.
Eysenck, H.J. (1992c). A reply to Costa and McCrae. P or A and C - The role of theory. *Personality and Individual Differences, 13*, 867-868.
Eysenck, H.J. and Eysenck, M.W. (1985). *Personality and individual differences: A Natural Science approach*. New York: Plenum Press.
Galton, F. (1884). Measurement of character. *Fortnightly Review, 42*, 179-185.
Guthrie, E.R. (1944). Personality in terms of associative learning. In J. McV. Hunt (Ed.), *Personality and the behavior disorders*, (Vol. 1). New York: Plenum Press, pp. 49-68.
Harris, N.H. and Levey, J.S. (Eds.) (1975). *The New Columbia Encyclopedia*, (4th ed.), New York: Wiley.
Hettema, J. (1989a). Principles of personality assessment. In P.J. Hettema (Ed.), *Personality and environment: Assessment of human adaptation*. Chichester: Wiley, pp. 31-45.
Hettema, J. (1989b). Towards a two-process conception of human adaptation. In P.J. Hettema (Ed.), *Personality and environment: Assessment of human adaptation*. Chichester: Wiley, pp. 203-221.
Hettema, J. and Kenrick, D.T. (1989). Biosocial interaction and individual adaptation. In P.J. Hettema (Ed.), *Personality and environment: Assessment of human adaptation*. Chichester: Wiley, pp. 3-29.

Hilgard, E.R. (1987). *Psychology in America: A historical survey*. San Diego: Harcourt Brace Jovanovich Publishers.

Hofstee, W.K.B. (1990). The use of everyday personality language for scientific purposes. *European Journal of Personality, 4*, 77-88.

Hofstee, W.K.B., de Raad, B. and Goldberg, L.R. (1992). Integration of the big five and circumplex approaches to trait structure. *Journal of Personality and Social Psychology, 63*, 146-163.

John, O.P. (1990). The "big-five" factor taxonomy: Dimensions of personality, in the natural language and in questionnaires. In L.A. Pervin (Ed.), *Handbook of personality: Theory and research*. New York: Guilford, pp. 66-100.

Kenrick, D.T. (1987). Gender, genes, and the social environment. In P.C. Shaver and C. Hendrick (Eds.), *Review of Personality and Social Psychology*, (Vol. 7, pp. 14-43). Beverly Hills, CA: Sage.

Kenrick, D.T. and Funder, D.C. (1988). Profiting from controversy: Lessons from the person-situation debate. *American Psychologist, 43*, 23-34.

Krout, M.H. (1931). Personality testing in the light of the situational approach. *American Journal of Psychiatry, 87*, 839-854.

Lader, M. (1983). Anxiety and depression. In A. Gale and J.A. Edwards (Eds.), *Physiological correlates of human behaviour. Vol. 3. Individual differences and psychopathology* (pp. 155-167). London: Academic Press.

Magnusson, D. (1978). On the psychological situation. *Reports from the Department of Psychology*, University of Stockholm, No. 544.

Magnusson, D. (1986). Antisocial conduct of boys and autonomic activity/reactivity. *Reports from the Department of Psychology*, University of Stockholm, No. 652.

Magnusson, D. (1990a). Personality development from an interactional perspective. In L. Pervin (Ed.), *Handbook of personality theory and research*. New York: Guilford, pp. 193-222.

Magnusson, D. (1990b). Personality research - Challenges for the future. *European Journal of Personality, 4*, 1-17.

Magnusson, D. (1992a). Back to the phenomena: Theory, methods and statistics in psychological research. *European Journal of Personality, 6*, 1-14.

Magnusson, D. (1992b). Individual development: A longitudinal perspective. *European Journal of Personality, 6*, 119-138.

McCrae, R.R. (1989). Why I advocate the five-factor model: Joint factor analysis of the NEO-PI with other instruments. In D.M. Buss and N. Cantor (Eds.), *Personality psychology: Recent trends and emerging directions*. New York: Springer-Verlag, pp. 237-245.

McCrae, R.R. and Costa, P.T. (1986). Clinical assessment can benefit from recent advances in personality psychology. *American Psychologist, 41*, 1001-1002.

Mead, G.H. (1934). *Mind, self and society*. Chicago: University of Chicago Press.

Miller, N.E. and Dollard, J. (1941). *Social learning and imitation*. New Haven, CT.: Yale University Press.

Mischel, W. (1968). *Personality and assessment*. New York: Wiley.

Mischel, W. (1973). Toward a cognitive social learning reconceptualization of personality. *Psychological Review, 80*, 252-283.

Nunnally, J.C. (1984). Self report measures of personality traits. In N.S. Endler and J. McV. Hunt (Eds.), Personality and the behavior disorders, (2nd ed., Vol. 1). New York: John Wiley and Sons, pp. 221-259.

Olweus, D. (1977). A critical analysis of the "modern" interactionist perspective. In D. Magnusson and N.S. Endler (Eds.), *Personality at the crossroads: Current issues in interactional psychology*. Hillsdale, NJ: Lawrence Erlbaum Associates, pp. 221-233.

Overton, W.F. and Reese, H.W. (1973). Models of development: Methodological implications. In J.R. Nesselrode and H.W. Reese (Eds.), *Life-span developmental psychology: Methodological issues*. New York: Academic Press, pp. 65-86.

Parker, J.D.A. (1990). From the intellectual to the non-intellectual traits: The origins of American personality research. Paper presented at the annual meeting of the Canadian Psychological Association, Ottawa, Ontario, Canada, June.

Parker, J.D.A. (1991). The search for the person: The historical development of American personality psychology. Ph.D. Dissertation, York University, Toronto, Canada.

Peabody, D. (1987). Selecting representative trait adjectives. *Journal of Personality and Social Psychology*, *52*, 59-71.

Pervin, L.A. (1978). Definitions, measurements and classifications of stimuli, situations, and environments. *Human Ecology*, *6*, 71-105.

Plomin, R. (1990). The role of inheritance in behavior. *Science*, *248*, 183-188.

Rorer, L.G. and Widiger, T.A. (1983). Personality structure and assessment. *Annual Review of Psychology*, *34*, 431-463.

Rotter, J.B. (1954). *Social learning and clinical psychology*. Englewood Cliffs, NJ: Prentice-Hall.

Rotter, J.B. (1975). Some problems and misconceptions related to the construct of internal versus external control of reinforcement. *Journal of Consulting and Clinical Psychology*, *43*, 56-67.

Sechrest, L. (1976). Personality. *Annual Review of Psychology*, *27*, 1-28.

Shute, C. (1973). Aristotle's interactionism and its transformations by some 20th century writers. *Psychological Record*, *23*, 283-293.

Stagner, R. (1937). *Psychology of personality*. New York: McGraw-Hill.

Stelmack, R.M. (1990). Biological bases of extraversion: Psychophysiological evidence. *Journal of Personality*, *58*, 293-311.

Symonds, P. (1924). The present state of character measurement. *Journal of Educational Psychology*, *15*, 484-498.

Thorndike, E.L. (1940). *Human nature and the social order*. New York: Macmillan.

Tolman, E.C. (1932). *Purposive behavior in animals and men*. New York: Century Co.

Van Heck, G.L. (1991). Temperament and the person-situation debate. In J. Strelau and A. Angleitner (Eds.), *Explorations in temperament: International perspectives on theory and measurement*. London: Plenum Press, pp. 163-175.

Vernon, P.E. (1933). The American vs. the German methods of approach to the study of temperament and personality. *British Journal of Psychology*, *24*, 156-177.

Wiggins, J.S. (1980). Circumplex models of interpersonal behavior. In L. Wheeler (Ed.), *A Review of Personality and Social Psychology*, (Vol. 1). Beverly Hills, CA: Sage, pp. 265-294.

Woodworth, R.S. (1918). *Personal data sheet*. Chicago: Stoelting.

Zuckerman, M. (1991). *Psychobiology of personality*. Cambridge: Cambridge University Press.

HAVE TRAIT PSYCHOLOGISTS BEEN IGNORANT OF CONTEXT?
Commentary on the chapter by N.S. Endler

Boele de Raad
Department op Psychology
University of Groningen,
Grote Kruisstraat 2/1
9712 TS Groningen,The Netherlands

Behavior in Context

All behavior takes place within a context. This is in a nutshell the distinctive leading motive of the interactional approach to personality. No behavior can be independent of either biological or social factors. Psychologists in general, and personality psychologists in particular, do not really call these views in question. From the early history of personality psychology, students of personality have sought to describe the characteristic psychological differences between people in terms of how they might be expected to react under particular circumstances. In Freud's conception the ego deals with reality and the superego represents the internalization of important features of the environment. Adler, Horney and others all explicitly emphasized the contribution of the social world in the development of personality and the production of behavior. Allport's definition of personality comprises the expression "unique adjustment to the environment". But also Sheldon, the proponent of constitutional psychology, was well aware of environmental determinants of behavior, witness items from his Temperament scale like 'love of being with others', 'fear of social involvement' and 'need of people when troubled'. Moreover, even the ancient prerunners of personality psychology focused on behavior in situ. The 30 characters of Theophrast, for example, without exception consist of striking contextual portrayals of types of people. Cattell and Eysenck, the two pioneers in the factor/trait approach, both strive for the improvement of predictions of what individuals will do in given situations. I believe that the history of personality psychology can be reconstructed as having been consequently aware of the contextual nature of behavior.

The major quest of the trait approach during the last few decades has been the taxonomizing of the differentiating psychological characteristics of people. Because the trait approach is the dominant one within personality, Endler has implicitly misnamed and misclassified this persistent and longtime endeavour, by stating (see also Endler, 1982) that personality research has been and still is "basically piecemeal, and ad hoc, rather than programmatic". Within the taxonomic program, lively controversies such as that between the Big Five and the Eysenck's P.E.N. model sometimes turn out to be biting events, though, it must be admitted, the recent discussions in this regard reflected more rapprochement than ever before. In no way, however, such a debate endangers the program. Quite the contrary, they form the confirmation of an enterprise at work.

J. Hettema and I. J. Deary (eds.), Foundations of Personality, 269–273.

What does it mean to say, for example, that a person is impulsive? According to Endler, trait theorists would view impulsivity as an internal factor accounting for consistently occurring impulsive behaviors. This definition, it should not be forgotten is, like any definition, a summarizing description of the characteristics of a concept. What is summarized, in my view, is the way people are consistently like themselves and different from others (cf. Peterson, 1992). If this description is somewhat vague, it better remains that way. It is meant to express the adherence to an observation that is smashingly evident to most people: although visibly changing in time, people remain recognizably the same in their essential characteristics, even across years. Definitions can easily be changed, and they should be, in the light of improved insights. The summary conception of traits (Hampson, 1984) seems most apt to the definition above. Calling a person impulsive thus means to summarize descriptions of behaviors of a varying nature taking place in divergent situations: 'instead of saying Nancy drives through red lights, buys the first pair of jeans she sees and says things without thinking, we say that Nancy is "impulsive"'. Traits, according to Hampson (1989), are shorthand terms for behaviors *plus their ascribed social significance*.

The Crisis

Personality psychology is said to have gone through a crisis by the mid 1960s, and according to Endler the crisis is not over yet. What are the characteristics of a crisis? One possible meaning of crisis is that of a serious disturbance or persistent disease. Another meaning is that of a decisive stage or turningpoint. Common to these two readings of a crisis in personality psychology is that, given the state of the art, the answers are considered to fall short of expectations, in that they are judged inadequate for coping with the contemporary demands. Was personality psychology seriously ill? Had personality psychology been arrived at a stage where it had to demonstrate its ultimate competence? These questions are mainly rhetorical, but so is the presumed suggestion of a crisis. At times, personality psychology may have looked a bit palish, but that is far behind now. Also, there was never any need to put the competence of personality psychology to a final test. The conclusion is that the diagnosis of a crisis with the matching suggestions was wrong. With such a conclusion it becomes important to ask who did the diagnosis, which question in itself again emphasizes the rhetorical nature of the diagnosis.

The interactionist critique, as particularly expressed Mischel, was not a negation either of personality as a field or of individual differences as a phenomenon. The focus was on the unique interactions of each person with the specific contexts of his or her life (Mischel 1990). I will not deny here that persons can be said to be unique and incomparable. For example, Nancy may be said to be impulsive in a manner that nobody else is, because her impulsivity is embedded in a personality structure all her own; to call her impulsive is to reshape the meaning of impulsivity. Strictly speaking, the extension of the uniqueness argument makes all comparisons between people impossible. Eysenck's old shoe is unique. Of course, shoes are not persons. It is important to take into account the distinction between persons considered as objects and persons considered as discourse partner (Kouwer, 1973). As an object the individual is not systematically different from other objects. Humans, however, are distinct from other entities because they are the ones we talk with. The unique person is the one to talk with, not to describe and treat as an object. The conclusion is that a scientific approach to uniqueness is impossible (for a more extensive documentation of the argument, see Hofstee and De Raad, 1992).

Major Models

Endler enumerates traits, situationism, psychodynamics, and interactionism, as the four major models in personality psychology. Situationism, strictly spoken, denies personality as a determinant of behavior, and is therefore not a model of personality. The other three models have in common the general notion that individuals can be differentially characterized in such a way that particular distinct behaviors are to be expected from different individuals in certain contexts. Personality psychologists do differ in emphasis they give to personality characteristics, and in the ways they conceive of the essential features of personality. These different conceptualizations, more than the four mentioned above, are the proper alternative models of personality: traits, temperaments, adaptive mechanisms, projects, capabilities, and more. They all try to capture alternative, sometimes overlapping, features of the differentiating characteristics of people, all with the promise to account for different niches and levels of personality. What is basic here is that the mainstream personality approach, and I prefer to reserve the denominator "trait approach" for this, has been very consistent, with known drawbacks all right, and flourishing more than ever, witness the Big Five boom.

The Big-Five Model

Peabody and Goldberg (1989) relate the Big Five to five different realms of life: sets of contexts to which the Big Five are supposed to refer: Power (I), Love (II), Work (III), Affect (IV), and Intellect (V). This conclusion does not stand alone. Throughout the taxonomic tradition, investigators have consistently referred to particular contexts to which the trait dimensions refer, and have pursued to show the importance of the trait dimensions for those contexts. One of the most interesting developments in this regard is the evolutionary personality psychology (Buss, 1991), which approach not only confirms the contextual connectedness of the personality factors as 'most important dimensions of our social adaptive landscape', but also provides an explanatory framework for their differential role.

Trait descriptive expressions, personality adjectives, nouns, phrases, are not to be considered as concepts liberated from context. Quite the contrary, the majority of trait words, used in personality taxonomies, not only capture important personality characteristics, but also represent contextual information. According to Au (1986), an interpersonal verb is worth a thousand words, referring to the contextual meaning that is implicitly accounted for by the use of certain verbs. For example, in *John amazed Mary*, John gave rise to the amazement and is therefore the stimulus. But in *John feared Mary*, Mary gave rise to the fear and is therefore the stimulus. In the same vein, personality descriptors should be conceived of as carriers of implicit contextual meaning. For instance, the expression John is *impudent* means that a situation is captured in which John expresses himself to a superior. That situation is implicit in the word *impudent*.

In *John is easily frightened*, the situation is the stimulus, but in *John is shy*, John is the stimulus. In going through the Dutch AB5C scheme (De Raad, Hendriks and Hofstee, 1992), the interesting criterion is whether personality adjectives more ore less directly evoke a certain context or not. The I+I+ terms, *exuberant, spontaneous, open*, probably do not. *Tolerant, peaceful, forgiving* (II+II+), give rise to certain context, and so do *careful, prompt, meticulous* (III+III+). The personality descriptors resonate a differential aptness for use in particular contexts. Terms referring to accuracy, precision, etc. belong to the context of WORK, many Factor I terms have meaning, particularly in the context of DOMINANCE, and so on. Actually, the use of certain descriptors in a *different* environment, in turn may give presence to the matching implicit contexts of use. *Being precise, accurate, industrious* at a party (the context for extraverts), evokes the WORK connotation for that person. The complete test is not pursued here, but probably the majority of the adjectives spontaneously evoke certain contexts and a minority does not. The argument is that the taxonomic enterprise implicitly captures context. Because of its comprehensiveness, the taxonomy approach captures an enormous variety of contextual features of different levels and of varying relevance.

An interesting observation here is related to those terms from the AB5C scheme that do not spontaneously evoke contexts of application. Examples are *active, lively, quiet, gentle, dynamic*, etc. The terms are scattered over the scheme. I suggest that the distinction that comes to the fore here between context evoking and context free terms parallells that between Strelau's distinction between content and form, respectively (Strelau, 1987). The suggestion is that to the extent that adjectives do not evoke associated contexts of application, they are temperamental in nature. It would be an interesting task to scrutinize personality descriptors for their distinctive contextual features, and to systematically investigate the temperamental (form) characteristics of (a subset of?) personality descriptors.

References

Au, K.T. (1986). A verb is worth a thousand words: The causes and consequences of interpersonal events implicit in language. *Journal of Memory and Language, 25*, 104-122.

Buss, D.M. (1991). Evolutionary personality psychology. *Annual Review of Psychology, 42*, 459-491.

De Raad, B., Hendriks, A.A.J. and Hofstee, W.K.B. (1992). Towards a refined structure of personality traits. *European Journal of Personality, 6*, 301-319.

Endler, N.S. (1982). Interactionism omes of age. In Zanna, M.P., Higgins, E.T., and Herman, C.P. (Eds.), *Consistency in social behavior: The Ontario Symposium.* (Volume 2, pp.202-249). Hillsdale, N.J.: awrence Erlbaum Associates.

Hampson, S.E. (1984). The social construction of personality. In Bonarius, H., Van Heck, G. and Smid, N. (Eds.), *Personality Psychology in Europe: Theoretical and Empirical Developments.* Lisse: Swets & Zeitlinger.

Hampson, S.E. (1989). Using traits to construct personality. In Buss, D.M. and Cantor, N. (Eds.). *Personality Psychology: Recent Trends and Emerging Directions.* New York: Springer-Verlag.

Hofstee, W.K.B. and De Raad, B. (1992). Personality structure through traits. In Caprara, G.V. and Van Heck, G.L. (Eds.). *Modern Personality Psychology; Critical Reviews and New Directions.* New York: Harvester/Wheatsheaf.

Kouwer, B.J. (1973). *Existentiële Psychologie: Grondslagen van het Psychologische Gesprek* [Existential Psychology]. Meppel: Boom.

Mischel, W. (1990). Personality dispositions revisited and revised: A view after three decades. In Pervin, L.A. (Ed.), *Handbook of Personality; Theory and Research*, New York: The Guilford Press.

Peabody, D. and Goldberg, L.R. (1989). Some determinants of factor structures from personality-trait descriptors. *Journal of Personality and Social Psychology, 57*, 552-567.

Peterson, C. (1992). *Personality*. Fort Worth: Harcourt, Brace, Janovich, Inc.

Strelau, J. (1987). The concept of temperament in personality research. *European Journal of Personality, 1*, 107-117.

SUBJECT INDEX